Progress in Sleep Research

The book series provides an overview of the physiological processes involved in sleep regulation and the essential role of sleep in various vital physiological processes. It also summarizes the different stages of sleep and the clinical features of common sleep disorders. The individual books discuss the neurological and neuropsychological implications of sleep and highlight the relation between sleep and brain structure-functions. It explores the impact of sleep on circadian biology, neuroanatomy, neurophysiology, neuropharmacology, neuroendocrinology, neuroimmunology, neuropathology, basic neurology, biological psychiatry, and the behavioral sciences. It reviews the genetic factors associated with normal sleep and sleep disorders. It also describes the recent advances in research aiming to elucidate the neurochemical mechanisms regulating sleep and wakefulness.

Ahmed S. BaHammam • Mahadevappa Hunasikatti
Editors

Sleep Apnea Frontiers

Pathophysiology, Diagnosis, and Treatment Strategies

Springer

Editors
Ahmed S. BaHammam
The University Sleep Disorders Center, Department of Medicine, College of Medicine
King Saud University
Riyadh, Saudi Arabia

Mahadevappa Hunasikatti
Division of Respiratory, Sleep and Anesthesiology (DSRA), OHT1, Food and Drug Administration (FDA)
Silver Spring, MD, USA

ISSN 2730-5678 ISSN 2730-5686 (electronic)
Progress in Sleep Research
ISBN 978-981-99-7900-4 ISBN 978-981-99-7901-1 (eBook)
https://doi.org/10.1007/978-981-99-7901-1

This Springer imprint is published by the registered company Springer Nature Singapore Pte Ltd.
The registered company address is: 152 Beach Road, #21-01/04 Gateway East, Singapore 189721, Singapore

Paper in this product is recyclable.

Preface

"Sleep is the only medicine that gives ease."
—Sophocles (496–406 B.C.E.)

Getting the right amount and quality of sleep is not a luxury but a necessity that no one can ignore. Early, undifferentiated stages of a problem are much harder to diagnose than later stages when symptoms have advanced to the point where the diagnosis is clear. A ton of literature has been written about sleep apnea. In the modern world, scientists and researchers have access to vast and expanding informational resources that offer in-depth knowledge, instruction, and training to advance and improve research. This book differs from others in that it focuses on a specific area of research rather than being a traditional volume on sleep medicine. It embodies the authors' seriousness, sincerity, and passion as they attempt to lead and direct the reader in their pursuit of research to discover new information, recognize issues, and offer solutions.

This specialty volume covers a wide range of important topics related to OSA and sleep-disordered breathing, including OSA screening, the current clinical perspective of adult OSA, perioperative evaluation and management of patients with sleep-disordered breathing, central sleep apnea, OSA co-occurring with insomnia, sleep telemedicine during the COVID-19 pandemic, wearable technology, sleep disorders and traffic safety, the current clinical perspective of adult OSA, recent developments in surgical treatments for snoring and OSA, the use of neurostimulators in OSA, advances in PAP and nonsurgical therapies, and the significance of REM on sleep dysfunctions, and the impact of COVID-19 pandemic on OSA and other sleep disorders.

The importance of the disorders is discussed, and the availability of new data was considered while selecting the topics for this volume, though most authors emphasize the need for more study. Additionally, the authors featured in this volume come from different parts of the globe, each representing a specific area of expertise in studying sleep and its disorders.

The title of this book makes clear that it is written with sleep specialists and other healthcare professionals in mind. Therefore, this volume should be an excellent reference for sleep medicine practitioners and physicians-in-training, as well as students, primary care and general physicians, and diverse medical specialty interested in the field of sleep medicine.

Although the chapters were written and revised at the height of the COVID-19 pandemic, the planning for this edition started in 2020. We are, therefore, even more appreciative than usual of the many contributors who made time to work on the book despite the pandemic's demands on their personal and professional lives.

In light of the fact that our population is inexorably sleep-deprived and that sleep medicine must deal with an increasing number of patients with sleep disorders who have multiple, complicated problems, we have been able to assemble the brightest minds and industry pioneers to offer authoritative leadership.

The greatest strength of this volume is the representation of several authors' personal experiences. It is encouraging to see how many people took the initiative to use this platform to share their knowledge and breadth of experience regarding sleep research. In addition, the book is easy to read and comprehend. The topic of sleep health, in general, and sleep apnea, in particular, is discussed. This is a valuable book that every sleep researcher, practitioner, and trainee will want to keep handy for quick reference.

Chapters are written by eminent clinicians, educators, and researchers in the field of sleep medicine. The volume's text, tables, and illustrations are well balanced and provide readers with useful information. The main objective of this volume is to increase readers' knowledge of sleep apnea, which will directly improve patient-centered care across a range of clinical settings. Overall, our authors include a sizable number of subspecialists from various sleep disciplines as well as many sleep specialists. The authors also come from a wide range of medical specialties.

The editors of the book have more than two decades of experience in sleep medicine practice and research. Furthermore, they have extensive experience contributing, editing, and managing journals relevant to sleep medicine. Finally, we hope this book will convey some of our joy at having been given a chance to work in the quickly expanding field of sleep medicine. The field is dynamic, so this book aims to provoke thought rather than provide a solution. As a result, we anticipate that this volume will be even more educational and useful for the readers.

Riyadh, Saudi Arabia — Ahmed S. BaHammam
Fairfax, VA, USA — Mahadevappa Hunasikatti
09/01/2022

Acknowledgments

A project like this represents the collective scholarship of an academic community, not just one or two individuals, and this volume in particular. The majority of the information is the contributions of other people. Hence, there are a few thanks that must be given.

First and foremost, this volume's editing was made possible thanks in large part to Prof. Seithikurippu R. Pandi-Perumal and Springer publishers' trust and support.

We appreciate each and every one of our contributors for dedicating precious time and lending their knowledge to the chapters they wrote. We appreciate Springer's staff for their strict publishing guidelines and commitment to excellence.

Ahmed S. BaHammam
Mahadevappa Hunasikatti
09/01/2022

Contents

Editors and Contributors

About the Editors

Ahmed S. BaHammam, is a tenured pulmonary and sleep medicine professor at King Saud University (KSU) in Riyadh, Saudi Arabia. He was awarded the KSU Lifetime Achievement Award in 2016 for his significant contributions to sleep medicine research. Prof. BaHammam has published 350 scientific articles, 65 book chapters, and a few books on sleep medicine and technology. His research focuses on various sleep-related topics, including hypersomnia, obesity hypoventilation syndrome, central hypersomnolence disorders, the impact of intermittent fasting and mealtime on sleep and circadian rhythm, and sleep monitoring. Prof. BaHammam serves as the President-Elect of the Asian Society of Sleep Medicine (ASSM), the Editor-in-Chief of *Nature & Science of Sleep*, the Executive Director of Academic Affairs at the Medical City, King Saud University, and the Director of Prince Naif Health Research Center.

Mahadevappa Hunasikatti, works as a medical officer in the sleep team at the Food and Drug Administration in Silver Spring, MD, USA, and serves a medical director for Mahadeva Sleep Tele Medicine (MSTM). He completed his residency in psychiatry and obtained a diploma in psychological medicine from the prestigious National Institute of Mental Health and Neurosciences in Bangalore, India. After moving to the United States, he did an internship/residency in psychiatry and internal medicine at the Medical College of Ohio in Toledo. Later, he did a fellowship in critical care medicine at Shock Trauma Center, University of Maryland, Baltimore, MD, and a fellowship in pulmonary medicine at Sinai Hospital in Detroit, MI. He is board certified in internal medicine, psychiatry, pulmonary medicine, critical care medicine, sleep medicine, and hospital medicine. His main interests include sleep education, regulation of sleep medical devices, sleep tele medicine, sleep research, and policy.

Contributors

Mashni Alsaeed Respiratory Care Department and Sleep Disorders Center, King Faisal Specialists Hopsital and Research Center, Jeddah, Saudi Arabia

Ahmed S. BaHammam The University Sleep Disorders Center, Department of Medicine, College of Medicine, King Saud University, Riyadh, Saudi Arabia
King Saud University Medical City, King Saud University, Riyadh, Saudi Arabia

Ana Rita Brito Evelima Children's Hospital, Guy's and St Thomas NHS Foundation Trust, London, UK

Cláudio D'Elia Department of Pediatrics, Lusíadas Hospital, Lisbon, Portugal
European Sleep Center, Lisbon, Portugal

Brian Freeman Georgetown University School of Medicine, Washington, DC, USA

Ravi Gupta Division of Sleep Medicine, Department of Psychiatry, All India Institute of Medical Sciences, Rishikesh, Uttarakhand, India
Department of Psychiatry and Division of Sleep Medicine, All India Institute of Medical Sciences, Rishikesh, Uttarakhand, India

Arup Haldar Consultant Pulmonologist Woodlands Hospital & Institute of Sleep Sciences, Kolkata, West Bengal, India

Yousef MohamedRabaa Hawsawi Research Center Department, King Faisal Specialist Hospital and Research Center, Jeddah, Saudi Arabia

Kumudhini Hendrix Anesthesia Devices, CDRH, Food and Drug Administration, Silver Spring, MD, USA

Mahadevappa Hunasikatti Division of Respiratory, Sleep and Anesthesiology (DSRA), OHT1, Food and Drug Administration (FDA), Silver Spring, MD, USA

Anjum Khan ENT Division, US Food and Drug Administration, Silver Spring, MD, USA

Shravan Kooragayalu University of Pittsburgh Medical Center—Western Maryland Health System, Cumberland, MD, USA

Sreelakshmi Kooragayalu University of Pittsburgh Medical Center—Western Maryland Health System, Cumberland, MD, USA

Barry Krakow Maimonides Sleep Arts and Sciences, Savannah, GA, USA

Vijay Krishnan Department of Psychiatry and Division of Sleep Medicine, All India Institute of Medical Sciences, Rishikesh, Uttarakhand, India

Chris Landon Family Practice and Pediatrics, University of California, Los Angeles, CA, USA

Atul Malhotra Division of Pulmonary, Critical Care and Sleep Medicine, University of California, San Diego, CA, USA

Miguel Meira e Cruz Cardiovascular Center, University of Lisbon, Lisbon School of Medicine, Lisbon, Portugal
European Sleep Center, Lisbon, Portugal

Vinh Nguyen Anesthesiology Department, University of Minnesota, Minneapolis, MN, USA

Seithikurippu R. Pandi-Perumal Somnogen Canada Inc., College Street, Toronto, ON, Canada
Saveetha Medical College and Hospitals, Saveetha Institute of Medical and Technical Sciences, Saveetha University, Chennai, Tamil Nadu, India

Lokesh Kumar Saini Division of Sleep Medicine, Department of Pulmonary Medicine, All India Institute of Medical Sciences, Rishikesh, Uttarakhand, India

Christopher N. Schmickl Division of Pulmonary, Critical Care and Sleep Medicine, University of California, San Diego, CA, USA

Ashutosh Suresh J.J.M Medical College, Davanagere, Karnataka, India
Fortis Hospital, Bangalore, Karnataka, India

Andrew M. Vahabzadeh-Hagh Division of Otolaryngology—Head and Neck Surgery, University of California, San Diego, CA, USA

Pratyaksh P. Vaishnav J.J.M Medical College, Davanagere, Karnataka, India
Vasavi Hospital, Bangalore, Karnataka, India

Chapter 1
Neuro-Stimulator Use in Obstructive Sleep Apnea—Past, Present and Future

Christopher N. Schmickl, Andrew M. Vahabzadeh-Hagh, and Atul Malhotra

Abstract Obstructive sleep apnea (OSA) is a common disorder with major neurocognitive and cardiovascular sequelae. OSA affects up to 1 billion people worldwide. OSA is now recognized to be a heterogeneous disorder with multiple underlying mechanisms (endotypes) and varying in clinical expression (phenotypes). Nasal CPAP (continuous positive airway pressure) is the treatment of choice for OSA, but issues with adherence can limit its effectiveness. Hypoglossal nerve stimulation has now been FDA approved for treatment of OSA and has been shown to have sustained benefits in some patients. OSA patients who have issues with pharyngeal mechanics and pharyngeal dilator muscle dysfunction are likely the group that benefits from hypoglossal nerve stimulation. A number of technologies are being developed to activate the hypoglossal nerve although the optimal approach remains unclear. Hard outcome data are currently not available for nerve stimulation for OSA; however, most experts agree that hypoglossal stimulation is a viable alternative therapy for a subset of OSA patients.

Keywords Obstructive sleep apnea (OSA) · Hypoglossal nerve stimulation · Continuous positive airway pressure (CPAP) · Pharyngeal mechanics · Alternative therapies

C. N. Schmickl · A. Malhotra (✉)
Division of Pulmonary, Critical Care and Sleep Medicine, University of California, San Diego, CA, USA
e-mail: amalhotra@health.ucsd.edu

A. M. Vahabzadeh-Hagh
Division of Otolaryngology—Head and Neck Surgery, University of California, San Diego, CA, USA

A. S. BaHammam, M. Hunasikatti (eds.), *Sleep Apnea Frontiers*, Progress in Sleep Research, https://doi.org/10.1007/978-981-99-7901-1_1

1.1 The Role of Pharyngeal Muscle Function in Sleep Apnea Pathophysiology

Obstructive sleep apnea (OSA) is defined as a repetitive collapse of the upper airway during sleep causing oxygen desaturations, hypercapnia and arousals (Jordan et al. 2014). OSA is primarily caused by an anatomical predisposition to upper airway collapse (e.g. crowded airway in the setting of obesity), but in most individuals non-anatomical factors or traits play a critical role in OSA pathogenesis as well (Eckert et al. 2013; Schmickl et al. 2018a). One key non-anatomical trait is upper airway dilator muscle function: Following pharyngeal collapse there is a gradual increase in respiratory stimuli (e.g. increase in carbon dioxide and negative intra-thoracic pressure) resulting in increased reflexes augmenting upper airway dilator muscle activity. If dilator muscles respond vigorously and are able to restore patency before an arousal and/or desaturation occur then sleep is maintained, otherwise sleep fragmentation and/or intermittent hypoxia—the hallmarks of OSA—ensue (Eckert and Younes 2014). This concept illustrates how poor pharyngeal muscle function and a low arousal threshold (waking up easily, i.e., leaving little time for dilators to activate) are two non-anatomical traits that can substantially contribute to OSA pathophysiology. Importantly, during wakefulness OSA patients compensate for their collapsible upper airway anatomy through increased upper airway dilator activity compared with matched subjects without OSA (Mezzanotte et al. 1992). But, with sleep-onset there is loss of upper airway motor output leading to collapse of the vulnerable airway (Mezzanotte et al. 1996).

There are 23 pairs of upper airway dilator muscles, the most important of which is probably the genioglossus muscle. The genioglossus muscle makes up the bulk of the tongue thus protecting the retroglossal space, which is a common site of collapse in OSA. The genioglossus muscle has both tonic activity (i.e. present throughout the respiratory cycle) as well as phasic activity (i.e. bursts with each inspiration). It is innervated by the hypoglossal nerve, thus receiving neuronal input from the hypoglossal motor nucleus in the brainstem. Importantly, activation via median branches of the hypoglossal nerve results in tongue protrusion, and electrical stimulation of the hypoglossal nerve has been shown to treat OSA effectively in select patients (see Sect. 1.2.1). However, the genioglossus muscle has also been the target of other potential therapeutic interventions described below.

1.1.1 Pharmacological Interventions

A number of strategies have been pursued to develop a pharmacological therapy for OSA (Gaisl et al. 2019; Mason et al. 2013). Agents to raise the arousal threshold or to stabilize control of breathing have been discussed (Eckert et al. 2011; Schmickl et al. 2020; Wellman et al. 2008), but we focus here on approaches to increasing hypoglossal motor output. A number of projections to the hypoglossal motor nucleus have been well defined including histaminergic and orexinergic from the hypothalamus, adrenergic from the locus ceruleus (or specifically the A7 region), cholinergic from the lateral dorsal tegmentum/pediculopontine tegmentum (LDT/PPT) and periobex region, serotonergic from the raphe neurons. These inputs are generally state dependent (i.e. fire more during wakefulness than during sleep) and are excitatory. Based on monosynaptic projections from various monoaminergic centers, investigators have suggested that agents targeting these pathways may be effective in promoting hypoglossal motor output. Thus, pharmacological studies to treat OSA in pre-defined subsets of patients are being undertaken. To date, no definitive study has shown benefit of this approach, although some provocative data have been published regarding the combination of atomoxetine and oxybutynin. In a small physiological cross-over study, single-night administration of atomoxetine-oxybutynin improved the apnea-hypopnea index by 63% (median 28.5 vs 7.5 events/h) compared with placebo (Taranto-Montemurro et al. 2019). However, longer-term studies with hard outcomes will be required before such approaches can be recommended.

1.1.2 Exercise-Based Interventions

In general, exercise is advocated for OSA patients as it can be an important component of weight loss and can help to improve overall health. Even in the absence of weight loss, exercise has been associated with some improvement in OSA severity (Awad et al. 2012). For upper airway muscles, motor control abnormalities in OSA are thought to be state dependent (i.e. problematic during sleep rather than wakefulness) rather than generalized dysfunction of these muscles (Mezzanotte et al. 1996). However, a number of investigators have identified either sensory or motor deficits in OSA patients even during wakefulness (Saboisky et al. 2012). As a result, some studies have assessed the potential role of muscle training as a therapy for OSA in at least some patients. One prominent study in *The BMJ* used the didgeridoo and showed potential improvements in OSA patients ostensibly on the basis of muscle training (Puhan et al. 2006). A group in Brazil reported provocative data showing some improvement in OSA using a set of pharyngeal muscle exercises (Guimaraes et al. 2009). In aggregate, the data suggest that upper airway muscle training may have a role in some OSA patients depending on the underlying endotype (Schmickl et al. 2018a; Camacho et al. 2015). However, further data are clearly required examining hard outcomes in multicenter studies before this approached can be definitively endorsed.

1.1.3 Electrical Stimulation in the Past

The concept of electrical stimulation of upper airway muscles has been around for several decades. A number of groups acquired data in both rodents and humans to examine the impact of nerve stimulation on pharyngeal mechanics. In the 1980s, a group in Japan explored direct stimulation of tongue muscles (Miki et al. 1989, 1988), but it soon became clear that indirect stimulation via the hypoglossal nerve has many advantages including greater opening of the upper airway with lower electrical amplitudes. Much of the groundwork for this approach, using hypoglossal nerve stimulation (HGNS) to activate pharyngeal muscles and improve OSA was led by the group at Johns Hopkins (Schwartz et al. 2001, 1993). Early challenges with this approach included difficulty with measuring sleep during electrical stimulation. Indeed, some suggested that electric stimulation was leading to arousal as a mechanism to restore airway patency rather than improving the mechanics of the airway per se. Debate also occurred regarding how to optimize stimulation, for example, medial branch of the hypoglossal nerve is largely responsible for tongue protrudors whereas the lateral branch is responsible for retractors. Some data from Fregosi suggested that stimulating both protudor and retractors had major benefit for pharyngeal mechanics (Fuller et al. 1998). In addition, there was some debate regarding whether respiratory gating was required, that is, did electrical stimulation of the genioglossus need to be phased with inspiration, or could tonic stimulation be provided throughout the respiratory cycle. Considerable progress was made during the 1990s and 2000s culminating in several clinical trials evaluating HGNS devices from at least three different companies: One device by Apnex Medical stimulated tongue protrudors during inspiration and showed promising results in a phase II trial (Kezirian et al. 2014), but a subsequent phase III trial (NCT01446601) was halted by investors when it became clear that efficacy thresholds would not be met (Strohl et al. 2016). Another device by Inspire Medical Systems used a similar approach demonstrating efficacy in the STAR trial (Strollo Jr. et al. 2014) (see below, Sect. 1.2) and in 2014 became the first FDA approved HGNS system in the United States. A third device called Aura6000 and made by ImThera Medical (now owned by LivaNova) stimulates both protrudors and retractors throughout the respiratory cycle. In a small uncontrolled study of 13 patients mean AHI improved from 45 to 21 events/h over 1 year (53% reduction) (Mwenge et al. 2013). Results were similar in another small uncontrolled trial (Friedman et al. 2016) and it has been approved for use in Europe, but for the United States an FDA evaluation is pending further phase III data (NCT02263859) (Strohl et al. 2016).

1.2 The Present: Hypoglossal Nerve Stimulation

In April 2014, the FDA approved the Inspire device (Inspire Systems., Maple Grove, MN) as the first HGNS device for the treatment of select patients with OSA (PMA P130008 2014). Based on emerging evidence from observational studies the eligibility criteria were subsequently liberalized (PMA P130008/S090 2023), with current criteria being summarized in Table 1.1. The implanted device consists of three main components: First, a pressure sensor in the fourth or fifth intercostal space to detect respiratory effort. Second, a pulse generator located in a midclavicular subcutaneous pocket which creates an electrical impulse synchronous with inspiration. Third, an electrode with three contacts surrounding unilaterally the medial branches of the hypoglossal nerve thus stimulating the genioglossus during inspiration, resulting in a tongue protrusion and thus widening of the retroglossal space. Of note, in many patients—especially responders to HGNS—the tongue motion is coupled with an anterior movement of the soft palate thus also widening the retropalatal space which is a common site of collapse in OSA (Safiruddin et al. 2015).

Table 1.1 Eligibility criteria for Inspire therapy based on the approval by the US Food and Drug Administration. Note, the body mass index (BMI) is not part of the eligibility criteria set forth by the FDA, but the pivotal STAR trial excluded patients with a BMI >32 kg/m^2 and most insurances will deny therapy for patients with a BMI >40 kg/m^2

Indications
Age > 21 years[a]
AHI 15-100 events/h*
PAP failure[b] or intolerance[c]
Contraindications
Central + mixed apnea >25% of total AHI
Complete concentric collapse (CCC) during drug-induced sleep endoscopy (DISE)
Anatomical issues that compromise upper airway stimulation
Conditions or procedures that compromised neurological control of the upper airway
Inability to operate the sleep remote (to activate the device at bedtime)
Pregnancy (now or planned)
Requiring magnetic resonance imaging (MRI) other than what is specified in the MR Conditional labeling[d]
Another implanted device that may interact with the Inspire device

* For patients between age 13–18y who have Down's syndrome the approved range is 15–50 events/h

[a] In April 2020, the FDA also approved Inspire for patients between 18 and 21 years who meet all of the criteria listed and are not able to undergo, or not effectively treated by, an adenotonsillectomy. In June 2023, the FDA also approved Inspire for pediatric patients 13–18 years of age who have Down's syndrome, OSA with an AHI of 15–50 events/h, and who are i) contraindicated for or not effectively treated with adenotonsillectomy, ii) have been confirmed to fail, or cannot tolerate PAP therapy despite attempts to improve adherence, and iii) have followed standard of care in considering all other alternative/adjunct therapies

[b] AHI >20 events/h despite PAP usage

[c] Inability to use PAP for >4 h/night on >5 nights/week or unwillingness to use PAP

[d] With the newest generation device MRIs excluding the torso are generally possible (for details see the "MRI Guidelines for Inspire Therapy" at https://manuals.inspiresleep.com/)

1.2.1 Efficacy

Regulatory approval was primarily based on the pivotal STAR trial, an industry-sponsored prospective multicenter study in which 126 patients were followed for 1 year after HGN stimulator implantation (Strollo Jr. et al. 2014). Key eligibility criteria included an apnea hypopnea index (AHI, with hypopneas defined as a 30% airflow reduction and a 4% oxygen desaturation) between 20 and 50 events per hour (including <25% central/mixed events), body mass index (BMI) <32 kg/m^2, inability to accept or adhere to CPAP therapy and lack of complete concentric collapse during a drug induced sleep endoscopy. Enrolled subjects were mostly middle-aged (mean 55 years) men (83%) that were overweight (28.4 kg/m^2). At 1 year of follow up, the median AHI (primary outcome) improved by 68% (29.3/h to 9.0/h, $P < 0.001$) and similar improvements were noted for secondary outcomes including subjective measures of daytime sleepiness, quality of life and partner-reported snoring. Of note, 83 subjects (65%) were classified as responders based on prespecified criteria (AHI reduction by 50% or more to less than 20 events per hour). Forty-six consecutive responders were randomized to therapy withdrawal versus maintenance: after 1 week mean AHI returned close to baseline levels in the withdrawal group but remained essentially unchanged in the maintenance group (25.8/h vs 8.9/h; $P < 0.001$) supporting a causal effect. Treatment effects were maintained in the subset of 97 patients who returned for follow up at 5 years post-implantation (Woodson et al. 2018). Furthermore, data from 382 patients who were included in the multicenter, prospective ADHERE (Adherence and Outcome of Upper Airway Stimulation for OSA International) registry showed an improvement in OSA severity and subjective daytime sleepiness of similar magnitude as in the STAR trial (AHI reduction by ~70%; Epworth Score reduction by 5 points; see Table 1.2) (Thaler and Schwab 2020).

1.2.2 Side Effects

Data about side effects from the STAR trial and the ADHERE registry are summarized in Table 1.3 and suggest that fewer than 6% of patients experience serious adverse events requiring revision or removal during the 1–5 years of follow-up periods currently available (Woodson et al. 2018; Thaler and Schwab 2020). Reported events were related to lead dislodgements, device failures, discomfort or attempts to improve stimulation by changing the lead placement. In addition, the FDA MAUDE (Manufacturer and User Facility Device Experience) database lists a few cases of the device causing a pneumothorax, pleural effusion or sensor lead migration into the pleural space (Bellamkonda et al. 2021). The relative frequency of these events is unclear but probably low. Non-serious events during the post-operative period included discomfort related to the incisions or device, as well as some tongue weakness, virtually all of which appear to resolve over time. Similarly,

Table 1.2 Summary of key outcome data for the Inspire hypoglossal nerve stimulator. Data are summarized as median [interquartile range], mean (standard deviation), or percent. For all outcomes listed in the table, changes from baseline were reported as statistically significant ($P < 0.05$) in the original reports (Woodson et al. 2018; Thaler and Schwab 2020)

	STAR trial						ADHERE registry			
	Baseline ($n = 126$)		1-year follow up ($n = 124$)		5-year follow up ($n = 97^a$)		Baseline ($n = 382$)		1-year follow up ($n = 382$)	
AHI, h^{-1}	29.3	[24–39]	9.0	[4–23]	6.2	[NA]	32.8	[24–45]	9.5	[4–19]
– Responder[b]			66%		75%				69%	
ESS	11	[8–15]	6	[4–10]	6	[NA]	11	[7–16]	6	[3–10]
– Normal (<10)	33%				78%		37%		74%	
FOSQ	14.6	[12–17]	18.2	[16–20]	18.7	[NA]				
– Normal (>17.9)	15%				67%					
Snoring:										
– Very intense	54%				2%					
– Loud	29%				8%					
– No/soft	17%				90%					
Adherence:										
– %Nights (self-report)			86%		80%					
– h/night (objective)									5.7	[4.1–7.1]
% Subjects Reporting:										
– Overall satisfaction									93%	
– Preference HGNS>PAP									95%	
– Would choose HGNS again									94%	
– Would recommend HGNS									96%	
Arousal index	27.8	(11.7)			7.8	(9.7)				

Abbreviations: *AHI* Apnea hypopnea index, *ESS* Epworth sleepiness score, *FOSQ* Functional outcomes of sleep questionnaire

[a] From the original 126, 21 were lost to follow up, 5 died of unrelated causes, 3 had device explanted (2 due to non-response, 1 due to septic arthritis); Of the 97 subjects who followed up at 5 years, only 71 underwent a polysomnography to assess AHI (results were similar in various sensitivity analyses)

[b] Defined as AHI reduction by ≥50% to <20/h

Table 1.3 Summary of key side effects data for the Inspire hypoglossal nerve stimulator. Events are not mutually exclusive, and one subject may have reported more than one event

	STAR trial		ADHERE registry			
	0–12 months (subjects = 124)	4–5 years[a] (subjects = 97)	0–6 months (subjects = 640)		6–12 months (subjects = 382)	
	Events	Events	Events	(% subjects)	Events	(% subjects)
Device-related serious events						
– Total	2[a]	7[a]	1[b]		2[b]	
Non-serious events						
Procedure related						
– Discomfort related to incision	47	1	14	(4%)	8	(2%)
– Discomfort independent of incision	41	0				
– Tongue weakness	34	0	3	(<1%)	0	(0%)
Device related						
– Discomfort due to electrical stimulation	81	5	41	(12%)	28	(8%)
– Tongue abrasion	28	2	12	(3%)	14	(4%)
– Mechanical pain due to device presence	7	1	10	(3%)	5	(1%)
– Insomnia/ arousals			10	(3%)	17	(5%)

[a] Serious events are reported for the period between 2 and 5 years; during the entire 5-year period there were a total of 9 events in 8 patients: repositioning to resolve discomfort ($n = 2$), repositioning to improve tongue movement ($n = 1$), replacement due to device failures ($n = 4$), replacement due to accidentally cut stimulation lead ($n = 1$)
[b] Repositioning to fix electrode dislodgement ($n = 3$)

after activation some patients may experience discomfort related to the electrical stimulation or tongue abrasion, but this typically resolves with device adjustments over time. Lastly, about 3–5% of patients experience some insomnia during the first year, but it is possible that this may in part reflect a co-morbidity rather than an adverse event.

1.2.3 *Predictors of Response*

An early study suggested that patients with a complete concentric collapse (CCC) during drug-induced sleep endoscopy (DISE) are unlikely to benefit from HGNS (Vanderveken et al. 2013). Thus, CCC on DISE was an exclusion criterion for the pivotal STAR trial and is a contraindication for the Inspire HGNS device as per the FDA label (Strollo Jr. et al. 2014; PMA P130008 2014). Since the initial approval of this device, there have been several retrospective analyses attempting to identify patient characteristics that are associated with a favorable response (defined as a drop in AHI by 50% to <20 events/h) (Woodson et al. 2018; Thaler and Schwab 2020; Heiser et al. 2019). As summarized in Table 1.4, published results have been somewhat mixed, but overall data suggest that patients with a higher age, female sex and lower BMI may be most likely to benefit from HGNS therapy. A lower ODI at baseline, higher device usage may further predict a higher likelihood of response. Lastly, based on the raw data shown in one report (Woodson et al. 2018), in the STAR trial subjects with versus without prior UPPP had eight-fold higher odds of response (17/53 Responders vs 1/18 Non-Responders) which may suggest a role for combination therapy in some patients, but other studies did not find any clear evidence that prior upper airway surgery affects HGNS outcomes (Kezirian et al. 2019; Huntley et al. 2018; Mahmoud and Thaler 2018).

1.2.4 *Surgical Techniques and Challenges*

Proper patient selection for the currently available hypoglossal nerve stimulators cannot be emphasized enough. One of the final steps in selection is the drug-induced sleep endoscopy. This procedure is often performed in a procedural or operative suite. It begins with nasal decongestion and a room setup that is sleep conducive (e.g. low lighting and noise). Intravenous propofol is typically infused at a rate of 50 mcg/kg/min and increased by 25 mc/kg/min every couple of minutes until the patient no longer responds to verbal stimuli and begins to snore and obstruct. Flexible fiber laryngoscope is advanced through either nares and used for dynamic assessment of the nasopharynx, velopharynx, oropharynx, and supraglottis (Soares et al. 2013). Patients without circumferential collapse of the retropalatal airway, or less than 75% lateral wall collapse are acceptable candidates.

Implantation of the Inspire hypoglossal nerve stimulator is performed under general anesthesia using oral or nasal intubation. Nerve integrity monitor (NIM 3.0, medtronic) is used to monitor genioglossus (GG) and styloglossus (SG) muscles. A neck incision is made 5 cm midway between the hyoid and mandible, 1 cm off midline. Through this incision, the superior margin of the anterior belly of the digastric is followed down to the tendon. The tendon is retracted inferiorly with vessel loops. The submandibular gland is retracted posterosuperior and the mylohyoid anterior. The hypoglossal nerve is identified and the functional

Table 1.4 Clinical predictors of response to HGNS (drop in AHI by 50% to <20 events/h). Table shows variables reported as significant in univariable analyses. Multivariable analyses in the original reports were based on stepwise regression and only included significant variables

	STAR trial (5-year follow up, $N = 71$) (Woodson et al. 2018)				ADHERE registry (Initial report, $N = 227$) (Heiser et al. 2019)				ADHERE registry (Subsequent report, $N = 382$) (Thaler and Schwab 2020)			
	Univariable		Multivariable		Univariable		Multivariable		Univariable		Multivariable	
	OR	P	OR	P	OR	P	OR	P	OR	P	OR	P
Age, years	1.07	0.03*			1.04	0.01*	?[a]	<0.05*	1.01	0.19		
Female sex	1.16	0.83			2.6	0.08			1.94	0.046*	3.4	0.005*
BMI, kg/m^2	0.97	0.76			0.91	0.03*			0.92	0.003*	0.91	0.005*
ODI, h^{-1}	0.94	0.02*	?[b]	<0.05*	NA							
Prior UPPP	8.0[c]	0.052			NA							
Usage (h/week at 12 months)	NA				NA				1.02	0.04*		

* denotes $P < 0.05$

Abbreviations: *UPPP* Uvulopalatopharyngoplasty, *ODI* Oxygen desaturation index

[a] Per report, based on multivariable stepwise regression only a higher age was predictive of response

[b] Per report, based on multivariable stepwise regression only a lower ODI was predictive of response

[c] The article reports an OR of 0.13, but the raw data shown in Table 1.4 suggest this was an inversion error (17/36)/(1/17) = 8.0 and 1/8.0 = 0.13 (note: on 3/4/2021 lead author Dr Woodson confirmed via email communication that this was indeed an inversion error)

breakpoint along the superior, distal hypoglossal nerve is identified and confirmed with the NIM and observed tongue movement. The goal is to achieve unhindered protrusion of the stiffened tongue by selective stimulation of the tongue protrudors; for example, genioglossus and geniohyoid muscles. A window is created around the inclusion nerve branches and the stimulation lead cuff electrode is placed. Next a subdermal pocket is created for the generator through a 5 cm incision inferior and parallel to the clavicle. Dissection is carried down to the pectoralis major fascia. Next is placement of the sensor lead. This is done through another 5 cm incision within the first intercostal space inferior to the pectoralis major and 5 cm lateral to the nipple line. Dissection is performed through the anterior serratus and external intercostal muscles. The internal intercostals are then encountered, noting a change in muscle fiber direction. A 4 cm pocket is created with a malleable retractor. The sensor lead is then placed within this pocket facing the internal intercostals. Tunneling for the sense lead is then performed superficial to the pectoralis major fascia from the generator pocket to the sense lead incision. The lead is then passed into the generator pocket ensuring adequate excess lead at both ends of the tunnel. Stimulation lead is then tunneled in the subplatysmal plane from the neck to the generator pocket superficial to the clavicle. The lead is then passed into the generator pocket. The leads are then secured into the generator which is then secured to prevent wandering. The system is then validated intraoperatively confirming tongue motion and accurate sensing. Wounds are closed in a multilayer fashion. Post-operative X-rays are performed to document implant position and integrity. The generator is activated at 4 weeks and a fine-tuning sleep study is performed at 12 weeks (Woodson et al. 2018).

The Genio system® (Nyxoah S.A., Belgium) is a bilateral hypoglossal nerve stimulator that is currently in a pivotal, Investigational Device Exemption (IDE) study (DREAM) aimed to support market authorization in the United States. It recently completed the BLAST OSA study which demonstrated its safety and efficacy (Eastwood et al. 2020). This system is comprised of an implantable stimulator (IS) that is activated by wireless energy transmission from an external wearable activation chip. The implantation is performed under general anesthesia via oral or nasal intubation. The NIM system is also used. A single 6 cm transverse midline incision is made midway between the mentum and the hyoid bone. Mylohyoid muscle is divided and the vertical fibers of the geniohyoid (GH) are then separated and retracted. The lateral border of the GG is identified and the hypoglossal nerve is then found superomedially intersecting the GG at a 60° angle. Nerve is confirmed with the NIM system. A deep pocket is created superior to the nerve to accommodate the legs of the IS. NIM stimulator is used to probe the nerve once more at low amplitude, 0.1 mA, to ensure no retrograde stimulation that might activate the SG electrode. This process is repeated on the opposite side. Once both stimulator legs are in place, an external stimulator is used to activate the IS. This approach should generate strong GG potentials with a low reading in the SG electrode. The legs are then secured to the GG muscle with suture. Flexible laryngoscope is placed transnasally to verify favorable tongue, epiglottis and palate movement. This is a critical step in confirming optimal placement of the IS on the hypoglossal nerve branches. Stimulation is uniquely cyclical with a stimulation window that has a preset percentage ON time. Distance between the IS and skin should be less than 3.5 cm. As such, some submental fat can be removed to accomplish this. The device is then activated at 6 weeks followed by a titration polysomnogram at 8 weeks (Lewis et al. 2019).

Any surgical procedure carries the possibility of complications. The Inspire system is a more involved procedure involving lead tunneling, multiple incision sites and instrumentation of the chest wall. Serious device-related events have been reported in up to 6% of patients (Woodson et al. 2018) including device migration, infection and less commonly pneumothorax, pleural effusion, and lead migration into the pleural space (Bellamkonda et al. 2021; Bestourous et al. 2020). The Nyxoah system provides some potential advantages including single incision, no tunneling, no wires or battery, and bilateral selective hypoglossal nerve stimulation. However, more remains to be learned as we gain more and longer-term data on this device.

1.3 Future Directions

While much progress has been made in the last few decades to develop HGNS from bench to bedside, there are several gaps in our knowledge which need to be filled in order to further advance it as a therapy for OSA. In the following we will highlight a few key issues, most of which are currently under active investigation.

1.3.1 Optimal Stimulation Approach and Comparisons with Other Modalities

From a technical standpoint, there are many different ways how HGNS can be achieved including variations in the type and anatomical placement of the stimulation lead (e.g. cuff vs flat electrode, 3 vs 6 vs more contact points and medial vs distal placement), and whether stimulation is synchronized with respiration versus not. Of note, while all the devices discussed above (Sect. 1.1.3) stimulate the hypoglossal nerve unilaterally (typically on the right side), there may be some theoretical benefits from bilateral stimulation. This was recently explored in a small uncontrolled pilot study of the Genio system® which stimulated both (bilateral) hypoglossal nerves at a rate close to patients' respiratory rate (but without actual respiratory synchronization) achieving similar results as reported for other devices (Eastwood et al. 2020). Eventually, head-to-head comparisons of different strategies in clinical trials will be required to establish the optimal approach. Similarly, comparative effectiveness studies are needed to assess better the benefit of HGNS versus other approaches such as upper airway surgery, taking into account that—unlike surgery—HGNS requires nightly adherence with therapy in order to be effective (Schmickl et al. 2018b).

1.3.2 Patient Selection

As outlined above (Sect. 1.2.3), there have been some attempts to identify likely responders based on clinical characteristics yielding limited success. However, OSA being increasingly recognized as a heterogeneous condition, and taking into account the mechanism (trait) causing sleep apnea in a given patient holds the promise of a true precision medicine approach for OSA. HGNS aims to improve the upper airway muscle function, thus a patient whose OSA is primarily caused by an unstable ventilatory control or a low arousal threshold (i.e. waking up too easily) is not expected to benefit much from HGNS therapy. Indeed, a secondary analysis of the STAR trial in which traits were estimated from baseline polysomnography supported this notion (Op de Beeck et al. 2021), but prospective validation is needed before a physiology-based approach for patient selection can be considered for clinical practice.

1.3.3 Rescue Strategies and Combination Therapies

While HGNS improves OSA substantially in many patients, strategies are needed to help those who experience only partial response despite adjustments in HGNS device settings. Based on a small retrospective study, patients who had residual obstruction at the oro-/velopharynx with HGNS may benefit from additional upper airway surgery (Steffen et al. 2019), but other data on combination of upper airway surgery with HGNS failed to show clear benefits and prospective randomized trials are lacking. Another interesting approach deserving further study is the addition of an oral appliance (using a model with sufficient space for tongue protrusion) to HGNS to improve response (Lee et al. 2015). This hybrid approach may allow treatment with less mandibular advancement and lower HGNS stimulation amplitude than with either therapy alone, thus potentially also alleviating issues related to patient discomfort. Further, we believe that a deeper understanding of individuals' underlying pathophysiological mechanisms may not just improve upfront patient selection discussed above (Sect. 1.3.2), but may also provide insights on potential pharmacological interventions to rescue non-responders (e.g. acetazolamide (Schmickl et al. 2020) for patients who experience partial improvement with HGNS and have high loop gain).

1.3.4 Long-Term Data on Hard Outcomes

HGNS is still a relatively new technology, and while efficacy appears to be maintained over time, data beyond 5 years of follow up are currently lacking. Furthermore, much of the data demonstrating efficacy of HGNS for OSA have been focused on the AHI as the outcome. However, while commonly used to measure OSA severity, it is important to remember that the AHI is just a proxy outcome and correlates only moderately with the various clinically important sequelae of OSA. It is somewhat reassuring that subjective daytime sleepiness and quality of life have been assessed as secondary outcomes and showed improvements in major studies. However, due to the lack of participant blinding, effects for subjective outcomes are likely biased away from the null (i.e. exaggerated). Further, data on objective daytime sleepiness or hard cardiovascular outcomes are generally lacking, although a trial focusing on the latter is currently ongoing (NCT03359096) (Dedhia et al. 2018).

References

Awad KM, Malhotra A, Barnet JH, Quan SF, Peppard PE (2012) Exercise is associated with a reduced incidence of sleep-disordered breathing. Am J Med 125(5):485–490. (1555-7162 (Electronic))

Bellamkonda N, Shiba T, Mendelsohn AH (2021) Adverse events in hypoglossal nerve stimulator implantation: 5-year analysis of the FDA MAUDE database. Otolaryngol Head Neck Surg 164(2):443–447. (1097-6817 (Electronic))

Bestourous DE, Pasick LJ, Benito DA, Zapanta PE (2020) Adverse events associated with the Inspire implantable hypoglossal nerve stimulator: a MAUDE database review. Am J Otolaryngol 41(6):102616. https://doi.org/10.1016/j.amjoto.2020.102616

Camacho M, Certal V, Abdullatif J et al (2015) Myofunctional therapy to treat obstructive sleep Apnea: a systematic review and meta-analysis. Sleep 38(5):669–675. https://doi.org/10.5665/sleep.4652

Dedhia RA-O, Quyyumi AA, Park JA-O, Shah AJ, Strollo PJ, Bliwise DL (2018) Cardiovascular endpoints for obstructive sleep apnea with twelfth cranial nerve stimulation (CARDIOSA-12): rationale and methods. Laryngoscope 128(11):2635–2643. (1531-4995 (Electronic))

Eastwood PR, Barnes M, MacKay SG et al (2020) Bilateral hypoglossal nerve stimulation for treatment of adult obstructive sleep apnoea. Eur Respir J 55(1):1901320. https://doi.org/10.1183/13993003.01320-2019

Eckert DJ, Younes MK (2014) Arousal from sleep: implications for obstructive sleep apnea pathogenesis and treatment. J Appl Physiol (Bethesda, Md: 1985) 116(3):302–313. https://doi.org/10.1152/japplphysiol.00649.2013

Eckert DJ, Owens RL, Kehlmann GB et al (2011) Eszopiclone increases the respiratory arousal threshold and lowers the apnoea/hypopnoea index in obstructive sleep apnoea patients with a low arousal threshold. Respirology (Carlton, VIC) 120(12):505–514. https://doi.org/10.1111/resp.13044.10.1042/cs20100588

Eckert DJ, White DP, Jordan AS, Malhotra A, Wellman A (2013) Defining phenotypic causes of obstructive sleep apnea. Identification of novel therapeutic targets. Am J Respir Crit Care Med 188(8):996–1004. https://doi.org/10.1164/rccm.201303-0448OC

Friedman M, Jacobowitz O, Hwang MS et al (2016) Targeted hypoglossal nerve stimulation for the treatment of obstructive sleep apnea: six-month results. Laryngoscope 126(11):2618–2623. https://doi.org/10.1002/lary.25909

Fuller D, Mateika JH, Fregosi RF (1998) Co-activation of tongue protrudor and retractor muscles during chemoreceptor stimulation in the rat. J Physiol 507(Pt 1):265–276. https://doi.org/10.1111/j.1469-7793.1998.265bu.x

Gaisl T, Haile SR, Thiel S, Osswald M, Kohler M (2019) Efficacy of pharmacotherapy for OSA in adults: a systematic review and network meta-analysis. Sleep Med Rev 46:74–86. https://doi.org/10.1016/j.smrv.2019.04.009

Guimaraes KC, Drager LF, Genta PR, Marcondes BF, Lorenzi-Filho G (2009) Effects of oropharyngeal exercises on patients with moderate obstructive sleep apnea syndrome. Am J Respir Crit Care Med 179(10):962–966. https://doi.org/10.1164/rccm.200806-981OC

Heiser C, Steffen A, Boon M et al (2019) Post-approval upper airway stimulation predictors of treatment effectiveness in the ADHERE registry. Eur Respir J 53(1):1801405. https://doi.org/10.1183/13993003.01405-2018

Huntley C, Vasconcellos A, Doghramji K, Hofauer B, Heiser C, Boon M (2018) Upper airway stimulation in patients who have undergone unsuccessful prior palate surgery: an initial evaluation. Otolaryngol Head Neck Surg 159(5):938–940. https://doi.org/10.1177/0194599818792191

Jordan AS, McSharry DG, Malhotra A (2014) Adult obstructive sleep apnoea. Lancet (London, England) 383(9918):736–747. https://doi.org/10.1016/s0140-6736(13)60734-5

Kezirian EJ, Goding GS Jr, Malhotra A et al (2014) Hypoglossal nerve stimulation improves obstructive sleep apnea: 12-month outcomes. J Sleep Res 23(1):77–83. https://doi.org/10.1111/jsr.12079

Kezirian EJ, Heiser C, Steffen A et al (2019) Previous surgery and hypoglossal nerve stimulation for obstructive sleep apnea. Otolaryngol Head Neck Surg 161(5):897–903. https://doi.org/10.1177/0194599819856339

Lee JJ, Sahu N, Rogers R, Soose RJ (2015) Severe obstructive sleep apnea treated with combination hypoglossal nerve stimulation and oral appliance therapy. J Dent Sleep Med 2(4):185–186

Lewis R, Pételle B, Campbell MC et al (2019) Implantation of the nyxoah bilateral hypoglossal nerve stimulator for obstructive sleep apnea. Laryngoscope Investig Otolaryngol 4(6):703–707. https://doi.org/10.1002/lio2.312

Mahmoud AF, Thaler ER (2018) Upper airway stimulation therapy and prior airway surgery for obstructive sleep apnea. Laryngoscope 128(6):1486–1489. https://doi.org/10.1002/lary.26956

Mason M, Welsh EJ, Smith I (2013) Drug therapy for obstructive sleep apnoea in adults. Cochrane Database Syst Rev 31(5):Cd003002. https://doi.org/10.1002/14651858.CD003002.pub3

Mezzanotte WS, Tangel DJ, White DP (1992) Waking genioglossal electromyogram in sleep apnea patients versus normal controls (a neuromuscular compensatory mechanism). J Clin Invest 89(5):1571–1579. https://doi.org/10.1172/jci115751

Mezzanotte WS, Tangel DJ, White DP (1996) Influence of sleep onset on upper-airway muscle activity in apnea patients versus normal controls. Am J Respir Crit Care Med 153(6 Pt 1):1880–1887. https://doi.org/10.1164/ajrccm.153.6.8665050

Miki H, Hida W, Inoue H, Takishima T (1988) A new treatment for obstructive sleep apnea syndrome by electrical stimulation of submental region. Tohoku J Exp Med 154(1):91–92. https://doi.org/10.1620/tjem.154.91

Miki H, Hida W, Chonan T, Kikuchi Y, Takishima T (1989) Effects of submental electrical stimulation during sleep on upper airway patency in patients with obstructive sleep apnea. Am Rev Respir Dis 140(5):1285–1289. https://doi.org/10.1164/ajrccm/140.5.1285

Mwenge GB, Rombaux P, Dury M, Lengelé B, Rodenstein D (2013) Targeted hypoglossal neurostimulation for obstructive sleep apnoea: a 1-year pilot study. Eur Respir J 41(2):360–367. https://doi.org/10.1183/09031936.00042412

Op de Beeck S, Wellman A, Dieltjens M et al (2021) Endotypic mechanisms of successful hypoglossal nerve stimulation for obstructive sleep Apnea. Am J Respir Crit Care Med 203(6):746–755. https://doi.org/10.1164/rccm.202006-2176OC

PMA P130008: FDA Summary of Safety and Effectiveness Data for the Inspire II Upper Airway Stimulator. 4/30/2014. Available for download at: http://www.accessdata.fda.gov/cdrh_docs/pdf13/P130008B.pdf. Last accessed 3/4/2020

Puhan MA, Suarez A, Lo Cascio C, Zahn A, Heitz M, Braendli O (2006) Didgeridoo playing as alternative treatment for obstructive sleep apnoea syndrome: randomised controlled trial. BMJ (Clinical research ed). 332(7536):266–270. https://doi.org/10.1136/bmj.38705.470590.55

Saboisky JP, Butler JE, Gandevia SC, Eckert DJ (2012) Functional role of neural injury in obstructive sleep apnea. Front Neurol 3:95. https://doi.org/10.3389/fneur.2012.00095

Safiruddin F, Vanderveken OM, de Vries N et al (2015) Effect of upper-airway stimulation for obstructive sleep apnoea on airway dimensions. Eur Respir J 45(1):129–138. https://doi.org/10.1183/09031936.00059414

Schmickl CN, Owens RL, Edwards BA, Malhotra A (2018a) OSA endotypes: what are they and what are their potential clinical implications? Curr Sleep Med Rep 4(3):231–242. https://doi.org/10.1007/s40675-018-0121-8. Available online at https://rdcu.be/2nNf

Schmickl CN, Chalise S, Borker PV, Strang A, Thomas RJ (2018b) High residual apnea burden may contribute to frequent outcome failure in randomized controlled trials of positive airway pressure—a proof-of-concept study. Sleep 41(Suppl 1):A213–A213. https://doi.org/10.1093/sleep/zsy061.571

Schmickl CN, Landry S, Orr JE, Chin K, Murase K, Verbraecken J, Javaheri S, Edwards B, Owens RL, Malhotra A (2020) Acetazolamide for obstructive and central sleep apnea: a comprehensive systematic review and meta-analysis. Chest 158:2632–2645

Schwartz AR, Thut DC, Russ B et al (1993) Effect of electrical stimulation of the hypoglossal nerve on airflow mechanics in the isolated upper airway. Am Rev Respir Dis 147(5):1144–1150. https://doi.org/10.1164/ajrccm/147.5.1144

Schwartz AR, Bennett ML, Smith PL et al (2001) Therapeutic electrical stimulation of the hypoglossal nerve in obstructive sleep apnea. Arch Otolaryngol Head Neck Surg 127(10): 1216–1223

Soares D, Folbe AJ, Yoo G, Badr MS, Rowley JA, Lin HS (2013) Drug-induced sleep endoscopy vs awake Müller's maneuver in the diagnosis of severe upper airway obstruction. Otolaryngol Head Neck Surg 148(1):151–156. https://doi.org/10.1177/0194599812460505

Steffen A, Abrams N, Suurna MV, Wollenberg B, Hasselbacher K (2019) Upper-airway stimulation before, after, or without uvulopalatopharyngoplasty: a two-year perspective. Laryngoscope 129(2):514–518. https://doi.org/10.1002/lary.27357

Strohl MDK, Baskin MDJ, Lance MDC et al (2016) Origins of and implementation concepts for upper airway stimulation therapy for obstructive sleep apnea. Respir Investig 54(4):241–249. https://doi.org/10.1016/j.resinv.2016.01.006

Strollo PJ Jr, Soose RJ, Maurer JT et al (2014) Upper-airway stimulation for obstructive sleep apnea. N Engl J Med 370(2):139–149. https://doi.org/10.1056/NEJMoa1308659

Taranto-Montemurro L, Messineo L, Sands SA et al (2019) The combination of atomoxetine and oxybutynin greatly reduces obstructive sleep apnea severity. A randomized, placebo-controlled, double-blind crossover trial. Am J Respir Crit Care Med 199(10):1267–1276. https://doi.org/10.1164/rccm.201808-1493OC

Thaler EA-O, Schwab R, Maurer J et al (2020) Results of the ADHERE upper airway stimulation registry and predictors of therapy efficacy. Laryngoscope 130(5):1333–1338. (1531-4995 (Electronic))

Vanderveken OM, Maurer JT, Hohenhorst W et al (2013) Evaluation of drug-induced sleep endoscopy as a patient selection tool for implanted upper airway stimulation for obstructive sleep apnea. J Clin Sleep Med 9(5):433–438. https://doi.org/10.5664/jcsm.2658

Wellman A, Malhotra A, Jordan AS, Stevenson KE, Gautam S, White DP (2008) Effect of oxygen in obstructive sleep apnea: role of loop gain. Respir Physiol Neurobiol 162(2):144–151. https://doi.org/10.1016/j.resp.2008.05.019

Woodson BT, Strohl KP, Soose RJ et al (2018) Upper airway stimulation for obstructive sleep apnea: 5-year outcomes. Otolaryngol Head Neck Surg 159(1):194–202. https://doi.org/10.1177/0194599818762383

Chapter 2
Positive Airway Pressure Therapies: Advances of the Decade

Mashni Alsaeed, Yousef MohamedRabaa Hawsawi, and Ahmed S. BaHammam

Abstract Positive airway pressure (PAP) devices are becoming the first-choice therapy for obstructive sleep apnea (OSA) and other sleep-disordered breathing. Neglecting OSA treatments may result in fatigue, sleepiness, headaches, cerebrovascular and cardiovascular diseases, decreased life quality, and increased risk of motor vehicle accidents. OSA therapy plan includes reducing weight, dental options such as oral appliance therapy, surgical procedures, and continuous positive airway pressure (CPAP).

PAP therapy has a significant role in eliminating more complex respiratory obstructions during sleep, such as Cheyne–Stokes respiration, central sleep apnea associated with heart failure, and chronic hypoventilation syndromes that require the use of advanced PAP therapy, such as adaptive servo-ventilation (ASV) and volume targeted limited pressure ventilation. Identifying patients with complex breathing disorders during sleep who would benefit from advanced treatment remains an active research area. Randomized controlled trials indicated that the PAP device could effectively reduce the apnea and hypopnea index and improve subjective and objective sleepiness. PAP therapy aims to reach the optimal treatment method to eliminate recurrent respiratory obstructions that negatively affect patients' health. The methods of titrating the device differ according to the severity of respiratory obstructions or their association with poor pulmonary ventilation. CPAP functions by a direct pump pushing air into the nasal and oral passageway via a specific mask

M. Alsaeed
Respiratory Care Department and Sleep Disorders Center, King Faisal Specialists Hopsital and Research Center, Jeddah, Saudi Arabia

Y. M. Hawsawi
Research Center Department, King Faisal Specialist Hospital and Research Center, Jeddah, Saudi Arabia

A. S. BaHammam (✉)
The University Sleep Disorders Center, Department of Medicine, College of Medicine, King Saud University, Riyadh, Saudi Arabia

King Saud University Medical City, King Saud University, Riyadh, Saudi Arabia
e-mail: ashammam@ksu.edu.sa

A. S. BaHammam, M. Hunasikatti (eds.), *Sleep Apnea Frontiers*, Progress in Sleep Research, https://doi.org/10.1007/978-981-99-7901-1_2

during sleep. The airflow creates a pressure that works in splinting the airway and restricting the soft tissue from collapsing. To achieve the goal of using the CPAP device, the patient must accept the device's daily use, and the patient's noncompliance with the device is one of the obstacles that must be avoided by introducing techniques that improve the acceptance of the device.

However, alternative methods for determining a PAP device's optimal pressure—such as autotitrating PAP (APAP)—have seen significant development over the past decade. This chapter will discuss the latest developments in the selection of PAP therapy modes over the past decade, emphasizing the guidelines for their implementation to achieve significant improvement in respiratory ventilation and clinically positive outcomes. Furthermore, it will describe the state-of-the-art technology underpinning contemporary PAP devices.

Keywords Sleep disorders · Noninvasive ventilation · Sleep apnea · Obesity hypoventilation syndrome · Adaptive servo-ventilation · Auto-CPAP

2.1 Introduction

With the continuous positive airway pressure (CPAP) machine's discovery by Colin Sullivan in the early 1980s (Sullivan et al. 1981), sleep medicine has entered a new era, especially in treating obstructive sleep apnea (OSA). As the device's effectiveness has led to an acceleration of interest in this new type of home treatment, companies have diversified in making devices and developing their capabilities by introducing the latest technology methods. (BaHammam et al. 2017).

The described air pressure must be adequate to secure the upper airway opening and improve or remove obstructions in the upper airway to benefit from the PAP device. To do this, we need to send patients to the sleep disorders laboratory. For the device to be calibrated according to the presence of obstruction, the sleep study usually ends with reaching the optimal pressure that normalizes breathing during sleep (Ahmed and Parthasarathy 2010).

Several PAP device manufacturers work with specific algorithms and equations to respond to the respiratory system's airflow changes. Nevertheless, patients usually respond differently to the respiratory system's specific breathing patterns, which may affect tolerance to the device and hence, adherence. Some systems quickly adapt to the prescribed treatment. Therefore, companies are looking to improve the airflow mechanism that is fitted to the patient's breathing.

Since the discovery of PAP devices, various device modes have been developed to deal with respiratory obstructions or hypoventilation, which have contributed to improving tolerance and sleep quality.

This chapter discusses new developments in PAP therapy for sleep-related breathing disorders in the past decade.

2.1.1 Blowers and Control of Interface Pressure

The invention led by Sullivan and colleagues expanded the use of PAP devices for OSA treatment. This success has catalyzed the creation of a growing market, with an increasing number of manufacturers engaged in producing PAP devices and their related accessories. This activity drove the evolution of significant advances in PAP technology.

PAP algorithms are designed to perform multiple functions: they utilize a flow-generating element (blower) to deliver air, regulate the level of positive pressure administered to the patient, measure the patient's airflow, detect respiratory events during sleep, and estimate the volume of air exchanged per minute as well as any leakage. The collected information is used to set the patient's needs for proper ventilation and normal breathing pattern.

To prevent upper airway collapse during sleep, positive airway pressure devices apply positive pressure to the airways' wall during inhalation and exhalation. Indeed, steady CPAP levels affect airway circuit resistance and gas leakage, especially tidal volume and breathing frequency.

The main challenge in using a PAP device is to maintain synchronization with the patient's spontaneous breathing and, hence adherence. This is done using a sensitive pressure sensor, motor, and blower with good performance, fast response time, and lower noise levels.

2.1.2 Flow Generator and Transducers

A PAP device has a microprocessor, pressure sensor, integrated DC motor with blower, motor controller, flexible tube, and different interface masks (Chen et al. 2012). It also includes transformers that control engine speed, airflow, and pressure at a fixed point in the flow generator. There is a sensor inside the PAP, not the mask, so it measures the mask's expected pressure according to the flow measurements at different points in the system. The idea behind using sensors is to help control airflow, pressure, humidity, and temperature and provide a smooth output. During exhalation, the patient's breathing is monitored, and outputs that reduce the device's internal blower fan flow are sent. Reducing resistance prevents the patient from feeling a fight against the PAP device while breathing. Thus, the airflow sensor's importance is to determine the degree of inhaling and exhaling during a patient's breathing cycle.

2.1.3 Advanced Blower Technology

An electronically controlled blower draws in ambient air through a filter and delivers it to the device outlet port. It is essential to choose a sensitive pressure sensor, motor, and blower to do a good job, keeping this by making the response fast and less noisy; the blower speed in the CPAP machine adjusts dynamically in response to the patient's breathing requirements.

The air flows through the air filter located at the back (side) of the device by the activated blower and then is compressed. Lastly, the therapeutic airway pressure is delivered into the lungs through a tube (an interface mask). The pressure signal is detected near the mask by the pressure sensor and then entered into the microprocessor.

2.1.4 Ramping

The ramp is a feature in the PAP devices that allows the device to deliver pressure at a low point set during a specific period, as the device aims to achieve the set pressure. For example, if the described and optimal pressure is 12 cmH_2O, the ramp pressure may start at 5 cmH_2O and "escalate" or slowly increase over a preset time. The start-up pressure and slope time can be controlled. According to the patient's comfort and clinical condition, the ramp may be enabled or disabled on the device.

Ramp generator: This operating mode allows the PAP device unit to operate independently of patient effort and provides an increase in PAP pressure. The technologist initiates this function by starting a switch to bring the function generator to the circuit. Alternatively, the ramp generator provision can be adjusted to be directed automatically to the summing amplifier if the patient fails to breathe automatically for a specified time.

Usually, patients face difficulties when the PAP devices deliver high pressure; therefore, the ramping system helps the patient to feel more comfortable during sleep. A new feature by some manufacturers is an auto-ramp, which allows the system to start at low air pressure while the patient is awake. Once it detects that the patient has fallen asleep, the ramping will automatically begin to allow pressure to build up to the prescribed level.

Less noise: Sleep using the PAP is difficult because of the loud noise emitted from the device, so auto-ramp's low pressure will provide comfort by maintaining a soft sound from the device while the patients and bed partner are still awake.

2.1.5 Expiratory Pressure Relief Mechanisms

With the development of PAP devices, manufacturers and developers continue to add more technological developments that have led to the improvement of technologies that contribute to pressure relief (such as C-flex® and EPR®) and automatic adjustment machines (APAP). This technique works by relieving pressure during early exhalation. At the same time, APAP may reduce the average optimal pressure required to treat sleep-disordered breathing (SDB), thus reducing the side effects and discomfort associated with the high pressure of the device. Nevertheless, despite these technological advances and theoretical advantages of these devices for controlling SDB compared to constant pressure CPAP, studies have not conclusively shown an improvement in adherence (Ip et al. 2012).

As the technology advances for PAP devices, manufacturers continue to add further improvements that have led to the development of technologies that contribute to pressure relief (such as C-flex® and EPR®) and automatic adjustment machines such as APAP. This technique operates by decreasing pressure during early exhalation, while APAP may reduce the average optimal pressure required to treat SDB. Despite the technological advances and the effectiveness of these devices to control SDB related to continuous positive airway pressure, studies conclusively show no improvement in the patient's compliance.

The transformers monitor the motor speed, airflow, and pressure at a fixed point downstream of the flow generator. The sensor is located inside the device and not in the mask, so the device must calculate the mask's expected pressure based on the flow measurements at different points in the system. The pressure varies at two points with the flow that can be calculated, and the length and diameter of the tube and the characteristics of the mask affect the continuous value (BaHammam et al. 2017), so the technical specialist must specify some measurements and enter them into the device settings such as the type of mask and tube for the microprocessor to perform the correct calculations.

The microprocessor modifies the blower speed in response to pressure variations secondary to leaks or regular air pressure changes due to breathing. The flow signal is sent through low- and high-pass filters to separate the artifacts' respiratory flow signal. Low-pass filters exclude the flow of large rapid deflections (e.g., coughing or sneezing), and high-pass filters exclude cardiac fluctuations. The feedback limits determine whether the flow or pressure exceeds the expected range of flow change at a particular motor speed (e.g., tube fracture) and prevent the device from delivering too much or too little pressure (Chen et al. 2012). As the altitude increases, the fan speed needs to be increased to maintain the same pressure (Patil et al. 2019, Patz et al. 2010). Several devices are automatically adjusted to the height.

2.1.5.1 Leak Compensation

While using noninvasive ventilation; air leakage is one of the main factors that must be compensated for. Air leakage significantly impacts PAP devices' performance, affects the pressure to be delivered to the patient, and influences the cycle and operating thresholds and respiratory events identification.

Leakage Compensation Mechanism

The leakage compensation is done through the flow control and continuously searches for deviations from the expected flow of breathing, and therefore, there are differences in engine speed to compensate for air leakage. Because there are common variations in a patient's breathing cycle, the average expected leakage is usually calculated over several breaths. If the leakage is high, the automatic devices compensate by lowering the pressure, and this will seal the mask and reduce leakage.

Leakage can be managed from the flow rate at the end of exhalation (Stieglitz et al. 2020). Intentional leakage includes the exhalation ports on the mask, which vary according to the type of mask and pressure level, and unintentional leakage includes leakage from the mouth or around the mask (Al Otair and BaHammam 2020). Generally, the leakage should be less than 24 L/min in nasal masks and less than 36 L/min in full-face masks, but this may vary depending on the used interface.

The lips and tongue sometimes work as a one-way valve opening during exhalation, when pharyngeal pressure is at its highest. This procedure is called valve leakage or exhalation. Exhalation puffs may incorrectly indicate a flow restriction, which could cause the robotic devices to increase the pressure unnecessarily (Jaffuel et al. 2020).

2.1.5.2 Respiratory Cycle Determination

It is essential to define inhalation and exhalation cycles to provide a two-level PAP, relieve expiratory pressure, and resolve automatic algorithms' inhalation flow limitations. The critical mark is the beginning of the breathing from the negative flow to the positive flow—the flow signal switches from positive to negative flow at the start of the expiration.

Servo-ventilators and some auto devices use a continuous definition of the breathing cycle. The ResMed servo-ventilator microprocessor uses vague inference rules that look at the direction of flow rate and volume to determine the respiratory cycle stage. Using the Respironics ventilator servo device divides breaths into short periods (64 ms) to determine the average inspiration expected at other points of the cycle is determined by the length of exhale to predict the duration of breathing (Johnson and Johnson 2015).

2.1.5.3 Other PAP Features

Built-in Software

Several PAP devices demonstrate standard therapy statistics on the onboard display, and others have optional advanced data tracking available through an optional software program. Some of these programs are available on physical disks and require download cables from the machine to a computer, while others are downloadable programs and can be used with the data cable added to the device.

Recent home PAP devices are equipped with built-in software that provides data such as compliance, estimates of leaks, tidal volumes, minute ventilation, respiratory rates, and apnea-hypopnea indexes.

2.1.5.4 Data Collection and Display

Manufacturers have implemented new technology for PAP therapy data downloads. In addition to added ease, new transmission systems help deliver more reliable interaction between the patient, the treatment provider, the insurance company, and the home medical equipment (HME) company. In different cases, they also encourage patients with OSA to play an active role in monitoring their progress.

New software and apps that sync with CPAP devices offer unique insights into sleeping patterns, potential mask leaks, and any overnight pressure fluctuations, allowing suitable therapy to accommodate individual needs, data that were previously only obtainable for sleep specialists to evaluate. Many machines can view minute-by-minute reports; other machines can create day-by-day reports.

2.1.5.5 Noninvasive Positive Airway Pressure (PAP) Ventilatory Modes

The first generation of PAP machines was simple continuous pressure. However, the need to expand technology development enabled manufacturers to add bilevel flow generator devices to increase PAP devices' proper utilization.

2.1.5.6 Continuous Positive Airway (CPAP)

OSA is a partial or complete blockage of the upper airway that causes frequent and reversible oxygen saturation in the blood and leads to sleep interruption. OSA affects 9% of middle-aged men and 3% of women in North America. If left untreated, it leads to fatigue, drowsiness, headaches, heart disease, blood vessels, lower quality of life, and increased risk of car accidents. OSA is determined by several measurements that are diagnosed by polysomnography, such as the apnea-hypopnea index (AHI), arousal index (AI), minimal oxygen saturation (SpO_2), desaturation index, and time spent with $SpO_2 < 90\%$. Neurobehavioral and OSA are linked with disease states indicated by a reduced Epworth sleepiness score (ESS), quality of life (QoL), and increased cardiovascular disease (Patil et al. 2019).

The OSA treatment plan includes a wide range of decisions, such as diet and lifestyle changes, drug therapy, constant air pressure (CPAP), and various oral devices to ensure an upper airway splint and surgical treatment.

CPAP is the standard treatment for OSA and includes a pump to deliver air into the nose or mouth and a mask during sleep. Positive pressure is created from the airflow, which opens the airway and limits soft tissues from collapse. CPAP's effectiveness is limited by low adherence rates, which underscores the importance of determining adherence boundaries and dealing with optimal pressure to reinforce adherence.

Most devices apply CPAP in pressure settings between 4 and 20 cmH_2O. CPAP pressure of 4 cmH_2O is the lowest pressure required to provide adequate flow to clear the device's dead space, tubes, and airways to prevent rebreathing of exhaled air. CPAP aims to increase upper airway pressure to produce an airway splint to open up an airway that might collapse during inhalation. Usually, the pressure is adjusted to eliminate impaired breathing, obstructive respiratory events, snoring, reduced flow, and insomnia. By providing positive end-expiratory pressure with the required pressure, CPAP may recruit the alveoli and improve ventilation.

2.1.5.7 Bilevel Positive Airway Pressure (BPAP)

BPAP affords two ways of pressure delivery, high pressure in inhalation and low pressure in exhalation. A theoretical advantage of BPAP (Patil et al., 2019) was that the lower set of expiratory PAP (EPAP) is tolerable and may increase PAP treatment acceptance. With advances in technology, flow sensors have been integrated into PAP devices to detect obstructive events. BPAP pressures typically fluctuate from a minimum EPAP of 4 cmH_2O to a maximum inspiratory PAP (IPAP) of 25–-30 cmH_2O. Most devices use a streaming engine to determine when to shift to IPAP. The trigger is set above zero flow to sense efforts from the patient. Various technologies are used, including flow, shape, and volume, to navigate EPAP to reduce desynchronization. The flow cycle algorithm switches to EPAP when the flow falls below a percentage of the peak flow; Thus, the patient will not experience exhalation of resistance. Airflow starts high to approach the targeted pressure rapidly, then gradually decreases as the airway pressure inside the alveoli builds up. Once a predefined percentage of the maximum inspiratory flow is reached (usually 25%), the expiratory valve opens, and the inspiratory cycle is terminated. Several parameters can be adjusted with pressure support ventilation, including the trigger-on threshold, the inspiratory rise time, the pressure level, and in some ventilators, the cycling-off airflow threshold (Al Otair and BaHammam 2020).

There is significant variation among PAP devices regarding how quickly pressure levels are reached and whether the device becomes a late or early cycle, particularly in leakage cases. Patients may experience discomfort if the device's cycle control is not synchronized with their respiratory cycle.

In conventional BPAP devices, the motor alternates between higher and lower pressure settings to match the patient's breathing cycle. Engine acceleration as the device moved from lower pressure to higher pressure, changing synchronization, and bearing. One way to enhance comfort is allowing the blower motor to rotate when moving between inspiratory and finish smoothly. More modern BPAP devices allow for a smoother change in pressure, and the waveform can be square, exponential, sloping, or sinusoidal. Generally, BPAP offers a standing wave of pressure support, but manual or automatic adjustments can make the pressure change more smoothly, helping with comfort (Fig. 2.1). Many BPAP devices allow changing the elevation time (angle of the extent of pressure change).

Some BPAP devices include cycle settings, trigger, inspiration time, and elevation time, which can be used to improve efficacy and patient comfort, especially for patients with COPD or neuromuscular disorders. Consider the ascent time should be

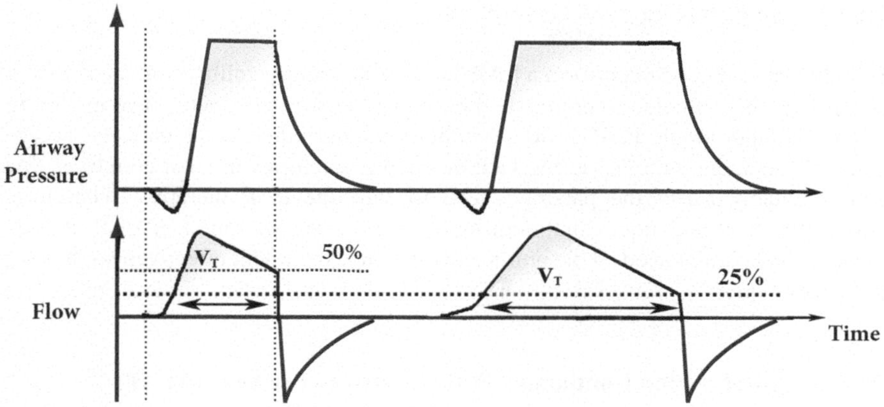

Fig. 2.1 This illustration shows the pressure and flow waves of pressure support. Cycling-off starts when airflow reaches a preset threshold, which is a preset percentage of maximal airflow (usually 25%). At this point, the ventilator stops delivering inspiratory flow, the expiratory valve opens to allow passive exhalation, and pressure goes down to the set positive end-expiratory pressure. Increasing the cycling-off preset percentage of maximal airflow results in decreasing the inspiratory cycle. The illustration demonstrates two cycles; one of them has a cycling-off preset percentage of maximal airflow of 25%, and the other is 50%. The inspiratory time is shorter by 50%, and hence the delivered tidal volume

shorter in COPD patients to allow the lungs to fill more quickly and enough time to exhale. The prolonged rise time in patients with neuromuscular impairment helps ensure proper tidal volume and gas exchange.

2.1.6 Advanced PAP Modes

One of the most challenging approaches to maintaining patients' compliance and adherence while using PAP therapy is getting PAP more comfortable. The technology underpinning PAP devices has advanced progressively, aiming to increase patient comfort and potentially boost adherence to treatment.

Technologies related to PAP systems have evolved from a simple pump that pushes air into the respiratory tract during inhalation and exhalation to develop technical additions that improve the device's performance in providing comfort to the patient and treating complex sleep-related breathing disorders.

It introduced features that control the amount of air and the appropriate pressures and mechanisms, such as:

2.1.6.1 Autotitrating Flow Generators

With the critical need to perform a PAP device's laboratory calibration, the device's air pressure is calibrated to eliminate the frequent airway obstructive events during sleep. The appropriate PAP devices' calibration contributes to the patients' acceptance of treatment, and due to the lack of sleep laboratories in most countries, this will inevitably affect the patients' need to start the PAP therapy. Therefore, developing new techniques that contribute to the auto-calibration of PAP devices to individual patient needs will significantly enhance treatment effectiveness, among these techniques:

2.1.6.2 Autotitrating Continuous Positive Airway Pressure (APAP)

AutoCPAP uses a unique technology that differs from one company to another. With so many manufacturers, the principle remains the same by sensing the amount of air that passes through the upper respiratory tract. Wherever there is resistance arising from the respiratory airway, the device senses the urgent need to increase the airflow flow and thus increase the air pressure to remove the obstructive events. Such differences and the change in air pressure affect the patient's tolerance to devices and clinical efficacy. If a patient enters REM sleep or changes the sleep position, there will be a sudden increase in the upper airway obstruction degree. The device will start to detect and act to eliminate respiratory-related events. The device can accommodate the required pressure needed by the patient and may develop desaturations or arousals. Studies describing the equivalence of AutoCPAP to in-lab titration recommend changing the EPAP minimum to the device's pressure at or below 90–95% of the time so that fixed CPAP settings should be tried if patients do not do well with AutoCPAP. Keeping AutoCPAP at a pressure of 4–20 cmH_2O is not recommended. Patients left on AutoCPAP 4–20 cmH_2O are undertreated and may exhibit frequent awakenings, residual symptoms and apneas, or difficulty tolerating PAP. Some patients are sensitive to pressure changes.

Earlier versions of AutoCPAP devices used to detect snoring vibration changes, making them unresponsive to many critical events. However, most of the new AutoCPAP systems use snore detection in conjunction with flow limitation detection. The flow is inspected multiple times per second, measured with a low-pass filter to eliminate defects, and then the average flow can be determined for any length of time. Peak flow can poorly measure breathing volume, and that could cause an overestimation or underestimation in case of apnea or shortness of breath. Respironics AutoCPAP uses the weighted peak flow (WPF) method to estimate ventilation. In contrast, ResMed AutoCPAP uses a low-pass filtered absolute value for respiratory flow and uses the square root method (RMS) to vary the medium's flow to match one moving period to another. DeVilbiss Intellipap Auto Adjust uses a graduated peak amplitude, while Auto Adjust 2 uses filtering techniques to fine-tune defects in peak flow.

There are some potential risks of using AutoCPAP therapy; if the pressure range is too broad, the device may take too long to change to a therapeutic setting, which may cause additional sleep apnea events and associated symptoms.

Reaching the APAP pressure setup too high may produce breath-holding episodes labeled as central apnea events (treatment-emergent central sleep apnea) (BaHammam 2017). Therefore, the settings become more critical when permitting the device to work through a range of options.

Taking into consideration that in some cases, the usage of autotitrating devices may be complicated. For instance, the use of AutoBPAP devices is not advised in obesity hypoventilation syndrome. Finally, an AutoCPAP may be slightly more expensive than a standard CPAP machine.

2.1.6.3 Autotitrating Bilevel Positive Airway Pressure (AUTO-BPAP)

AutoBPAP delivers spontaneous bilevel therapy, automatically adjusting EPAP and IPAP levels to meet the patient's needs (Carlucci et al., 2015). Just like AutoCPAP, not all AutoBPAP devices operate similarly. Some machines provide only stable PS, others specify a maximum PS (Pmax), while others allow a minimum PS (Pmin) and Pmax. Thus, AutoBPAP may not provide adequate ventilator support if some PSmin related upper respiratory tract obstruction cannot be detected. AutoBPAP devices do not usually have spontaneous timed (ST) mode backup breathing options, so they are not recommended for use in people with central apnea.

ResMed's VPAP-AutoBPAP and DeVilbiss AutoBPAP include a static set of PS. Respironics Series 50 AutoBPAP fixes PSmin at two and allows setting for PSmax, while Series 60 AutoBPAP allows setting for PSmin and Pmax. Within PSmin and PSmax limits, Respironics AutoBPAP changes EPAP in response to apnea (two pauses in breathing or one-time pauses in breathing and shortness of breath) and IPAP snoring response to hypopnea (hypopnea) and flow restriction, using algorithms similar to REMstar AutoCPAP (Johnson and Johnson 2015).

2.1.7 Adaptive Servo-Ventilation (Asv)

Adaptive servo-ventilation (ASV) is a BPAP therapy design that is frequently used to manage sleep-related breathing disorders, particularly central sleep apnea (CSA). Like BPAP and CPAP, ASV provides a set to control obstructive-related events by adjusting EPAP. Nevertheless, there is a difference between ASV therapy and CPAP or BPAP, where ASV implements dynamic adjustment to support IPS and uses an auto-backup rate to normalize breathing rate relevant to a predetermined target. Because elevated PAP pressure and elevated PS may lead to periodic breathing and CSR, devices have been advanced to detect and correct breathing patterns. Many companies have introduced these progressive changes, such as ResMed VPAP AdaptSV (equivalent to Teijin AutoSet CS) with ASV setting, AirCurve ASV, renamed AirCurve CS PaceWave with ASV, and ASVAuto settings, and Respironics BPAP AutoSV Advanced. Another device, SOMNOvent CR.

The most widely recognized manufacturers currently offering ASV devices include ResMed and Philips Respironics.

CSA is linked to increased mortality in heart failure patients (HF). A few randomized controlled trials have reported a statistically significant and clinically meaningful reduction in sympathetic activity in heart failure patients after treating CSR (Javaheri et al. 2020). Nevertheless, the two largest PAP trials in HF patients were negative with respect to their primary outcomes, and both trials were linked with excess mortality (Javaheri et al. 2020).

Positive airway pressure therapy, mostly automatic ASV, contributes to overcoming abnormal breathing patterns. It also has a role in enhancing cardiac function in HF patients with CSA. Studies have indicated the efficacy of ASV in treating primary heart impairment in patients with HF with CSA.

ResMed's standard ASV 8 uses a fixed EPAP, samples flow 50 times per second and change IPAP throughout inspiration to reach target minute ventilation. The PS range can be 0–20 cmH_2O, a PSmin of three, and a PSmax of 15 cmH_2O. The device determines a target minute ventilation based on the minute ventilation, calculated for the last 3 min. ASV uses fuzzy logic to determine the respiratory cycle part and whether the current ventilation is below or above the desired target and then adjusts the PS throughout the cycle to achieve that target, thus avoiding abrupt pressure changes. The change in pressure is calculated by multiplying a gain of 0.3 cm H_2O/L/min/s by the difference between the target minute ventilation and the actual minute ventilation. The auto-backup rate starts at the current auto rate based on a moving average calculated in several breaths. It gradually adapts during apnea to 15 beats per minute. ASVAuto also adjusts EPAP level similar to AutoSet's algorithm in response to obstructive apnea, flow restriction, and snoring.

When the breathing is stabilized, the periodic breathing pattern will subside. If the device is cycling the pressures frequently, it indicates an underlying periodic pattern, and often, the patient will not tolerate the device, or there will be a suboptimal clinical response.

Advanced Respironics BiPAP AutoSV is tuned with minimum and maximum EPAP, PSmin, and PSmax, maximum pressure, and automatic or fixed rate. The inspiratory rise time (Ti) and Bi-Flex features can also be customized to achieve patients' comfort. The PS level is targeted based on mean inspiratory flow, which is the sum of inspiratory flows over a period divided by the number of samples to set spurious values. The target peak flow is generally set at 90–95% of the mean peak flow of the last 4 min. In an attempt to stabilize breathing in the setting of obstructive features (apnea [<20% flow], hypopnea [20–60% flow], and snoring), the target peak flow increases from 95% of the mean inspiratory peak flow over the last 4 min to the 60th percentile of the inspiratory peak flow values. The target peak flow is also increased to the 60th percentile if CSR features are noted based on a CSR index, which measures similarity to a CSR pattern or a flow index that measures fluctuations in breath size (Johnson and Johnson 2015).

The respiratory cycle of AutoSV Advanced is determined using the average length of the respiratory cycle at the last breaths and calculating the expected midpoint of inspiration. Associated with multiple small modifications of PS during the breathing cycle with the ASV of ResMed, the inspiratory PS is determined based on the PS plus for the last breath based on an average moved for the required pressure in the last 30 breaths multiplied by the difference between the peak target inspiratory flow and the peak current inspiratory flow. PS is given within the breath if the actual flow is less than the target flow at 100 ms before the expected midpoint to inspire the current breathing. If the actual flow is greater than the target flow, PS will decrease in the next breath.

The EPAP adjustment algorithm of AutoSV Advanced is similar to the REMstar Auto algorithm. However, it uses a triggered breath rather than forced oscillation to differentiate obstructive from open-airway events (Johnson and Johnson 2015). The automatic backup rate of AutoSV Advanced is determined based on a moving window of the last 12 spontaneous breaths. A mandatory breath is given if the breath does not occur within specific parameters, with a minimum breath rate of 8–10 bpm.

Older versions of the BiPAP AutoSV devices did not provide automatic expiratory pressure calibration, and the automatic backup rate algorithm was not proportional to the baseline respiratory rate. However, it will give a breath if no spontaneous breath occurs within 8 s of the end of the expected breathing length or 4 s if there has been recent breathing.

2.1.7.1 Volume-Targeted Pressure Support (VTPS)

Worsening of chronic hypoventilation in conditions such as obesity hypoventilation syndrome (OHS), chronic obstructive pulmonary disease (COPD), and neuromuscular disease (NMD) may not respond well to PAP therapy. Those types of conditions have primary adverse outcomes on healthcare utilization, quality of life, and mortality (Bahammam and Al-Jawder 2012). Patients with chronic hypoventilation are being treated with noninvasive ventilation (NIV). Several studies have described improvements in gas exchange, hospitalization rates, quality of life, and mortality in patients with OHS when treated with NIV (Piper et al. 2017).

Setting patients on optimal NIV needs specialized expertise, thoughtful review, and close monitoring. Attended titration polysomnography (PSG) or a PSG review in those initiated on NIV empirically is recommended to optimize therapy. PSG is also essential to optimize the treatment of coexisting OSA, as described by current recommendations (Mokhlesi et al. 2019). Effective treatment of OSA is crucial not only to alleviate sleep disruption but also because it is linked to reduced alveolar ventilation, which can affect oxygenation.

While traditional treatments often involve constant settings for EPAP and PS, newer NIV modes are capable of automatically adjusting these pressures to meet the patient's varying ventilation requirements and ensure adequate breathing.

A theoretical drawback of fixed-level BPAP is the variability of delivered tidal volume (V_T) due to respiratory effort variations, respiratory system elastance, and

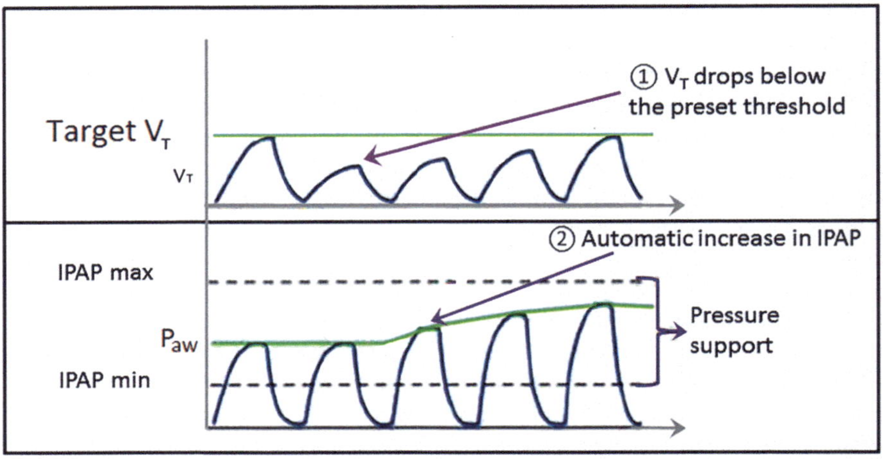

Fig. 2.2 An illustration of the operation of the VTPS ventilatory mode. When the tidal volume drops to less than the set tidal volume, the system discovers the reduction in tidal volume and automatically raises IPAP to rebuild tidal volume to the target tidal volume

sleep position in OHS patients. Volume-targeted pressure support (VTPS) combines the advantages of pressure-limited and volume-limited ventilation modes into one mode to attain sufficient V_T and minute ventilation throughout a sleep period and long term (Piper et al. 2017). Built-in pneumotachograph examines delivered V_T, and when V_T falls below a preset level, the system automatically reacts by increasing IPAP to increase V_T to approach the target V_T (Fig. 2.2).

Volume-targeted (VT) modes of positive airway pressure noninvasive ventilation (NIV) devices are hybrid modes to combine the advantages of more stable minute ventilation produced by traditional volume-limited (VL) ventilator modes with the comfort and superior leak compensation associated with pressure-limited (PL) ventilator modes. Tuning algorithms are also available with EPAP mechanisms to keep upper airway patency and control OSA (AutoEPAP) in some VTPS devices. These devices respond in real-time to changing pressure requirements during sleep and may have clinical benefits such as reducing air leakage and improving patient comfort. Besides, they can reduce the need for PSG monitoring and reduce health care costs.

A potential benefit of this mode over fixed BPAP is the ability to adjust inspiratory support to maintain ventilation despite changes in lung or chest wall compliance that can arise from changes in sleep stage or body position.

In comparison to BPAP ST, clinical studies have demonstrated that VTPS resulted in higher IPAP levels with higher VT or minute ventilation, yielding more significant reductions in overnight CO_2 (Piper et al. 2017). Nevertheless, the improved nocturnal ventilation with VTPS did not offer any other health-related quality of life benefits (Piper et al. 2017).

2.2 Conclusion

A broader range of PAP device modes uses different algorithms and ancillary features, allowing the treating specialist to choose the suitable one. That will provide the opportunity to select a mode of therapy that best fits the patient's ventilatory needs, and it also requires a clear understanding of the chosen mode's principles and indications. Automatically adjusting modes of ventilation can both simplify and optimize therapy in patients with complex sleep breathing abnormalities. Nonetheless, it must also be acknowledged that robust evidence displays these newer modes of ventilation provide more significant clinical benefits than fixed pressure modes are lacking.

Improvements in technology are also increasing our opportunities for continuous long-term monitoring of a patient's response to and acceptance of therapy. Simultaneously, these developments are a very welcomed addition to our clinical practice but do not replace informed clinical judgment and are considered manual interventions.

References

Ahmed O, Parthasarathy S (2010) APAP and alternative titration methods. Sleep Med Clin 5:361–368

Al Otair HA, BaHammam AS (2020) Ventilator- and interface-related factors influencing patient-ventilator asynchrony during noninvasive ventilation. Ann Thorac Med 15:1–8

BaHammam AS (2017) Treatment-Emergent Central Sleep Apnea (Complex Sleep Apnea). Sleep Vigilance 1:53–56

BaHammam AS, Al-Jawder SE (2012) Managing acute respiratory decompensation in the morbidly obese. Respirology 17:759–771

BaHammam AS, Singh T, George S, Acosta KL, Barataman K, Gacuan DE (2017) Choosing the right interface for positive airway pressure therapy in patients with obstructive sleep apnea. Sleep Breath 21:569–575

Carlucci A, Ceriana P, Mancini M, Cirio S, Pierucci P, D'artavilla Lupo N, Gadaleta F, Morrone E, Fanfulla F (2015) Efficacy of bilevel-auto treatment in patients with obstructive sleep apnea not responsive to or intolerant of continuous positive airway pressure ventilation. J Clin Sleep Med 11:981–985

Chen ZL, Hu ZY, Dai HD (2012) Control system design for a continuous positive airway pressure ventilator. Biomed Eng Online 11:5

Ip S, D'ambrosio C, Patel K, Obadan N, Kitsios GD, Chung M, Balk EM (2012) Auto-titrating versus fixed continuous positive airway pressure for the treatment of obstructive sleep apnea: a systematic review with meta-analyses. Syst Rev 1:20

Jaffuel D, Rabec C, Philippe C, Mallet JP, Georges M, Redolfi S, Palot A, Suehs CM, Nogue E, Molinari N, Bourdin A (2020) Patterns of adaptive servo-ventilation settings in a real-life multicenter study: pay attention to volume!: Adaptive servo-ventilation settings in real-life conditions. Respir Res 21:243

Javaheri S, Brown LK, Khayat RN (2020) Update on apneas of heart failure with reduced ejection fraction: emphasis on the physiology of treatment: part 2: central sleep Apnea. Chest 157:1637–1646

Johnson KG, Johnson DC (2015) Treatment of sleep-disordered breathing with positive airway pressure devices: technology update. Med Devices (Auckl) 8:425–437

Mokhlesi B, Masa JF, Brozek JL, Gurubhagavatula I, Murphy PB, Piper AJ, Tulaimat A, Afshar M, Balachandran JS, Dweik RA, Grunstein RR, Hart N, Kaw R, Lorenzi-Filho G, Pamidi S, Patel BK, Patil SP, Pepin JL, Soghier I, Tamae Kakazu M, Teodorescu M (2019) Evaluation and management of obesity hypoventilation syndrome. An official American Thoracic Society clinical practice guideline. Am J Respir Crit Care Med 200:e6–e24

Patil SP, Ayappa IA, Caples SM, Kimoff RJ, Patel SR, Harrod CG (2019) Treatment of adult obstructive sleep apnea with positive airway pressure: an American Academy of Sleep Medicine clinical practice guideline. J Clin Sleep Med 15:335–343

Patz DS, Swihart B, White DP (2010) CPAP pressure requirements for obstructive sleep apnea patients at varying altitudes. Sleep 33:715–718

Piper AJ, BaHammam AS, Javaheri S (2017) Obesity hypoventilation syndrome: choosing the appropriate treatment of a heterogeneous disorder. Sleep Med Clin 12:587–596

Stieglitz S, Galetke W, Esquinas A (2020) Treatment of Cheyne-Stokes respiration with adaptive servoventilation-analysis of patients with regard to therapy restriction. Somnologie (Berl):1–6

Sullivan CE, Issa FG, Berthon-Jones M, Eves L (1981) Reversal of obstructive sleep apnoea by continuous positive airway pressure applied through the nares. Lancet 1:862–865

Chapter 3
Advances in Alternate, Non-PAP, Non-surgical Therapies in the Treatment of Obstructive Sleep Apnea

Mahadevappa Hunasikatti, Seithikurippu R. Pandi-Perumal, and Ahmed S. BaHammam

Abstract Positive pressure therapy (PAP) is widely recognized as the gold standard treatment for OSA. Long-term adherence to PAP therapy, on the other hand, is typically low, ranging from 40 to 60%. Lifestyle modifications (weight loss, avoidance of alcohol or other drugs that reduce upper airway patency), positional therapy, continuous or bi-level positive airway pressure (BPAP), oral appliance therapy, and surgery are all options for treating OSA. It is especially important to see the non-PAP = non-surgical therapy alternatives for these patients who refuse or are intolerant of CPAP therapy.

There are multiple varieties of oral appliances broadly classified into regular and custom-fitted oral appliances. Custom-fitted oral appliances have higher performance regarding the effectiveness to treat OSA. It is also especially important for these patients to have good and close follow-up with dentists and may need to repeat the sleep test to document the control of the disease.

Positional therapy devices range from simple improvised devices mimicking the popular tennis ball technique to specially made apparatus, of which there are several types of devices available. Examples include ZZoma, Nightshift, and SONA Pillow. Many such devices are coming to the market with varying performance data. Sleep medicine physicians are overwhelmed with several choices in the market. The authors have tried to simplify the management of positional sleep apnea.

M. Hunasikatti (✉)
Division of Respiratory, Sleep and Anesthesiology (DSRA), OHT1, Food and Drug Administration (FDA), Silver Spring, MD, USA

S. R. Pandi-Perumal
Somnogen Canada Inc., College Street, Toronto, ON, Canada

Saveetha Medical College and Hospitals, Saveetha Institute of Medical and Technical Sciences, Saveetha University, Chennai, Tamil Nadu, India

A. S. BaHammam
The University Sleep Disorders Center, Department of Medicine, College of Medicine, King Saud University, Riyadh, Saudi Arabia

King Saud University Medical City, King Saud University, Riyadh, Saudi Arabia

A. S. BaHammam, M. Hunasikatti (eds.), *Sleep Apnea Frontiers*, Progress in Sleep Research, https://doi.org/10.1007/978-981-99-7901-1_3

In addition, additional novel therapies have been added to the armamentarium of sleep devices, like oral pressure therapy, Provent therapy, which have varying success in real use in the market due to non-clinical business factors. Even though not yet approved, novel negative pressure therapy for the management of these patients seems interesting and may result in another weapon in controlling the disease.

Keywords Oral appliance therapy · Positional therapy devices · Lifestyle modifications · Novel OSA therapies · Adherence to PAP therapy

3.1 Introduction

These are exciting times for the sleep physician taking care of patients with obstructive sleep apnea (OSA). Positive pressure therapy (PAP) is widely recognized as the gold standard treatment for OSA. Long-term adherence to PAP therapy, on the other hand, is typically low ranging from 40 to 60% (Qaseem et al. 2013). Lifestyle modifications (weight loss, avoidance of alcohol or other drugs that reduce upper airway patency), positional therapy, continuous or bi-level positive airway pressure (BPAP), oral appliance therapy, and surgery are all options for treating OSA.

3.2 Oral Appliance (OA) therapy

Oral appliance therapy has been used to treat OSA for the last three decades. However, only in the last decade have there been well-controlled trials that have proved their place in OSA treatment. It is accepted that dental appliances are effective in mild to moderate OSA.

There is good evidence of research that demonstrates the efficacy of OA (Mandibular Advancement Devices) in terms of reducing snoring and obstructive breathing events, as well as showing beneficial effects on associated health outcomes such as daytime sleepiness.

Well-controlled randomized controlled studies have established OA efficacy by comparison to placebo or inactive appliance (does not provide mandibular advancement). Four parallel-group randomized controlled trials have compared a monobloc appliance (75% of maximum mandibular advancement) to a control device over treatment periods from 2 weeks to 3 months. Three other studies with cross-over design also confirm OSA improvement with mandibular advancement device (MAD) with a significant reduction in apnea–hypopnea index (AHI) (both REM + NREM), arousal index, and oxygen desaturation index (ODI) (Naismith et al. 2005).

Recently, a group of experts collaborated and drafted a definition for an effective OA in the treatment of OSA (Scherr et al. 2014). The most important concepts are as follows:

Oral appliances work best for mild to moderate OSA patients and primary snoring. They also may be used for patients with severe OSA who do not respond to or are unwilling or unable to tolerate PAP therapies. We do not live in a one-size-fits-all society. Patients deserve therapeutic choices. However, the efficacy of oral appliances in treating severe OSA is not proven.

Custom fabrication improves treatment efficacy. Johal et al., conducted a randomized crossover trial comparing custom-made (MRDc) vs ready-made mandibular repositioning devices (MRDr) on 25 patients with a mild degree of OSA (apnea–hypopnea index of 13.3 events/h) and daytime sleepiness (Epworth Sleepiness Scale of 11). The MRDc achieved a complete treatment response in 64% of participants, compared with 24% with the MRDr. A significant difference was observed in treatment failures, when comparing the MRDr (36%) with the MRDc (4%). Excessive daytime sleepiness (Epworth Sleepiness Scale ≥ 10) persisted in 33% (MRDc) and 66% (MRDr) of OSA subjects, following treatment. A statistically significant improvement was observed in quality-of-life scales following MRDc therapy only. Significant differences were observed in relation to both the number of nights per week and hours per night between the two different designs of the device. Their study demonstrates the significant clinical effectiveness of a custom-made mandibular repositioning device, particularly in terms of patient compliance and tolerance, in the treatment of OSA (Johal et al. 2017).

Dual-arch design and a mechanism that permits gradual mandibular advancement over a minimum 5 mm protrusive range. The mechanism setting must be both verifiable and reversible. This ensures the treatment position can be optimized, reproduced nightly, and readily altered as needed to maintain patient comfort.

Advancement of the mandible helps maintain a patent upper airway during sleep: The principal mechanisms underlying the effect of MAD are expansion of the lateral diameter of the velopharynx, the enlargement of the velopharyngeal area, the reduction of velopharyngeal and glossopharyngeal collapsibility, and the stabilization of the upper airway. MAD results in an anterior and inferior movement of the jaw, generating anatomical variations in the UA that enable an increase in the pharyngeal area (Rodriguez-Lozano et al. 2008; Luo et al. 2016; Jayesh and Bhat 2015). Even as new advancements in appliance design evolve, support of free breathing during sleep remains the primary therapeutic focus of oral appliance therapy.

3.2.1 Is It Clinically Significant?

Data are accumulating to indicate the health benefits of MAD in OSA patients (Liu et al. 2020; Serra-Torres et al. 2016; Zhu et al. 2020; Yamamoto et al. 2019). Epworth Sleepiness Score (ESS) has improved with MAD compared to placebo devices as well as objective parameters like multiple sleep latency test (MSLT) in these patients (Gotsopoulos et al. 2002).

The effects on blood pressure have been reported. A crossover study monitored 24-h ambulatory blood pressure after 4 weeks of OAm and inactive appliance wear in 61 patients and found a reduction in 24-h diastolic but not systolic blood pressure. Awake blood pressure was reduced on average by 3.3 mmHg, although there was no effect on blood pressure measurements during sleep. In contrast, a recent report of the meta-analysis showed that therapy with CPAP significantly reduces BP in patients with OSA but with a small effect size. Patients with frequent episodes may benefit most from CPAP. The mean difference in systolic and diastolic BP was, respectively, 2.2 ± 0.7 and 1.9 ± 0.6 mmHg during daytime and 3.8 ± 0.8 and 1.8 ± 0.6 mmHg during nighttime (Fava et al. 2014).

3.2.2 What Are the Factors for Success in Dental Appliance Therapy?

The best approach to predict success is doing a titration sleep study with mandibular advancement, just like doing a CPAP titration sleep study. A single-night titration study has shown promise in predicting the success of a few prototype devices (Dort et al. 2006). Recently, commercially available remotely controlled mandibular protrusion devices have made this more feasible (Remmers et al. 2013).

The device connects to upper and lower dental trays containing impressions of the patient's dentition and advances by moving forward the lower tray with polysomnographic monitoring. This has been evaluated in 67 patients. The only exclusion criteria were BMI > 40 kg/m^2 and any applicable contraindications to oral appliances. During the titration study, the technologist forwards the lower dental tray by 0.2–0.6 mm increments in response to any observed apneas or hypopneas. The advancement is halted in any event of arousal. The protrusion is continued until apnea or hypopneas are eliminated in both REM and NREM stages as per the AASM criteria (Kushida et al. 2008). The set prediction rule was ≤1 respiratory event/5 min supine REM sleep, and it predicted success. Respiratory events of >1/5 min predicted failures. The definition of success was AHI < 10/h plus a 50% reduction in the AHI baseline. The mandibular titration predicted therapeutic success in 30 of 23 patients. Only five of the predicted failures of the 29 were found to be treatment responders. Overall, the initial study shows this titration method is a good prediction tool and shows promise in sleep physician armamentarium (Remmers et al. 2013).

3.2.3 Compliance Is the Key

Since the monitoring of the adherence to the CPAP treatment has been possible (Krieger et al. 1988), using the Medicare guidelines, many studies have shown that less than half of the patients will be using CPAP <4/h per night on <70% of the nights. Until now, adherence to OA therapy has been based on the self-report of the patient's usage. However, with recent objective monitoring capabilities in clinical settings, measuring the use of the patient's is possible (Sutherland et al. 2011). The objective data is especially important in both research and clinical settings as well as it will also address the commercial driver's fitness to drive with the attestation by the monitoring technology that they are using the device as they were supposed to. This also has medico-legal implications in North America and some European countries.

This is also important as this is covered by entities like Medicare. A custom fabricated mandibular advancement oral appliance (E0486) used to treat obstructive sleep apnea (OSA) is covered if criteria A–D are met.

A. The beneficiary has an in-person clinical evaluation by the treating practitioner prior to the sleep test to assess the beneficiary for obstructive sleep apnea testing. Refer to the ICD-10 code list in the LCD-related Policy Article for applicable diagnoses.
B. The beneficiary has a Medicare-covered sleep test that meets one of the following criteria (1–3):
   1. The apnea–hypopnea index (AHI) or respiratory disturbance index (RDI) is greater than or equal to 15 events per hour with a minimum of 30 events: or,
   2. The AHI or RDI is greater than or equal to 5 and less than or equal to 14 events per hour with a minimum of ten events and documentation of:
      (a) Excessive daytime sleepiness, impaired cognition, mood disorders, or insomnia; or,
      (b) Hypertension, ischemic heart disease, or history of stroke; or,
   3. If the AHI > 30 or the RDI > 30 and meets either of the following (a or b):
      (a) The beneficiary is not able to tolerate a positive airway pressure (PAP) device; or,
      (b) The treating practitioner determines that the use of a PAP device is contraindicated.
C. The device is ordered by the treating practitioner following a review of the report of the sleep test. (The practitioner who provides the order for the oral appliance could be different from the one who performed the clinical evaluation in criterion A.)
D. The device is provided and billed for by a licensed dentist (DDS or DMD) (LCD – Oral Appliances for Obstructive Sleep Apnea (L33611) (cms.gov) n.d.).

Key Point

Comparative effectiveness studies have shown that health outcomes from using oral appliances are nearly comparable to those of Continuous Positive Airway Pressure (CPAP) therapy, even though CPAP remains the most effective treatment for Obstructive Sleep Apnea (OSA). However, for patients who refuse or cannot tolerate CPAP, oral appliances present an alternative option. The higher compliance rates associated with oral appliances are a significant factor to consider when planning therapy. The potential to predict treatment success through mandibular titration is promising. Furthermore, monitoring patient adherence to oral devices through remote technology could benefit not only the patient and clinician but also regulatory authorities

(continued)

in industries such as heavy motor vehicle operation, mass transportation, and aviation. This could help reduce work related accidents.

3.3 Position Therapy

Positional therapy devices range from simple improvised devices mimicking the popular tennis ball technique to specially made apparatus, of which there are several types of devices available.

ZZoma
Nightshift
SONA Pillow

De Vries et al. conducted a study on 40 patients using a tennis ball as a positional therapy. PT (positional therapy) was successful in 27 out of 40 patients (68%). Overall, AHI reduced significantly from a median (interquartile range [IQR]) AHI of 14.5 (10.7–19.6) to 5.9 (3.1–8.5), $p < 0.001$. In the short term, PT using the tennis ball technique is an easy method to treat most patients with positional OSA, showing significant reductions in AHI. However, after a mean of 13 ± 5 months, 26 patients (65%) reported they no longer used PT, especially patients with moderate positional OSA (89%) (de Vries et al. 2015). Taking all of this into account, PT using the tennis ball technique might not be the best solution in treating patients with OSA.

Examples include devices like ZZoma. The ZZoma Positional device is an FDA-cleared Class II (Prescription only) Positional medical device intended for the treatment of mild to moderate positional OSA. Permut et al. studied 38 patients with positional OSA (non-supine AHI < 5 events/h) identified on a baseline polysomnogram. Patients were randomly assigned to a night with a positional device (PD) and a night on CPAP (10 ± 3 cm H2O). Positional therapy was equivalent to CPAP, decreasing the AHI to less than five events per hour (92% and 97%, respectively). Sleep efficiency, spontaneous arousal index, and sleep architecture were unchanged with both therapies. The authors concluded that positional therapy is equivalent to CPAP at normalizing the AHI in patients with positional OSA, with similar effects on sleep quality and nocturnal oxygenation (Permut et al. 2010).

3.3.1 Home | Zzoma (zzomaosa.com)

Based on the current body of research, the AASM endorses positional therapy as a secondary treatment option for OSA. This therapy involves techniques that prevent patients from sleeping on their back, thus maintaining a non-supine position. It is particularly recommended as an adjunct to primary OSA therapies for patients who exhibit a lower AHI when not lying on their back compared to when they are in a supine position. According to AASM guidelines, patients who show improvement in AHI when sleeping non-supine typically present with a less severe form of OSA, tend to be less overweight, and are often younger. The recommendation is supported by three Level II studies, one directly contrasting the supine position with an upright one. The AASM also advises that the efficacy of positional therapy in reducing AHI to normal levels should be confirmed through a proper sleep study, as not all patients may experience a normalization of AHI when sleeping non-supine. (Morgenthaler et al. 2006).

3.3.2 *Night Shift*

The battery powered NPTD (Night Shift, Advanced Brain Monitoring, Carlsbad, CA) weighs 44 g and includes electronics housed in a 5.5 (l) × 3.8 (w) × 1.6 (h) cm enclosure affixed on the back of the neck with an adjustable non-latex silicone rubber strap secured by a magnetic clasp.

Levendowski et al. (2014), 30 patients between ages 18 and 75 years, BMI ≤ 35 with an overall apnea–hypopnea index (AHI) ≥ 5 and an overall AHI ≥ 1.5 times the non-supine AHI, and an Epworth score ≥ 5 were prospectively studied. Subjective reports and polysomnography were used to assess efficacy resulting from 4 weeks of in-home supine-avoidance therapy and to measure device accuracy. From 363 polysomnography reports, 209 provided sufficient positional data to estimate one site's prevalence of positional OSA.

The study revealed that 83% of participants demonstrated a reduction in their overall AHI by more than half, with the mean and median reductions being 69% and 79%, respectively. This significant decrease in AHI was a result of substantial reductions in the overall and supine AHI, apnea index, the duration of time with blood oxygen saturation (SpO2) below 90%, and snoring. These improvements led to notable enhancements in the lighter stages of non-REM sleep, known as stage N1 and N2 sleep. There was also a decrease in cortical arousals and awakenings, which are known to disrupt sleep and reduce its restorative quality. Additionally, the study observed an improvement in depression scores, indicating potential mental health benefits from the intervention. The study also found that the supine position was under-detected by more than 5% in a mere 3% of cases, demonstrating the device's accuracy in detecting sleep position. Moreover, when position feedback was provided, sleep efficiency as measured by neck actigraphy was within 10% of polysomnography in 87% of the studies, indicating the device's reliability in monitoring sleep efficiency. The prevalence of Positional OSA (POSA) remained consistently above 70% when the overall AHI was less than 60, suggesting that the device effectively limits supine sleep, which is known to worsen sleep apnea. The neck position therapy device was found to be both accurate and effective in improving AHI, enhancing sleep architecture and continuity, and in monitoring treatment outcomes. However, the study highlighted that patient tolerance to the device is key. If patients can withstand the discomfort and adapt to the vibrotactile feedback, the device could be a worthwhile alternative for those who are unable to tolerate CPAP therapy Levendowski et al. (2014).

3.3.3 Night Shift Positional Therapy (nightshifttherapy.com)

3.4 Provent Sleep Apnea Therapy

Provent sleep therapy is a clinically proven treatment for obstructive sleep apnea. Provent therapy is an FDA-cleared, a prescription-only device that utilizes nasal EPAP (expiratory positive airway pressure) to treat all levels of severity of obstructive sleep apnea (OSA). It may be an ideal treatment option for your patient who has rejected or is non-compliant with CPAP.

Berry et al. studied in 19 sites, including both academic and private sleep disorder centers; the patients selected were OSA patients with a pre-study AHI $\geq$ 10/h. Interventions included treatment with a nasal EPAP device ($N = 127$) or a similarly appearing sham device ($N = 123$) for 3 months. Polysomnography (PSG) was performed on two non-consecutive nights (random order: device-on, device-off) at week one and after 3 months of treatment. Analysis of an intention-to-treat group (ITT) (patients completing week 1 PSGs) (EPAP $N = 119$, sham $N = 110$) was performed. At week one, the median AHI value (device-on versus device-off) was significantly lower with EPAP (5.0 versus 13.8 events/h) but not sham (11.6 versus 11.1 events/h); the decrease in the AHI (median) was greater (−52.7% vs. −7.3%) for the ITT group. At month 3, the percentage decrease in the AHI was 42.7% (EPAP) and 10.1% (sham), $P < 0.0001$. Over 3 months of EPAP treatment, the Epworth Sleepiness Scale decreased (9.9 ± 4.7 to 7.2 ± 4.2), and the median percentage of reported nights used (entire night) was 88.2%. This indicates that the nasal EPAP device significantly reduced the AHI and improved subjective daytime sleepiness compared to the sham treatment in patients with mild to severe OSA with excellent adherence (Berry et al. 2011; Riaz et al. 2015).

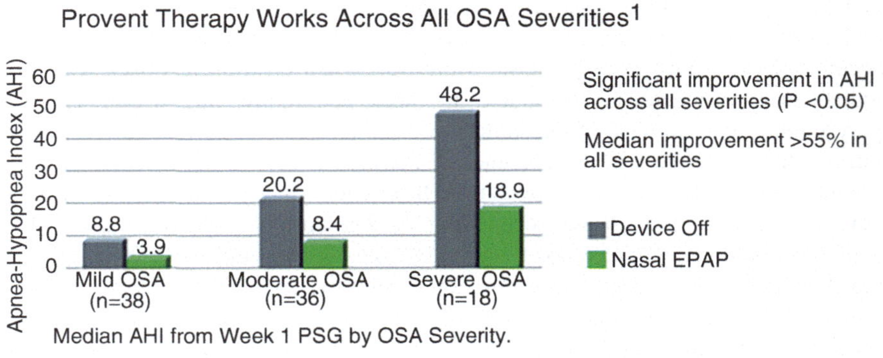

(Berry RB; Kryger MH; Massie CA. A novel nasal expiratory positive airway pressure (EPAP) device for the treatment of obstructive sleep apnea: a randomized controlled trial. SLEEP 2011;34 (4):479–485. Data showed on file, Theravent, Inc.)

We are not sure if this is available in the market at present. We are waiting for further information. Sleep review reports the manufacturer's statement, "Our current inventory of Provent has been depleted and so it is with heavy hearts that we are announcing an end to our company, and the end of the manufacturing life for the Provent Sleep Apnea Therapy product" (Petri et al. 2008).

3.5 BONGO

The Bongo intranasal appliance is a simple device that is placed just inside the nostrils. The device directs expiratory flow through small vent ports, which increases intranasal pressure like the expiratory portion of the breathing cycle during the use of the predicate device. The Bongo consists of valve assemblies coupled with nasal inserts. The Bongo has soft nasal inserts to form a seal within the nasal openings. The nasal inserts are offered in many sizes. Bongo is indicated for use in the treatment of mild to moderate obstructive sleep apnea (OSA) in adults.

The FDA approval was based on the clinical study. A prospective, non-randomized, open-label, single-center clinical study was performed. The clinical study had ten ($N = 10$) patients who completed the study, with seven patients responding with AHI reduction using the Bongo. For the patients ($N = 10$): The mean diagnostic AHI was 15.7, and the mean AHI for the final PSG with Bongo was 7.1. The median diagnostic AHI was 15.4, and the median AHI for the final PSG with Bongo was 7.0. The results demonstrated that AHI was significantly reduced with the Bongo compared to the baseline diagnostic. There were no serious adverse events (PREMARKET NOTIFICATION (fda.gov) n.d.).

Gay et al. evaluated both Bongo and Provent devices to determine inspiratory and expiratory resistance to flow (RTF) and expiratory work of breathing (WOB) in the laboratory. At flow rates of 10/20/30 LPM: Average inspiratory RTF was 1.0/1.3/1.7 vs 2.2/2.9/3.6 cmH2O/l/s for Bongo Rx vs Provent, average expiratory RTF was 17.4/34.1/49.1 vs 59.0/76.9/83.1 cmH2O/l/s for Bongo Rx vs Provent. They concluded that Bongo Rx may result in a more comfortable OSA therapy experience by providing a significantly lower expiratory WOB, the result of significantly lower average inspiratory/expiratory RTF (Gay n.d.; Srijithesh et al. 2019).

3.6 Oral Pressure Therapy

Oral pressure therapy (OPT) is a treatment for obstructive sleep apnea (OSA) that uses negative pressure in the mouth to shift the soft palate and tongue forward. The negative pressure is created by a bedside console connected by a small tube to a mouthpiece worn inside the mouth during sleep. The device is successful in between a quarter and a third of people.

Nigam et al. performed a systematic review evaluating oral pressure therapy as a treatment for OSA OPT. Treatment success rates varied widely between 25 and 79% depending on the residual apnea–hypopnea index (AHI) cut-off defined for treatment success. When using standard definitions (at least 50% reduction from the baseline AHI and the post-OPT treatment residual AHI less than or equal to 10), the success rate of OPT treatment varied between 25 and 37%. Although there was a significant reduction in AHI with OPT, the baseline AHI was not normal in most patients to the extent to of calling these patients optimally treated. Successful treatment with OPT did not correlate with the severity of OSA; however, patients with retropalatal collapse responded better to treatment.

Cochrane review revealed that CPAP is better than positional therapy in improving AHI in people with OSA. However, participants' self-reported adherence was better with positional therapy than CPAP. Studies comparing positional therapy with inactive control therapy showed lower AHI and ESS in favor of positional therapy. The mean difference for AHI noted in the review for positional therapy versus inactive control is higher than the minimal clinically significant difference (MCID) of five, while that for the ESS is less than the MCID of 2 to 3. It is reasonable to try these devices in selected patients (Gay n.d.; Patel et al. 2017; Kim et al. 2017; Nigam et al. 2015; Srijithesh et al. 2019).

3.7 Medications

Physostigmine/Mirtazapine: Smith et al. meta-analysis evaluated the effect of multiple drugs on OSA. Physostigmine at a dose of 0.12 μg/min/kg reduced the AHI from 54 (placebo) events/h to 41 (intervention) events/h in ten patients. Common side effects of physostigmine include excessive sweating, nausea/vomiting, salivation, diarrhea, and stomach cramps.

Mirtazapine at a dose of 15 mg reduced the by 50%, and the placebo reduced the AHI by 29%. Common side effects of mirtazapine include dizziness, constipation, dry mouth, increased appetite, and weight gain. Protriptyline led to symptomatic improvement (improved versus not improved) in two out of three crossover trials (13 participants), but there was no change in the apnea frequency. In one trial, naltrexone did reduce AHI, but total sleep time favored placebo. There is insufficient evidence to recommend the use of drug therapy in the treatment of OSA (Smith et al. 2006).

Paroxetine: Paroxetine reduced the AHI to 23.3 events/h compared to 30.3 events/h for the placebo. Acetazolamide reduced the AHI by 24 events/h, but patients tolerated it poorly in the long term.

Nonbenzodiazepines: Zhang et al. conducted a meta-analysis (Eight relevant placebo-controlled clinical trials involving 448 patients were included) on the effect of nonbenzodiazepines on OSA. The inclusion criteria included adult patients with OSA (AHI ≥5 events/h) based on polysomnography, at least one nonbenzodiazepine medication was studied (eszopiclone, zaleplon, zolpidem), and the study was a controlled clinical trial with pre- and posttreatment data available. The authors found that the weighted estimate had no significant effect on the AHI or the LSAT. Medications evaluated included zolpidem 20 mg, zolpidem 10 mg, flurazepam 30 mg, modafinil 200 mg, eszopiclone 3 mg, and zaleplon 10 mg. The wakefulness after sleep onset was shortened significantly in the OSA patients who were treated with nonbenzodiazepines. The administration of non-BZDs at the commonly recommended dose has been shown to improve objective sleep quality in OSA patients without worsening sleep apnea. It suggests that OSA patients with a complaint of insomnia symptoms may benefit from taking non-BZDs (Zhang et al. 2014). Recent studies have identified mechanisms by which clinical polysomnography can be used to determine physiologic phenotypes. Certainly, the determination of low arousal threshold phenotype can be readily accomplished for measures routinely reported during diagnostic polysomnography. Personalized treatment of OSA based on physiologic phenotype like low respiratory arousal threshold may be treated better with hypnotics like eszopiclone (Abreu et al. 2021).

Hypnotics/Sedatives: Mason et al. evaluated 14 studies, looking at the effect of ten drugs with a total of 293 participants. One study showed that a single administration of eszopiclone 3 mg significantly decreased AHI compared with placebo (24 ± 4 vs. 31 ± 5; P-value <0.05), and the second study of sodium oxybate 4.5 g showed a significant decrease in AHI compared with placebo (mean difference (MD) -7.41, 95% confidence interval (CI) -14.17 to -0.65; $N = 48$). Most studies were small and of short duration, with some methodological issues. Larger, longer trials involving patients across a broader spectrum of OSA severity are needed to clarify these results (Mason et al. 2015).

Oropharyngeal muscle exercise: Verma et al. studied 20 patients in their center with oropharyngeal muscles exercise. Oropharyngeal exercises are a new, non-invasive, cost-effective treatment modality for the treatment of mild to moderate obstructive sleep apnea. It acts by increasing the tone of pharyngeal muscles, is more physiological, and effects may be long-lasting. A total of 20 patients with mild to moderate obstructive sleep apnea syndrome (OSAS) were given oropharyngeal exercise therapy for 3 months, divided into three phases in the graded level of difficulty. Each exercise had to be repeated ten times, five sets per day at their home. Oropharyngeal exercises were derived from speech-language pathology and included soft palate, tongue, and facial muscle exercises. Anthropometric measurements, snoring frequency, intensity, Epworth daytime sleepiness, Berlin sleep questionnaire, and full polysomnography were performed at baseline and study conclusion. There was a significant reduction in the neck circumference (38.4 ± 1.3 to 37.8 ± 1.6) at the end of the study. Significant improvement was seen in symptoms of daytime sleepiness, witnessed apnea, and snoring intensity. Significant improvement was also seen in sleep indices like minimum oxygen saturation, time duration of $Sao2 < 90\%$, sleep efficiency, and arousal index. Graded oropharyngeal exercise therapy increases compliance and reduces the severity of mild to moderate OSA (Verma et al. 2016). As the sample size was small, this may need to be repeated in other centers with a larger sample to recommend a change in management at present.

3.8 New Novel Intraoral Negative Airway Pressure Therapy

Even though this is not currently marketed/approved product, new novel intraoral negative airway pressure therapy for treatment of obstructive sleep apnea. Intermittent negative airway pressure (iNAP) therapy, the latest form of OPT, is characterized by a lower oral tissue discomfort rate and more favorable compliance. By providing intermittent negative pressure to the oral cavity, the iNAP device moves both the soft palate and tongue in an anterior direction, thereby increasing retropalatal and retroglossal areas and reducing airway obstruction during sleep. The device is battery-powered, and the airflow is designed to stop when the targeted negative oral pressure is achieved and restart again only when airflow is required to prevent a decrease in target pressure. Through this design, iNAP enhances the patency of the pharyngeal upper airway without a mask and forced air, thereby preventing local tissue discomfort and systematic dryness of the mouth. In the first month of usage, the average sealing percentage is 60–70%, which increases to 80–90% after 2–3 months. The device may be indicated for patients with OSA who intolerant to CPAP therapy are. No definite contraindication has been indicated in the current literature. Kuo et al. demonstrated that in TCI-DISE, iNAP therapy relieved velar, oropharyngeal, and tongue base obstruction in the supine position in some patients. Moreover, iNAP therapy can be combined with positional therapy to alleviate velar, tongue base, and epiglottic obstruction in some patients. TCI-DISE can also be used to screen the possible responders for iNAP therapy because it is less time-consuming than PSG. Both the design and mode of operation are novel and their application to CPAP intolerant patients is interesting and may need to wait for the formal review by regulatory authorities after their clinical studies is complete (Kuo et al. 2021).

References

Abreu AR, Ramos AR, Dib SI et al (2021) Pathophysiology of obstructive sleep apnea: the role of physiologic phenotypes. Int J Head Neck Surg 12(2):71–73

Berry RB, Kryger MH, Massie CA (2011) A novel nasal expiratory positive airway pressure (EPAP) device for the treatment of obstructive sleep apnea: a randomized controlled trial. Sleep 34(4):479–485

de Vries GE, Hoekema A, Doff MH, Kerstjens HA, Meijer PM, van der Hoeven JH, Wijkstra PJ (2015) Usage of positional therapy in adults with obstructive sleep apnea. J Clin Sleep Med 11(2):131–137

Dort LC, Hadjuk E, Remmers JE et al (2006) Mandibular advancement and obstructive sleep apnoea: a method for determining effective mandibular protrusion. Eur Respir J 27:1003–1009. [PubMed]

Fava C, Dorigoni S, Vedove FD, Danese E, Montagnana M, Guidi GC, Narkiewicz K, Minuz P (2014) Effect of CPAP on blood pressure in patients with OSA/hypopnea: a systematic review and meta-analysis. Chest 145(4):762–771. https://doi.org/10.1378/chest.13-1115

Gay P. A comparison of two nasal expiratory PAP devices for the treatment of OSA (chestnet.org)
Gotsopoulos H, Chen C, Qian J, Cistulli PA (2002) Oral appliance therapy improves symptoms in obstructive sleep apnea: a randomized, controlled trial. Am J Respir Crit Care Med 166:743–748
Jayesh SR, Bhat WM (2015) Mandibular advancement device for obstructive sleep apnea: an overview. J Pharm Bioallied Sci 7(Suppl 1):S223–S225
Johal A, Haria P, Manek S, Joury E, Riha R (2017) Ready-made versus custom-made mandibular repositioning devices in sleep apnea: a randomized clinical trial. J Clin Sleep Med 13(2): 175–182
Kim J, Tran K, Seal K, Almeida F, Ross G, Messier MR et al (2017) Interventions for the treatment of obstructive sleep apnea in adults: a health technology assessment. Canadian Agency for Drugs and Technologies in Health; 2017 Mar; CADTH Optimal Use Report, No. 6.1b (accessed prior to 8 April 2019). [ISSN: 1927-0127]
Krieger J, Kurtz D et al (1988) Objective measurement of compliance with nasal CPAP treatment for obstructive sleep apnoea syndrome. Eur Respir J 1:436–438. [PubMed]
Kuo YH, Liu TJ, Chiu FH, Chang Y, Lin CM, Jacobowitz O et al (2021) Novel intraoral negative airway pressure in drug-induced sleep endoscopy with target-controlled infusion. Nat Sci Sleep 13:2087–2099. https://doi.org/10.2147/NSS.S327770
Kushida CA, Chediak A, Berry RB, Brown LK, Gozal D, Iber C, Parthasarathy S, Quan SF, Rowley JA, Positive Airway Pressure Titration Task Force of the American Academy of Sleep Medicine (2008) Clinical guidelines for the manual titration of positive airway pressure in patients with obstructive sleep apnea. J Clin Sleep Med 4(2):157–171
LCD – Oral Appliances for Obstructive Sleep Apnea (L33611) (cms.gov) (n.d.)
Levendowski DJ, Seagraves S, Popovic D, Westbrook PR (2014) Assessment of a neck-based treatment and monitoring device for positional obstructive sleep apnea. J Clin Sleep Med 10(8): 863–871
Liu C, Kang W, Zhang S, Qiao X, Yang X, Zhou Z et al (2020) Mandibular advancement devices prevent the adverse cardiac effects of obstructive sleep apnea-hypopnea syndrome (OSAHS). Sci Rep 10(1):3394
Luo H, Tang X, Xiong Y, Meng L, Yi H, Yin S (2016) Efficacy, and mechanism of mandibular advancement devices for persistent sleep apnea after surgery: a prospective study. J Otolaryngol Head Neck Surg 45(1):56
Mason M, Cates CJ, Smith I (2015) Effects of opioid, hypnotic, and sedating medications on sleep-disordered breathing in adults with obstructive sleep apnoea. Cochrane Database Syst Rev (7): CD011090. https://doi.org/10.1002/14651858.CD011090.pub2. PMID: 26171909
Morgenthaler TI, Kapen S, Lee-Chiong T et al (2006) Practice parameters for the medical therapy of obstructive sleep apnea. Sleep 29(8):1031–1035
Naismith SL, Winter VR, Hickie IB, Cistulli PA (2005) Effect of oral appliance therapy on neurobehavioral functioning in obstructive sleep apnea: a randomized controlled trial. J Clin Sleep Med 1:374–380
Nigam G, Pathak C, Riaz M (2015) Effectiveness of oral pressure therapy in obstructive sleep apnea: a systematic analysis. Sleep Breath. [PubMed] [Google Scholar]
Patel S, Kon SS, Nolan CM, Simonds AK, Morrell MJ, Man WD et al (2017) Minimum clinically significant difference of the Epworth Sleepiness Scale. Eur Respir J 50(Suppl 61):PA330. https://doi.org/10.1183/1393003.congress-2017.PA330
Permut I, Diaz-Abad M, Chatila W, Crocetti J, Gaughan JP, D'Alonzo GE, Krachman SL (2010) Comparison of positional therapy to CPAP in patients with positional obstructive sleep apnea. J Clin Sleep Med 6(3):238–243
Petri N, Svanholt P, Solow B, Wildschiodtz G, Winkel P (2008) Mandibular advancement Appliance for obstructive sleep apnoea: results of a randomised placebo-controlled trial using parallel group design. J Sleep Res 17:221–229
PREMARKET NOTIFICATION (fda.gov)

Qaseem A, Holty JE, Owens DK, Dallas P, Starkey M, Shekelle P (2013) Clinical Guidelines Committee of the American College of Physicians. Management of obstructive sleep apnea in adults: a clinical practice guideline from the American College of Physicians. Ann Intern Med 159(7):471–483. https://doi.org/10.7326/0003-4819-159-7-201310010-00704

Remmers J, Charkhandeh S, Grosse J et al (2013) Remotely controlled mandibular protrusion during sleep predicts therapeutic success with oral appliances in patients with obstructive sleep apnea. Sleep 36:1517–1525. [PubMed]

Riaz M, Certal V, Nigam G et al (2015) Nasal expiratory positive airway pressure devices (provent) for OSA: a systematic review and meta-analysis. Sleep Disord 2015:734798. https://doi.org/10.1155/2015/734798

Rodriguez-Lozano FJ, Saez-Yuguero Mdel R, Linares Tovar E, Bermejo Fenoll A (2008) Sleep apnea and mandibular advancement device. Revision of the literature. Med Oral Patol Oral Cir Bucal 13(9):E549–E554

Scherr SC, Dort LC, Almeida FR, Bennett KM, Blumenstock NT, Demko BG, Essick GK, Katz SG, McLornan PM, Phillips KS, Prehn RS, Rogers RR, Schell TG, Sheats RD, Sreshta FP (2014) Definition of an effective oral appliance for the treatment of obstructive sleep apnea and snoring: a report of the American Academy of Dental Sleep Medicine. J Dent Sleep Med 1(1): 39–50

Serra-Torres S, Bellot-Arcis C, Montiel-Company JM, Marco-Algarra J, Almerich-Silla JM (2016) Effectiveness of mandibular advancement appliances in treating obstructive sleep apnea syndrome: a systematic review. Laryngoscope 126(2):507–514

Smith I, Lasserson TJ, Wright J (2006) Drug therapy for obstructive sleep apnoea in adults. Cochrane Database Syst Rev 19(2):Cd003002. pub2. PubMed PMID: 16625567; eng

Srijithesh PR, Aghoram R, Goel A, Dhanya J (2019) Positional therapy for obstructive sleep apnoea. Cochrane Database Syst Rev (5):CD010990. https://doi.org/10.1002/14651858.CD010990.pub2. Accessed 27 Nov 2021

Sutherland K, Cistulli P, authors. (2011) Mandibular advancement splints for the treatment of sleep apnea syndrome. Swiss Med Wkly 141:w13276. [PubMed]

Verma RK, Johnson JR, Goyal M et al (2016) Oropharyngeal exercises in the treatment of obstructive sleep apnoea: our experience. Sleep Breath 20:1193–1201. https://doi.org/10.1007/s11325-016-1332-1

Yamamoto U, Nishizaka M, Tsuda H, Tsutsui H, Ando SI (2019) Crossover comparison between CPAP and mandibular advancement device with adherence monitor about the effects on endothelial function, blood pressure and symptoms in patients with obstructive sleep apnea. Heart Vessel 34(10):1692–1702

Zhang XJ, Li QY, Wang Y, Xu HJ, Lin YN (2014) The effect of nonbenzodiazepine hypnotics on sleep quality and severity in patients with OSA: a meta-analysis. Sleep Breath 18(4):781–789. https://doi.org/10.1007/s11325-014-0943-7. Epub 2014 Jan 29. PMID: 24474447

Zhu D, Kang W, Zhang S, Qiao X, Liu J, Liu C et al (2020) Effect of mandibular advancement device treatment on HIF-1alpha, EPO and VEGF in the myocardium of obstructive sleep apnea-hypopnea syndrome rabbits. Sci Rep 10(1):13261

Chapter 4
Perioperative Assessment and Management of Patients with Sleep Disorders

Vinh Nguyen, Brian Freeman, and Kumudhini Hendrix

Abstract In the United States, about 60,000 patients receive general anesthesia for surgery per day. The most important postoperative cardiac, neurologic, and psychiatric complications occur frequently around the middle of the first postoperative week—between postoperative day 2 and 5 (Lazic et al., Physiol Behav 168:41–54, 2017). In addition to acute myocardial infarction, myocardial ischemia, strokes, and delirium, postoperative sleep disorders also have significant adverse impact on postoperative outcomes. Animal models show that sleep deprivation leads to depletion of glycogen stores and increases of oxidative stress and free radical formation with production of pro-inflammatory cytokines all leading to poor postoperative recovery. Importantly, sleep-disordered breathing (SDB) is exacerbated, peaking on postoperative day 3 and returning to baseline only on postoperative night 7 (Chung et al., Anesthesiology 120:299–311, 2014).

Though sleep and general anesthesia share physiological common ground, they are both different states of unconsciousness. Sleep is a state of decreased arousal generated by nuclei in the hypothalamus, brain stem, and basal forebrain. Normal human sleep cycles between two states: rapid eye movement (REM) sleep and non-REM sleep at approximately 90 min intervals. REM sleep is characterized by active high-frequency, low-amplitude rhythms with concomitant rapid eye movements, dreaming, irregularities of respiration and heart rate, penile and clitoral erection, and airway and skeletal muscle hypotonia. Non-REM sleep has three distinct EEG stages—with higher-amplitude, lower frequency rhythms and is accompanied by waxing and waning muscle tone, decreased body temperature and decreased heart rate.

V. Nguyen
Anesthesiology Department, University of Minnesota, Minneapolis, MN, USA

B. Freeman
Georgetown University School of Medicine, Washington, DC, USA

K. Hendrix (✉)
Anesthesia Devices, CDRH, Food and Drug Administration, Silver Spring, MD, USA
e-mail: Kumudhini.hendrix@fda.hhs.gov

A. S. BaHammam, M. Hunasikatti (eds.), *Sleep Apnea Frontiers*, Progress in Sleep Research, https://doi.org/10.1007/978-981-99-7901-1_4

General anesthesia is a drug-induced reversible state that is characterized by unconsciousness, amnesia, analgesia, and akinesia. If a procedure does not require full general anesthesia, then lower doses of sedatives or hypnotics are used to produce diminished cortical function or activity while maintaining intact respiratory and cardiovascular function. In vitro and in vivo studies have identified GABA$_{\lambda}$ and N-methyl-D-aspartate (NMDA) receptors in the cortex, thalamus, brain stem, and striatum as two important targets of hypnotic drugs (Brown et al., NEJM 363:27, 2010). A bolus of hypnotic drugs rapidly reaching the brain stem arousal centers will contribute to unconsciousness. Clinical signs are consistent with the actions of the brain stem. Impaired brain stem function due to actions of hypnotic agents on oculomotor, trochlear, abducens, trigeminal and facial nuclei in the midbrain and the pons produce loss of oculocephalic and corneal reflexes. Apnea ensues due to actions of hypnotic agents on GABA$_{\lambda}$ interneurons in the respiratory control network in the ventral medulla and pons (Brown et al., NEJM 363:27, 2010). Rapid atonia that ensues induction is most likely due to the actions of hypnotic agents in the spinal cord, pontine, and medullary reticular nuclei that control antigravity muscles.

Not surprisingly, sleep disorders have many implications, especially respiratory, for the surgical patient during the perioperative period. Unfortunately, many sleep disorders are often not diagnosed or remain misdiagnosed prior to surgery—especially in the pediatric populations. Type of surgery, anesthesia, type of sleep disorders, type and severity of sleep disorder, comorbidities, and use of opioids have significant effects on patient outcomes.

Keywords Perioperative sleep disorders · Sleep-disordered breathing (SDB) · Postoperative complications · General anesthesia and sleep · Preoperative assessment and management

4.1 Preoperative Concerns

Obstructive sleep apnea (OSA) is characterized by recurrent episodes of upper airway obstruction and desaturation leading to fragmentation of sleep and daytime fatigue and sleepiness (Raveendran et al. 2017). The severity of OSA is determined by apnea hypopnea index (AHI). Apnea is a cessation of airflow for 10 s or more, while hypopnea is reduced airflow with a concomitant desaturation of at least 4%. Prevalence of OSA is 13% in men and 6% in women when OSA is defined as AHI more than 15 events/h. In morbidly obese patients scheduled for bariatric surgery, prevalence of OSA is about 78%. Unfortunately, a significant portion of patients are undiagnosed when scheduled for surgery. Furthermore, OSA is associated with many serious comorbidities such as heart failure, myocardial ischemia, hypertension, arrhythmias, insulin resistance, gastroesophageal reflux, and obesity. Analgesic and anesthetic drugs can impact these patients' consciousness and sleep, upper airway anatomy and physiology, arousal responses, muscle activation and ventilatory drive and thereby potentially increasing perioperative risks in patients with

OSA. OSA patients remain at high risk for respiratory depression even for cases for mere procedural sedation. Though these patients remain at risk for opioid-induced respiratory depression, fragmentation and sleep deprivation heighten sensitivity to pain. Importantly, there is a positive association between OSA and difficult mask ventilation and intubation (Cozowicz et al. 2018).

Though the importance of obstructive sleep apnea in surgical patients is widely acknowledged and codified within professional guidelines (Society of Anesthesia and Sleep Medicine Guideline on Intraoperative Management of Adult Patients with OSA 2018), the importance of other disorders is less clear. Restless leg syndrome (RLS), a sensorimotor sleep disorder, resulting in nocturnal leg discomfort and sleep disruption was found to have an incidence of 9% in patients undergoing spinal anesthesia (Hogl et al. 2002). Though this could not be reproduced, the perioperative period of prolonged immobility, blood loss, iatrogenic sleep deprivation, temporary hold on symptomatic treatment, and use of medications such as metoclopramide can worsen RLS.

Similarly, very little is known about the anesthetic implications for patients with central disorders of hypersomnolence (CDH). This group of disorders include narcolepsy type 1 (with cataplexy), narcolepsy 2 (without cataplexy), and idiopathic hypersomnia. CDH result in persistent and pathologic daytime sleepiness despite adequate amount and quality of nocturnal sleep. Kleine–Levin syndrome, a CHD, is characterized by bouts of recurrent hypersomnia punctuated by periods of relative normalcy. Though only supported by case reports and small case series, CHD raises following concerns for surgical patients in the perioperative period: (1) perioperative cataplexy or sleepiness, (2) chronic use of psychotropic and/or vasoactive medications such as amphetamines, and importantly (3) delayed emergence form anesthesia.

Additionally, most anesthesiologists are not aware of obesity hypoventilation syndrome (OHS) and overlap syndrome (combination of OSA and chronic obstructive pulmonary disease). Both these syndromes are associated with significant comorbid conditions and perioperative morbidity than OSA alone (Raveendran et al. 2017). Daytime hypercapnia, defined as $PaCO_2$ of more than 45 mmHg during wakefulness, differentiates OHS from OSA. OHS is defined as a combination of BMI > 30 kg/m^2, daytime hypercapnia, and absence of an alternative explanation for hypoventilation such as neuromuscular, mechanical, or metabolic in origin. Malignant obesity hypoventilation syndrome (MOHS) is severe OHS characterized by multi-organ dysfunction in morbidly obese patients. Although 90% of patients with OHS have OSA, 10% of OHS patients do not. OHS prevalence is approximately 0.6% of general population and increasing to 50% for cohort with BMI > 50 kg/m^2. Unfortunately, OHS is often not diagnosed and morbidity is higher than patients with OSA. Most patients with OHS are diagnosed in their fifth decade of life after hypercapnic respiratory failure or in sleep clinic evaluations. OHS patients have a higher likelihood to develop heart failure, angina pectoris and/or cor pulmonale. MOHS is defined as a patient with BMI > 40 kg/m^2, awake hypercapnia, metabolic syndrome (central obesity, hypertension, hyperlipidemia, and insulin resistance), and multi-organ function related to obesity (Raveendran et al. 2017).

Patients with OHS have restrictive respiratory physiology impairing lung compliance by 60% compared to eucapnic obese counterparts who have 20% impaired lung compliance. This has implications for increased work of breathing leading to quicker fatigue of respiratory muscles. Typically, obese patients have higher carbon dioxide production that leads to a concomitant drive to increase minute ventilation. However, patients with OHS have blunting of ventilatory drive to carbon dioxide. This blunted response is worsened by increased hypercapnia and serum bicarbonate. Mortality for patients with OHS undergoing bariatric surgery is between 2% and 8%. Also, patients with OHS and overlap syndrome are more likely to experience postoperative respiratory failure, postoperative heart failure, prolonged intubation, postoperative ICU transfer, and longer stay in the hospital.

A multipronged approach, including lifestyle modification and positive airway pressure (PAP) and management of comorbid state is necessary for patients with OHS. Use of BPAP (bi-level PAP) less than 4 h/night is the strongest independent predictor of mortality among patients with OHS. Treatment of patients with OHS on PAP may have protective benefits during perioperative period.

Though there may not be a clear pathophysiological link, overlap syndrome is the combination of COPD and OSA. Overlap syndrome has a prevalence of 1% while COPD alone is 10% while OSA 5–10% of adult populations. However, in those patients with moderate to severe COPD, OSA has a prevalence rate of almost 66%. Nocturnal hypoxemia is an important sing of overlap syndrome. Smoking and obesity are two major predictors of overlap syndrome. Daytime hypercapnia, respiratory failure, and pulmonary hypertension are higher in cohort with overlap syndrome than in isolated OSA or COPD. Within COPD, those with chronic bronchitis than emphysema are more likely to have overlap syndrome.

In overlap syndrome, increased accessory muscles are used for breathing due to flattening of the diaphragm and increased dead space ventilation. In contrast to OHS patients, overlap patients have normal to enhanced respiratory drive to carbon dioxide. Five-year survival in these patients is higher with CPAP and oxygen compared to oxygen alone.

Children with SDB and OSA have been found to have increased perioperative respiratory related adverse events (Scalzitti and Sarb 2018). That is, they have a higher rate of significant oxygen desaturations, airway obstructions, laryngospasms, over-sedation, and need for escalation of care compared to their counterparts. Unfortunately, most children with SDB undergo general anesthesia without PSG or a diagnosis. One important clinical tool, the snoring, trouble breathing, un-refreshed sleep questionnaire (STBUR)—a five question tool, has been shown to be predictive of such adverse events. A score of three or more increased the risk of opioid related adverse events and perioperative respiratory events. Children with OSA who underwent adenotonsillectomy showed that hypoxemia reduced the anesthesia requirement while opioids doubled the level of respiratory suppression. Also, there is increasing evidence of codeine in children with SDB. Polymorphism of CYP 2D6, a metabolic enzyme, leads to slower or no metabolism in 10% of this cohort while increasing metabolism in 2% of these children converting codeine to morphine with production of higher blood levels of morphine than expected. This has led to restriction on use of codeine and tramadol for children by the FDA.

4.1.1 Preoperative Screening and Risk Assessment

Careful preoperative screening is the first and most important step in management of any morbidly obese patient with SDB. Most of these patients are often not diagnosed or have been misdiagnosed. A sensitive screening tool to identify daytime hypercapnia such as an arterial blood gas analysis is considered invasive. However, serum bicarbonate is a sensitive surrogate biomarker along with lower oxygen saturation. A serum HCO_3 of 27 mEq/L in absence of other causes such as diuretics demonstrates a 92% sensitivity in predicting hypercapnia in arterial blood gas. One unit increase in serum HCO_3 increases possibility of having OHS by 14%. End tidal CO_2 can also substitute for arterial blood gas to detect hypercarbia. A two-step screening process, first with serum HCO_3 to exclude patients without OHS and then AHI improves specificity. Additionally, hypoxemia defined as $SpO_2 < 90\%$ or $PaO_2 < 60$ mmHg during wakefulness should lead clinicians to suspect OHS in patients with OSA.

Since polysomnography is expensive and time consuming, the STOP-Bang Questionnaire can alternatively be used as a screening tool. STOP-Bang Questionnaire has the highest methodological validity and accuracy in predicting diagnosis of OSA while a STOP-Bang score of 5–8 identifies those with high probability of moderate to severe OSA. Therefore, in addition to HCO_3 of at least 28 mEq/L and a STOP-Bang score of 3 or more improves the specificity for preoperative OSA recognition. Patients with preoperative mean overnight SpO_2 of less than 93% or oxygen desaturation index of 29 events/h have been shown to have higher risk for postoperative adverse events.

Modified STOP-Bang score adds 1 additional point for BMI 35–40 kg/m^2, 2 points for BMI 40–45 kg/m^2, 3 points for greater than 45 kg/m^2, 1 point for HCO_3 26–28 mEq/L, and 2 points for greater than 28 mEq/L. Modified score of 6 or more has increased sensitivity with a decreased specificity (89.2% and 47.6%) compared to original score of 6 or more (71.6% and 59.1%). Additionally, preoperative clinicians should evaluate for clinical signs of pulmonary hypertension (pretibial edema, jugular venous distension, etc.). When patients have poor exercise tolerance, a preoperative echocardiogram may be needed.

Recently published guidelines from Society of Anesthesia and Sleep Medicine recommends OSA screening a part of routine preoperative evaluation. Preoperative cardiopulmonary evaluation will be necessary in those with hypoventilation syndrome, pulmonary hypertension, and hypoxemia of unknown origin. They may need preoperative sleep studies as well. ASA's perioperative management of patients with OSA is based on severity of OSA, invasiveness of surgery, type of anesthesia, and need for perioperative opioids. The Society of Ambulatory Anesthesia Consensus Statement recommends the STOP-Bang Questionnaire as a screening tool for all surgical patients. Ambulatory surgery may not be an option for patients with OHS or overlap syndrome due to significant postoperative complications and comorbidities.

4.2 Intraoperative Concerns

4.2.1 OSA

Adequate perioperative screening and reliable risk predictors help to devise an anesthetic plan and mitigate risks. Although PSG is the gold standard for suspected OSA patients, the time and resources to obtain this test may not suit all patients. Therefore, using specific OSA screening questionnaires preoperatively can help identify those at high risk for airway compromise (Chung et al. 2016). Intraoperative challenges with OSA patients include difficult airway after induction, appropriate dosage of hypnotic and sedative drugs, and neuromuscular blocking agents (NMBA). OSA is an independent risk factor for difficulty masking or intubation (Brodsky et al. 2002). Using AHI as a metric, those with AHI >40 had a significantly higher prevalence of a difficult airway. One study categorized three groups as AHI < 40, 40–70, AHI > 70. The incidence for a difficult airway was 3.3%, 19.3%, and 27.6%, respectively (Kim and Lee 2006). These patient populations are at higher risk for airway complications, including laryngeal injury, airway trauma, rapid desaturation, aspiration, unexpected ICU admission, and possible death. As a result, the American Society of Anesthesiology (ASA) has devised a practice guideline for possible difficulty airway management, including mask ventilation and intubation. This information allows anesthesia providers to be prepared for a potentially difficult airway. During the anesthesia induction, mask ventilation can be challenging and can lead to a higher incidence of inadequate oxygenation. It is imperative to have other airway adjuncts available and possibly two trained providers for mask ventilation. A recent analysis demonstrated that during intubation of OSA patients, poor visualization of the glottis was expected. The need for multiple intubations attempts increased by more than threefold compared to non-OSA patients (Nagappa et al. 2018). Thereby, it is imperative to have other intubation methods such as videolaryngoscopy, intubating laryngeal mask airway, or awake intubation.

The use of sedative or hypnotic can easily lead to airway collapsibility and depression of central respiratory drive. The upper airway collapse is dose-dependent with anesthetic agents. Anesthetic agents depress central respiratory output, which in turn relaxes the upper airway dilator muscles and suppresses upper airway reflexes in a dose-dependent manner. Occlusion is mainly at the soft palate level, but a higher level of obesity can be at multiple sites, causing total airway occlusion (Hartfield et al. 2023; Eastwood et al. 2005). Consequently, hypnotic agents should be titrated to the anesthetic depth level by using bispectral indexing (BIS) monitor or A-line autoregression. This, in turn, will eliminate the cumulative anesthetic dosages and help potential postoperative airway complications.

Opioid therapy is used to treat moderate to severe pain but can suppress ventilatory drive to breath. In general, the obese patient may have OSA or a component of airway collapsibility. These patients are associated with a 50% increase in opioid-induced respiratory depression. As well, chronically disturbed sleep continuity and intermittent hypoxia are associated with heightened pain sensitivity. The working theory is a hypoxia-triggered upregulation of opioid receptors, promoting sensitivity to narcotics. As a result, there is a growing concern for intraoperative opioid usage and appropriately safe opioid delivery in clinical practice. Studies point to a dose-dependent pattern with increased odds for respiratory complications at increasing opioid dose among OSA versus non-OSA patients (Subramani et al. 2017). These observational studies demonstrated a higher use of naloxone and positive airway pressure ventilation in the postoperative period (Weingarten et al. 2016; Lee et al. 2015). Opioid use in OSA patients presents a risk factor for critical postoperative respiratory events. To help combat the potential respiratory events, intraoperative opioids should be cautiously administered at a lower dose. They require far less than the non-OSA type patient. Furthermore, a multimodal analgesic approach is necessary to avoid post-extubation respiratory compromise, including non-opioids pharmacology therapy. If applicable, a regional or neuraxial anesthetic may be an appropriate anesthetic and potentially eliminate the use of narcotics.

The purpose of NMBA is used to improve intubation conditions and allow optimal surgical relaxation. The incidence of residual NMBA has been reported by up to 64% of patients in the PACU (Murphy and Brull 2010). The current evidence supports a relationship between OSA patients and higher postoperative events secondary to residual NMBA (Pereira et al. 2013). Therefore, a good clinical practice would include minimizing the NMBA dose given to OSA patients. During a surgical case, monitoring the depth of the NMBA is critical to prevent a deep neuromuscular block and prevent early NMB reversal with residual weakness. After completing the case, a complete reversal on NMBA before extubation is vital in the setting of any OSA patient. Two different pharmacological NMBA reversals are commonly used, neostigmine, a cholinesterase inhibitor, or a more recent drug—sugammadex, a cyclodextrin, used specifically for rocuronium. The evidence is limited, but two studies suggest sugammadex has fewer postoperative respiratory complications and fewer postoperative chest radiographic changes (Unal et al. 2015; Llaurado et al. 2014).

Besides the potential airway management issue on induction, premature extubation among OSA patients can lead to detrimental airway collapse. One group reported an incidence of 5% after OSA corrective surgery causing airway obstruction. Ultimately, this can lead to negative pressure pulmonary edema (Baluch et al. 2009). Ideally, all extubations would be performed when patients are fully awake and following commands. In turn, complete NMBA reversal is needed, and motor twitch verification is essential. Optimal body position is essential for adequate chest and diaphragm movement. Therefore, a 30-degree reverse Trendelenburg position will prevent abdominal compression against the diaphragm. The soft palate is highly collapsible and worsens with rostral fluid shift due to pooling and edema in the supine position. Airway adjuncts, either orally or nasally, should be applied before extubation to prevent airway collapse. Naturally, the neutral supine head position causes the head to be more flexed, worsening OSA. Lateral head rotation after extubation help alleviate the severity of OSA and maintain adequate ventilation (Zhu et al. 2017).

4.2.2 Restless Leg Syndrome

The perioperative environment can be linked to exacerbation of restless leg syndrome during surgery and ultimately affect postoperative outcomes. RLS can be associated with iron deficiency, neuropathy related to diabetes, pregnancy, chronic renal failure, and spinal stenosis (Klingelhoefer et al. 2016). Furthermore, sleep deprivation, immobility, and drug interference of the central nervous system's dopamine pathway can precipitate RLS (Goldstein 2015; Michaud et al. 2002; Hening et al. 1999). Certain medications, including dopaminergic agonist and alpha-2 delta calcium channel ligands, are known to improve RLS. Before intraoperative management, assessment of the patient's RLS symptoms and the treatment options are essential. Narcotic therapy effectively controls RLS to allow adequate pain control and sedation to allow less sleep disruption. Sequential compression devices (SCD) are important to prevent deep vein thrombosis but are equally important to control RLS exacerbations after surgery. SCD allows blood to circulate freely and avoid muscle immobility. Since RLS is linked to iron deficiency, clinicians should limit phlebotomy and aggressively treat blood loss intraoperatively. In suspected anemic patients, a dose of IV iron can help with RLS. Those patients with RLS under sedation anesthesia can benefit from a dose of physostigmine to eliminate periodic limb movement. Continuous dexmedetomidine infusion throughout the operation with a target dose rate of 0.4 μg/kg/h can benefit suspected RLS patients (Ohshita et al. 2020). The mechanism may target the spinal cord by counteracting increased spinal excitability. Dexmedetomidine specifically enhances the descending inhibition and decreases the descending facilitation to modulate pain and nociception (You et al. 2016). Finally, avoiding intraoperative medication including dopamine antagonists, first-generation antihistamine, metoclopramide, haloperidol, or selective serotonin reuptake inhibition can ameliorate RLS.

4.2.3 Narcolepsy

Narcolepsy is a sleep disorder characterized by excessive daytime sleepiness, cataplexy, hypnagogic or hypnopompic hallucinations, and sleep paralysis. This disorder has autonomic dysregulation, especially control of heart rate and body temperature. Intraoperative management is essential in these patients to prevent manifestations of narcoleptic symptoms. Narcolepsy is an incurable, lifelong disease that requires chronic pharmacologic treatment to prevent adverse events. As a result, narcoleptic medications should be continued preoperatively. Worse narcolepsy symptoms were seen in those who discontinued preoperatively or those who previously used medications for symptom control (Wang and Greenberg 2013). Medical therapy includes stimulants (amphetamine, methylphenidate, and modafinil) or sodium oxybate. Chronic use of stimulant downregulates sympathetic nervous system while increasing anesthetic requirement. This leads to profound hypotension. A few case reports reported serious adverse effects, including brief hypertension during laryngoscopy and intraoperative hypotension (Fischer et al. 2006; Burrow et al. 2005). Close hemodynamic monitoring is advised throughout the case, especially upon induction. Direct-acting vasoactive or vasodilator agents should be readily available to counteract the either hypotension or hypertension, respectively. Arterial line monitoring may be necessary if refractory hypotension persists despite a large dose of vasoactive agents. Chronic stimulants usage, especially modafinil, may carry the risk of intraoperative awareness. A common practice is to utilize BIS monitoring to guide the appropriate awareness depth and prevent delayed emergence (Morimoto et al. 2011). To limit the sedative properties of hypnotic and opioids drugs, short-acting narcotics such as remifentanil or fentanyl are ideal. The use of TIVA, utilizing propofol, has been described to benefit narcoleptic patients (Pelaez et al. 2004). Regional anesthesia, such as epidural, is effective and advantageous because patients can be monitored for narcolepsy signs. Obstetric populations are at a high risk for narcolepsy exacerbation; thus, epidural can provide the appropriate anesthetic while monitoring such symptoms.

4.2.4 Obesity Ventilation Syndrome or Malignant Obesity Ventilation Syndrome and Overlap Syndrome

OHS patients who present to the operating room are typically obese (BMI $>$ 30 kg/m^2) with daytime hypercapnia (elevated carbon dioxide level $>$ 45 mmHg). About 90% of OHS patients develop OSA. On the other hand, malignant obesity ventilation syndrome (MOHS) is a severe form of OHS involving multi-organ dysfunction. Intraoperative management for OHS or MOHS patients can be quite challenging. These patient populations are morbidly obese with high likelihood of airway collapse during anesthesia induction. Difficulty of mask ventilation and intubation is proportionate to patient's BMI. The airway management principle is similar to OSA patients. Critical respiratory depression is a concern upon anesthesia induction. One technique to limit the fall of oxygenation after induction is to position the patient at a 25-degree head-up position and applying positive end-expiratory pressure of 10 cm H_2O during preoxygenation. Preoxygenation should be greater than 3 min to prevent increase apnea tolerance time. After the loss of consciousness and apnea, use of nasopharyngeal oxygen insufflation or transnasal humidified rapid insufflation ventilatory exchange (THRIVE) technique can prevent desaturation (Patel and Nouraei 2015; Ramachandran et al. 2010). An array of airway tools should be readily available to manipulate the airway if masking or intubating proves to be challenging. The use of two-hand ventilation with oral or nasal airway adjuncts may be useful. Supraglottic airway devices can help bridge difficult ventilation or facilitate intubation. Videolaryngoscopic devices have made intubation more successful with morbidly obese patients. In suspected or known difficult masking or intubation, awake intubation is more prudent.

During the maintenance phase, ventilation strategies with low tidal volume, low plateau pressure, low driving pressure, and a moderate PEEP level are key to maintaining protective ventilation (Guldner et al. 2015). Atelectasis in these obese patients is common, and alveolar recruitment maneuvers can improve intraoperative oxygenation and compliance (Aldenkortt et al. 2012). OSA management principle, that is,careful titration of sedatives, hypnotics and NMBA, should be applied to OHS or MOHS populations. Extubation is a critical period, especially in the obese population with OHS or MOHS. They are at high risk for postoperative respiratory failure. Therefore, applying a positive airway pressure device may limit the amount of respiratory failure post-extubation. If possible, regional anesthesia can offer an advantage over general anesthesia. Less airway manipulation and avoidance or limited amount of respiratory depressed drugs have led to fewer airway complications.

MOHS anesthesia management is like OHS, but the syndrome has a multi-organ dysfunction component. It is vital to know what organ system MOHS has affected and tailored the intraoperative anesthetic based on the patient comorbidities. Most MOHS patients are diagnosed with nonalcoholic steatohepatitis. Focal segmental glomerulopathy is linked to morbidly obese patients, and the incidence of pulmonary hypertension can be as high as 50% of MOHS (Marik and Desai 2013; Serra et al. 2008). Each diseased organ's physiology must be considered to minimize organ failure and judicious anesthetic drug selection.

Overlap syndrome refers to the coexistence of OSA and COPD. These patients may have increased dead space ventilation with flattening diaphragm due to COPD. They tend to have more severe nocturnal desaturation due to the synergistic effect of both conditions as well, as pulmonary hypertension. Intraoperative anesthetic care should be like any OSA patient but with a focus on not aggravating COPD. Ventilation strategies include increasing expiratory time (to prevent air trapping) and limiting hyperoxia. Pulmonary hypertension physiology should be monitored so to prevent any exacerbations that may lead to right heart failure.

4.2.5 Pediatric Sleep Disorder

Sleep disordered breathing has become more prevalent in the pediatric population and thus bringing unique airway challenges to the anesthesiologists. These pediatric populations are at higher risk for significant oxygen desaturations due to airway obstruction, opioid intolerance, and laryngospasm. These disorders encompass mostly upper airway abnormalities leading to obstructive sleep apnea, central sleep apnea, and hypoventilation. Intraoperative management of these patients includes an adequate assessment of the severity of the patient sleep disorder. A PSG can help facilitate intraoperative considerations. Not all children can tolerate PSG; thus, the STBUR sleep questionnaire can help delineate which child is at an increased risk of opioid-related adverse events.

The pediatric airway anatomy is different than the typical adult airway. In general, their tongue is larger than the oral cavity. The occiput is more prominent, leading to a more flexed position when lying supine. This causes a more natural tendency to obstruct the airway. Their thyromental distance is short; thus, the tongue is closer to the soft palate leading to a propensity for airway obstruction. Given all these structural changes, enlarged pharyngeal tissues will cause a significant risk of airway collapse. It is essential to relieve obstructions using airway adjuncts during the induction phase. Two distinctive devices are fundamental to alleviate obstruction, an oropharyngeal and/or nasopharyngeal adjunct. Patients with SDB may have intermittent hypoxia. Studies show that hypoxia increases activation of opioid receptors in the brain. Therefore, in these populations, less opioids are needed for pain control. As a result, the usual weight-based opioid dosing can lead to overdosing in these children. It is good practice to reduce opioids by half for patients with obstructive sleep apnea during surgery. This has been adopted by the current American Society of Anesthesiologists practice guideline (ASA taskforce 2014). A multimodal approach is effective in the treatment of pain for children with SDB. Furthermore, nonsteroidal anti-inflammatory drugs (NSAIDs) reduce the requirement of opioids. However, NSAIDs inhibit platelet aggregation and prolong bleeding. Although some physicians are concerned about postoperative bleeding with NSAIDs administration, the Cochrane database has not shown an increased risk of bleeding postoperatively (Lewis et al. 2013).

Congenital central hypoventilation syndrome (CCHS) is characterized by a diminished drive to breathe during sleep despite progressive hypoxia or hypercapnia. CCHS can be linked to genetic condition, hypoxic brain injury, or seizure. Intraoperative ventilation management for these patients is challenging. It is a progressive disease where a perioperative patient is mechanically ventilated via tracheostomy. In contrast, another patient may present with a noninvasive ventilator, such as bilevel pressure ventilation or even a diaphragmatic pacemaker. Patients with diaphragmatic pacemakers should be evaluated for electrosurgical unit interference. A safe option would be to turn off the transmitter and disconnect itself from the electrodes during proximity cauterization (Venkatraghavan et al. 2009). Once the procedure is done, reconnecting the diaphragm pacemaker can facilitate extubation. There is no evidence that one anesthetic is superior to the other. It is judicious to avoid steroid NMBAs, but, if necessary, then vigilant neuromuscular monitoring and reversal with sugammadex should be performed. Using short-acting sedative-inhaled agents or drugs and restricting narcotic use would be ideal. Thus, total intravenous anesthesia with propofol and remifentanil have shown positive result while avoiding NMBAs (Niazi et al. 2011). There should be a low threshold to include a bridging post-extubation noninvasive ventilation with bilevel PAP before arriving in the recovery room. Anesthesia providers need to be knowledgeable of this condition, limit long-acting depressants, and monitor for postoperative apnea.

4.3 Postoperative Concerns

4.3.1 Obstructive Sleep Apnea

Postoperative complications are most likely to occur in OSA patients during the first 24 h following emergence from anesthesia. Obstructive sleep apnea renders patients at very high risk for postoperative respiratory depression. The specific problems which can occur in the post-anesthesia care unit include hypoxemia, apnea/respiratory arrest, acute respiratory failure, urgent reintubation, negative-pressure pulmonary edema, and overall worsening of upper airway obstruction (Kaw et al. 2012).

Due to the loss of airway patency, OSA patients carry a more significant risk for inadequate oxygenation and ventilation. The administration of continuous supplemental oxygen is routine immediately after emergence from anesthesia. Oxygen delivered at low flow rates should be carefully titrated to pulse oximetry with a goal to achieve at least 90% oxygen saturation. The rate should not be too high so as to miss detecting acute hypercapnia secondary to hypoventilation. Oxygen therapy may be lowered once the patient can maintain baseline oxygen saturation on room air. Compared to patients without OSA, monitoring of oxygenation via pulse oximetry should continue at least an extra hour once meeting discharge criteria from the post-anesthesia care unit (Gali et al. 2009).

Monitoring of ventilation is equally as critical as oxygenation for patients with known or suspected OSA in the immediate postoperative period (Lee and Weinger 2012). There are several methods to detect hypoventilation: clinical observation and evaluation, capnography (end-tidal carbon dioxide monitoring), and quantitative measurements of arterial blood gases. The presence of hypercapnia can help determine whether the OSA needs a lower rate of supplemental oxygen, non-invasive positive pressure ventilation, or urgent re-intubation and mechanical ventilation. Hypoventilation and airway obstruction in the postoperative period can be treated with non-invasive positive pressure ventilation (NIPPV) strategies such as continuous positive airway pressure (CPAP) or biphasic positive airway pressure (BiPAP) devices. In fact, unless contraindicated by the surgical procedure, these therapies should be used routinely for all patients with OSA who are compliant with NIPPV prior to surgery (ASA 2014). NIPPV not only treats respiratory insufficiency but prevents arterial desaturation and the risk of postoperative negative pressure pulmonary edema secondary to airway obstruction. NIPPV therapy should start in the PACU and continue through the rest of recovery. To improve compliance, OSA patients should bring their own equipment to the hospital.

Postoperative pain control can be challenging for patients with obstructive sleep apnea in the first 24 h after surgery. Opioids increase the risk of airway obstruction, hypoventilation, hypoxemia, and other respiratory complications. Patients with OSA often have lower pain thresholds leading to higher analgesic requirements as well as greater sensitivity to the respiratory depressant effects of opioids (Cozowicz et al. 2019). To minimize respiratory depression, multimodal analgesia with opioid-sparing techniques has been shown to be effective and is the preferred approach (Cozowicz et al. 2019). Non-opioid analgesics such as non-steroidal anti-inflammatory drugs, acetaminophen, gabapentin, cyclo-oxygenase-2 inhibitors, and ketamine should be considered. Regional anesthesia provided by epidural catheters or peripheral nerve block provides prolonged analgesia while reducing systemic opioid requirements. Regardless of the specific medications administered, vigilance for respiratory depression should be maintained throughout the patient's postoperative recovery period for respiratory depression. After leaving a closely monitored environment (e.g., PACU or intensive care unit), continuous infusions of opioids should be avoided when prescribing patient-controlled analgesia. Patients who received epidural or intrathecal bolus doses of hydrophilic opioids (morphine, hydromorphine) should be monitored for delayed respiratory depression for 24 h after the dose (ASA 2016).

In addition to the usual PACU discharge criteria, additional factors will determine the timing and disposition of the patient with OSA following surgery. These factors include the severity of the patient's obstructive disease, the nature of the surgery, preoperative use of CPAP, and postoperative analgesic requirements. Patients with mild OSA and low opioid requirements have a lower risk for respiratory depression and may be considered for discharge to an unmonitored setting (e.g., home or hospital ward). These patients should be able to maintain baseline oxygen saturation levels while breathing room air in an unstimulated setting such as sleeping (also known as the "room air challenge test"). In contrast, patients with higher risk for postoperative respiratory complications should be admitted to the hospital for at least 24 h of continuous pulse oximetry. The following factors typically necessitate additional monitoring: high pain scores and analgesic needs, frequent desaturation and apneic events in the PACU, moderate to severe OSA, major surgery particularly that involving the airway, significant medical comorbidities, and preoperative noncompliance with NIPPV (Hillman and Chung 2017).

4.3.2 Patients with Narcolepsy

Much is unknown regarding the postoperative outcomes of narcolepsy patients. In the immediate recovery period, theoretical adverse events include both cardiovascular (autonomic dysfunction leading to hemodynamic instability, dysrhythmias) or respiratory (depression, apnea) perturbations. Patients with narcolepsy demonstrate increased sensitivity to anesthetics which could lead to delayed emergence and hypersomnolence. One case report described a narcolepsy patient who required up to 12 h to emerge from general anesthesia after three operations (Mesa et al. 2000). However, the limited data document few postoperative complications. A retrospective review of 37 narcolepsy patients found no difference in emergence time, length of recovery, morbidity, or mortality in those patients who received their preoperative narcolepsy therapy (Burrow et al. 2005). A retrospective study of 76 patients found no difference in episodes of respiratory depression, time in recovery, or ICU transfers in patients with narcolepsy (Cavalcante et al. 2017). Increased postoperative risk may still exist since this study also found a higher frequency of emergency response team activation on the hospital wards primarily for the management of hemodynamic instability.

After discharge from the post-anesthesia care unit, long- and short-term postoperative concerns include an exacerbation of pre-procedure narcolepsy symptoms with worsening daytime sleepiness and increased sleep paralysis or hallucination episodes. The literature consists of mainly case reports and series and lacks prospective clinical trials. A recent systematic review described 49 cases of patients with narcolepsy undergoing surgery (Hu et al. 2018). Narcolepsy symptoms worsened in the postoperative period only for those patients for whom preoperative therapy was discontinued or decreased or for whom no therapy was prescribed for symptom control. Obstetric patients carried the highest risk. The review also noted that patients undergoing neurosurgery around the hypothalamus, pineal gland, and the third and fourth ventricles developed new-onset postoperative narcolepsy. This finding is perhaps due to the loss of hypothalamic neurons which secrete the neuropeptide hypocretin, a mediator within the arousal system.

4.3.3 Patients with Restless Leg Syndrome

Patients with restless leg syndrome carry a greater risk for disease exacerbation in the immediate postoperative period regardless of anesthetic technique (general, regional, or local/sedation). The exacerbation may be misdiagnosed as delirium or agitation. Acute exacerbation of RLS may occur in postoperative setting due to the use of neuroleptics and antihistamines, withdrawal of dopaminergic agents, and blood loss. The two most significant inciting factors were found to be immobility during the case and/or postoperative pain (Smith and White 2009). Postoperative management should include aggressive pain control, early ambulation, and re-initiation of the patient's preoperative dopamine agonist therapy (pramipexole, ropinirole). Antiemetic drugs that are dopaminergic antagonists, such as metoclopramide, prochlorperazine, and droperidol, are contraindicated for these patients with RLS. Selective serotonin reuptake inhibitors, tricyclic antidepressants, and some antihistamines may also have antidopaminergic activity. Naloxone could also worsen RLS symptoms. In one case of RLS exacerbation following bilateral sagittal split ramus osteotomy, symptoms resolved after initiation of a continuous infusion of dexmedotomidine, a central alpha-two receptor agonist (Ohshita et al. 2020).

4.4 Conclusion

History and physical exam data remain imperfect at predicting sleep disordered breathing. Surgeons and anesthesiologists should consider sleep disordered breathing as a comorbidity in the surgical patient. Non-opioid alternatives should be considered in the perioperative period. Postoperative monitoring is important in surgical patients with sleep disordered breathing. Determining which patients benefit from monitoring remains uncertain. Further research will help refine guidelines to make treatment in patients with sleep disordered breathing safer in the perioperative period.

References

Aldenkortt M, Lysakowski C, Elia N, Brochard L, Tramer MR (2012) Ventilation strategies in obese patients undergoing surgery: a quantitative systematic review and meta-analysis. Br J Anaesth 109(4):493–502

American Society of Anesthesiologists (2016) Practice guidelines for the prevention, detection, and management of respiratory depression associated with neuraxial opioid administration: an updated report by the American Society of Anesthesiologists Task Force on Neuraxial Opioids and the American Society of Regional Anesthesia and Pain Medicine. Anesthesiology 124:535–552

American Society of Anesthesiologists Task Force on Perioperative Management of Patients with Obstructive Sleep Apnea (2014) Practice guidelines for the perioperative management of patients with obstructive sleep apnea: an updated report by the American Society of Anesthesiologists task force on Perioperative Management of patients with obstructive sleep apnea. Anesthesiology 120(2):268–286

Baluch A, Mahbubani S, Al-Fadhli F, Kaye AD, Frost EA (2009) Anesthetic care of the patient with obstructive sleep apnea. Middle East J Anaesthesiol 20(2):143–152

Brodsky JB, Lemmens HJ, Brock-Utne JG, Vierra M, Saidman LJ (2002) Morbid obesity and tracheal intubation. Anesth Analg 94(3):732–736, table of contents

Brown E, Lydic R et al (2010) General anesthesia, sleep, and coma. NEJM 363:27

Burrow B, Burkle C, Warner DO, Chini EN (2005) Postoperative outcome of patients with narcolepsy: a retrospective analysis. J Clin Anesth 17(1):21–25

Cavalcante AN, Hofer RE, Tippmann-Peikert M, Sprung J, Weingarten TN (2017) Perioperative risks of narcolepsy in patients undergoing general anesthesia: a case-control study. J Clin Anesth 41:120–125

Chung F, Liao P et al (2014) Factors associated with postoperative exacerbation of sleep—disordered breathing. Anesthesiology 120:299–311

Chung F, Memtsoudis SG, Ramachandran SK, Nagappa M, Opperer M, Cozowicz C et al (2016) Society of Anesthesia and Sleep Medicine guidelines on preoperative screening and assessment of adult patients with obstructive sleep apnea. Anesth Analg 123(2):452–473

Cozowicz C, Chung F, Doufas AG et al (2018) Opioids for acute pain management in patients with obstructive sleep apnea: a systematic review. Anesth Analg 127:988

Cozowicz C, Poeran J, Zubizarreta N et al (2019) Non-opioid analgesic modes of pain management are associated with reduced postoperative complications and resource utilisation: a retrospective study of obstructive sleep apnoea patients undergoing elective joint arthroplasty. Br J Anaesth 122:131

Eastwood PR, Platt PR, Shepherd K, Maddison K, Hillman DR (2005) Collapsibility of the upper airway at different concentrations of propofol anesthesia. Anesthesiology 103(3):470–477

Fischer SP, Schmiesing CA, Guta CG, Brock-Utne JG (2006) General anesthesia and chronic amphetamine use: should the drug be stopped preoperatively? Anesth Analg 103(1):203–206, table of contents

Gali B, Whalen FX, Schroeder DR et al (2009) Identification of patients at risk for postoperative respiratory complications using a preoperative obstructive sleep apnea screening tool and postanesthesia care assessment. Anesthesiology 110:869

Goldstein C (2015) Management of restless legs syndrome/Willis-Ekbom disease in hospitalized and perioperative patients. Sleep Med Clin 10(3):303–310, xiv

Guldner A, Kiss T, Serpa Neto A, Hemmes SN, Canet J, Spieth PM et al (2015) Intraoperative protective mechanical ventilation for prevention of postoperative pulmonary complications: a comprehensive review of the role of tidal volume, positive end-expiratory pressure, and lung recruitment maneuvers. Anesthesiology 123(3):692–713

Hartfield PJ, Janczy J, Sharma A, Newsome HA, Sparapani RA, Rhee JS, Woodson BT, Garcia GJM (2023) Anatomical determinants of upper airway collapsibility in obstructive sleep apnea: a systematic review and meta-analysis. Sleep Med Rev 68:101741. https://doi.org/10.1016/j.smrv.2022.101741. Epub 2022 Dec 30. PMID: 36634409

Hening WA, Walters AS, Wagner M, Rosen R, Chen V, Kim S et al (1999) Circadian rhythm of motor restlessness and sensory symptoms in the idiopathic restless legs syndrome. Sleep 22(7): 901–912

Hillman DR, Chung F (2017) Anaesthetic management of sleep-disordered breathing in adults. Respirology 22:230

Högl B, Frauscher B, Seppi K, Ulmer H, Poewe W (2002) Transient restless legs syndrome after spinal anesthesia: a prospective study. Neurology 59:1705–1707. https://doi.org/10.1212/01.WNL.0000036606.56405.3D

Hu S, Singh M, Wong J et al (2018) Anesthetic management of narcolepsy patients during surgery: a systematic review. Anesth Analg 126(1):233–246

Kaw R, Chung F, Pasupuleti V et al (2012) Meta-analysis of the association between obstructive sleep apnoea and postoperative outcome. Br J Anaesth 109:897

Kim JA, Lee JJ (2006) Preoperative predictors of difficult intubation in patients with obstructive sleep apnea syndrome. Can J Anaesth 53(4):393–397

Klingelhoefer L, Bhattacharya K, Reichmann H (2016) Restless legs syndrome. Clin Med (Lond) 16(4):379–382

Lazic K, Pertovic J et al (2017) REM sleep disorder following general anesthesia in rats. Physiol Behav 168:41–54

Lee L, Weinger M (2012) Anesthesia Patient Safety Foundation consensus conference report: essential monitoring strategies to detect clinically significant drug-induced respiratory depression in the postoperative period. Anesth Analg 114:918

Lee LA, Caplan RA, Stephens LS, Posner KL, Terman GW, Voepel-Lewis T et al (2015) Postoperative opioid-induced respiratory depression: a closed claims analysis. Anesthesiology 122(3):659–665

Lewis SR, Nicholson A, Cardwell ME, Siviter G, Smith AF (2013) Nonsteroidal anti-inflammatory drugs and perioperative bleeding in paediatric tonsillectomy. Cochrane Database Syst Rev 2013(7):CD003591

Llaurado S, Sabate A, Ferreres E, Camprubi I, Cabrera A (2014) Postoperative respiratory outcomes in laparoscopic bariatric surgery: comparison of a prospective group of patients whose neuromuscular blockade was reverted with sugammadex and a historical one reverted with neostigmine. Rev Esp Anestesiol Reanim 61(10):565–570

Marik PE, Desai H (2013) Characteristics of patients with the "malignant obesity hypoventilation syndrome" admitted to an ICU. J Intensive Care Med 28(2):124–130

Memtsoudis SG, Cozowicz C, Nagappa M, Wong J, Joshi GP, Wong DT, Doufas AG, Yilmaz M, Stein MH, Krajewski ML, Singh M, Pichler L, Ramachandran SK, Chung F (2018) Society of

anesthesia and sleep medicine guideline on intraoperative management of adult patients with obstructive sleep apnea. Anesth Analg 127(4):967–987. https://doi.org/10.1213/ANE.0000000000003434. PMID: 29944522; PMCID: PMC6135479
Mesa A, Diaz AP, Frosth M (2000) Narcolepsy and anesthesia. Anesthesiology 92:1194–1196
Michaud M, Lavigne G, Desautels A, Poirier G, Montplaisir J (2002) Effects of immobility on sensory and motor symptoms of restless legs syndrome. Mov Disord 17(1):112–115
Morimoto Y, Nogami Y, Harada K, Shiramoto H, Moguchi T (2011) Anesthetic management of a patient with narcolepsy. J Anesth 25(3):435–437
Murphy GS, Brull SJ (2010) Residual neuromuscular block: lessons unlearned. Part I: definitions, incidence, and adverse physiologic effects of residual neuromuscular block. Anesth Analg 111(1):120–128
Nagappa M, Wong DT, Cozowicz C, Ramachandran SK, Memtsoudis SG, Chung F (2018) Is obstructive sleep apnea associated with difficult airway? Evidence from a systematic review and meta-analysis of prospective and retrospective cohort studies. PLoS One 13(10):e0204904
Niazi AU, Mocon A, Varadi RG, Chan VW, Okrainec A (2011) Ondine's curse: anesthesia for laparoscopic implantation of a diaphragm pacing stimulation system. Can J Anaesth 58(11):1034–1038
Ohshita N, Yamagata K, Himjima A et al (2020) Anesthetic management of a patient with restless legs syndrome: a case report. Anesth Prog 67(4):226–229
Patel A, Nouraei SA (2015) Transnasal Humidified Rapid-Insufflation ventilatory Exchange (THRIVE): a physiological method of increasing apnoea time in patients with difficult airways. Anaesthesia 70(3):323–329
Pelaez R, Hortal FJ, Bastida E, Barrio JM, Riesgo MJ (2004) Narcolepsy and cardiac surgery: can anesthesia with propofol and remifentanil be safe? J Cardiothorac Vasc Anesth 18(2):201–203
Pereira H, Xara D, Mendonca J, Santos A, Abelha FJ (2013) Patients with a high risk for obstructive sleep apnea syndrome: postoperative respiratory complications. Rev Port Pneumol 19(4):144–151
Ramachandran SK, Cosnowski A, Shanks A, Turner CR (2010) Apneic oxygenation during prolonged laryngoscopy in obese patients: a randomized, controlled trial of nasal oxygen administration. J Clin Anesth 22(3):164–168
Raveendran R, Wong J, Singh M, Wong DT, Chung F (2021) Obesity hypoventilation syndrome, sleep apnea, overlap syndrome: perioperative management to prevent complications. Curr Opin Anaesthesiol 30(1):146–155. https://doi.org/10.1097/ACO.0000000000000421
Scalzitti N, Sarb K (2018) Diagnosis and perioperative management in pediatric sleep-disordered breathing. Pediatr Anesth 28:940–946
Serra A, Romero R, Lopez D, Navarro M, Esteve A, Perez N et al (2008) Renal injury in the extremely obese patients with normal renal function. Kidney Int 73(8):947–955
Smith P, White S (2009) Anaesthesia and restless legs syndrome. Eur J Anaesthesiol 26(1):89–90
Subramani Y, Nagappa M, Wong J, Patra J, Chung F (2017) Death or near-death in patients with obstructive sleep apnoea: a compendium of case reports of critical complications. Br J Anaesth 119(5):885–899
Unal DY, Baran I, Mutlu M, Ural G, Akkaya T, Ozlu O (2015) Comparison of sugammadex versus neostigmine costs and respiratory complications in patients with obstructive sleep apnoea. Turk J Anaesthesiol Reanim 43(6):387–395
Venkatraghavan L, Chinnapa V, Peng P, Brull R (2009) Non-cardiac implantable electrical devices: brief review and implications for anesthesiologists. Can J Anaesth 56(4):320–326
Wang J, Greenberg H (2013) Status cataplecticus precipitated by abrupt withdrawal of venlafaxine. J Clin Sleep Med 9(7):715–716
Weingarten TN, Chong EY, Schroeder DR, Sprung J (2016) Predictors and outcomes following naloxone administration during phase I anesthesia recovery. J Anesth 30(1):116–122

You HJ, Lei J, Xiao Y, Ye G, Sun ZH, Yang L et al (2016) Pre-emptive analgesia and its supraspinal mechanisms: enhanced descending inhibition and decreased descending facilitation by dexmedetomidine. J Physiol 594(7):1875–1890

Zhu K, Bradley TD, Patel M, Alshaer H (2017) Influence of head position on obstructive sleep apnea severity. Sleep Breath 21(4):821–828

Chapter 5
Central Sleep Apnea: Advances in the Last Decade

Ahmed S. BaHammam, Mahadevappa Hunasikatti, and Seithikurippu R. Pandi-Perumal

Abstract Central sleep apnea (CSA) is characterized by recurrent apneic episodes with absent respiratory effort. This is not a common clinical condition like obstructive sleep apnea. Central sleep apnea is diagnosed when more than 50% of the events are central types.

Cheyne–Stokes respiration (CSR) is a breathing pattern characterized by cycles of crescendo-decrescendo changes in tidal volume followed by central sleep apneas. These changes are associated with fluctuations in hemodynamic, respiratory, neurologic, and acid-base parameters. Historical and current theories of causation strongly implicate an alteration of central respiratory control as the most key factor in the genesis of periodic respiration. Tissue perfusion and metabolically determined changes in acid-base balance have dominant roles in the production of CSR. At least ten consecutive minutes of these cycles should be recorded before CSR can be diagnosed. The *ICSD-3* specifies that at least ten central apneas and hypopneas per hour of sleep should occur, accompanied by arousals and derangement of sleep structure.

Other types of central sleep apnea are also important and have been discussed in this chapter by the authors. The management also differs from the underlying cause, which is clearly emphasized. The controversy about ASV versus other forms of PAP therapy in the management of CSA is delineated. This chapter discusses a range of

---

A. S. BaHammam
The University Sleep Disorders Center, Department of Medicine, College of Medicine, King Saud University, Riyadh, Saudi Arabia

King Saud University Medical City, King Saud University, Riyadh, Saudi Arabia

M. Hunasikatti (✉)
Division of Respiratory, Sleep and Anesthesiology (DSRA), OHT1, Food and Drug Administration (FDA), Silver Spring, MD, USA

S. R. Pandi-Perumal
Somnogen Canada Inc., College Street, Toronto, ON, Canada

Saveetha Medical College and Hospitals, Saveetha Institute of Medical and Technical Sciences, Saveetha University, Chennai, Tamil Nadu, India

A. S. BaHammam, M. Hunasikatti (eds.), *Sleep Apnea Frontiers*, Progress in Sleep Research, https://doi.org/10.1007/978-981-99-7901-1_5

interventions in the management of CSA, from acetazolamide and overdrive atrial pacing to the placement of implantable *remedē* systems.

Keywords Central sleep apnea (CSA) · Cheyne–Stokes respiration (CSR) · Central respiratory control · Acid-base balance · Adaptive servo-ventilation (ASV)

5.1 Introduction

Central sleep apnea (CSA) is characterized by recurrent apneic episodes with absent respiratory effort. This is not a common clinical condition like obstructive sleep apnea. Depending on the epidemiologic studies, this occurs only in 2–5% of sleep clinic populations. This is more commonly seen in patients with congestive heart failure, using narcotics, suffering from a stroke, or living in high altitudes.

5.2 Classification of Central Sleep Apnea

Central sleep apnea is diagnosed when more than 50% of the events are central types. Apnea in adults is scored when there is a drop in the peak signal excursion by ≥90% of the pre-event baseline using an oronasal thermal sensor (diagnostic study), PAP device flow (titration study), or an *alternative* apnea sensor, for ≥10 s (Berry et al. 2020). The respiratory inductance plethysmography (RIP) signals can be utilized as alternative sensors for the detection of apnea and hypopnea, as specified. The task force reached consensus on the use of the same sensors for adult and pediatric patients except for the following: (1) the end-tidal PCO_2 signal can be used as an *alternative* sensor for apnea detection in children only, and (2) polyvinylidene fluoride (PVDF) belts can be used to monitor respiratory effort (thoracoabdominal belts) and as an *alternative* sensor for detection of apnea and hypopnea (PVDFsum) only in adults. A more accurate determination of the absence of inspiratory effort would entail the use of an esophageal balloon catheter, which is not used in daily practice. Another option is diaphragmatic electromyography used only in research settings.

5.3 Classifications of Central Sleep Apnea

The diagnostic criteria for CSA vary according to the type of CSA (ICSD 2014) (AASM 2014):

5.3.1 Primary CSA Requires the Following Four Findings

Polysomnography (PSG) reveals ≥5 central apneas and/or central hypopneas per hour of sleep; the number of central apneas and/or central hypopneas is >50% of the total number of apneas and hypopneas, and there is no evidence of Cheyne–Stokes breathing.

The patient reports sleepiness, awakening with shortness of breath, snoring, witnessed apneas, or insomnia (difficulty initiating or maintaining sleep, frequent awakenings, or nonrestorative sleep).

There is no evidence of daytime or nocturnal hypoventilation.

The disorder is not better explained by another current sleep disorder, medical or neurological disorder, medication use, or substance use disorder.

Central sleep apnea disorders include a number of disorders, which according to the ICSD-3, can be divided into (AASM 2014):

- Central sleep apnea (CSA) with Cheyne–Stokes breathing (CSB)
- Central apnea due to a medical disorder without CSB
- Central sleep apnea due to high-altitude periodic breathing
- Central sleep apnea due to a medication or substance
- Primary central sleep apnea
- Primary central sleep apnea of infancy
- Primary central sleep apnea of prematurity
- Treatment-emergent central sleep apnea

5.3.2 CSA with Cheyne–Stokes Breathing Requires the Following Four Findings

CSB is observed chiefly in patients with a low left ventricular ejection fraction, but it has also been reported in patients with acute pulmonary edema, central nervous system lesions, and advanced renal failure (AlDabal and BaHammam 2010). In addition, the appearance of CSB in patients with heart failure is linked with increased morbidity and mortality and diminished quality of life (Carmona-Bernal et al. 2008).

The AASM describes CSB as a breathing disorder in which there are cyclical fluctuations in breathing, with periods of central apneas or hypopneas that substitute with periods of hyperpnea in a regular waxing and waning fashion. Many patients with left ventricular systolic impairment experience CSB and/or other sleep-related breathing disorders (AlDabal and BaHammam 2010). Patients with severe heart failure and an ejection fraction as low as 35% to 40% are predisposed to have pulmonary congestion and raised pulmonary venous pressure (AlDabal and BaHammam 2010). Pulmonary congestion may cause stimulation of pulmonary J receptors and finally results in hyperventilation (AlDabal and BaHammam 2010). When the $PaCO_2$ drops beneath the apnea threshold, central apneas are initiated. Afterward, this cycle of hyperventilation-apnea remains during sleep.

5.3.2.1 Symptoms

It is usually challenging to distinguish heart failure symptoms from breathing-related sleep disorders, as both may share symptoms such as poor sleep quality, daytime sleepiness or insomnia, paroxysmal nocturnal dyspnea, and fatigue. Hence, individual clinical assessment is needed to avert delays in diagnosing comorbid sleep disorders in heart failure patients. Commonly, patients with heart failure and CSB have a lower body mass index than patients with concurrent OSA (Sin et al. 1999). Moreover, CSB has been described to be more prevalent amongst men and patients with atrial fibrillation (AlDabal and BaHammam 2010).

5.3.2.2 Diagnosis

In general, patients with CSB usually have low or normal $PaCO_2$, and an enhanced ventilatory response to CO_2 (Xie et al. 2002). However, it is not uncommon to see patients with heart failure who have coexisting OSA and CSB (AlDabal and BaHammam 2010). The overnight polysomnography typically exhibits periodical changes in ventilation, with cycles of central apnea or hypopnea that alternate with cycles of hyperpnea that prevail during stages N1 and N2 of NREM sleep. Figure 5.1 shows an example of CSB.

5.3.3 CSA Due to High-Altitude Periodic Breathing Requires the Following Four Findings

Recent ascent to a high altitude (typically at least 2500 m, although some individuals may exhibit the disorder at altitudes as low as 1500 m) may develop CSB.

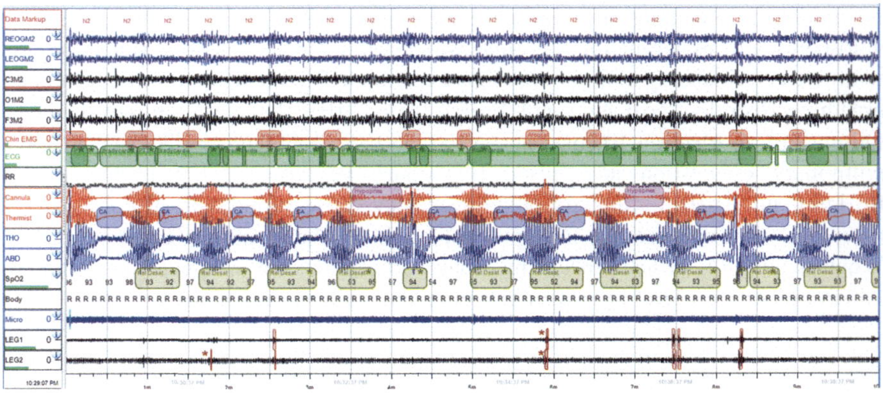

Fig. 5.1 A 9-minute window of a sleep study displaying Cheyne-Stokes Respiration (CSR) with a crescendo-decrescendo pattern, highlighting the characteristic fluctuations in breathing amplitude.

The patient reports sleepiness, awakening with shortness of breath, snoring, witnessed apneas, or insomnia (difficulty initiating or maintaining sleep, frequent awakenings, or nonrestorative sleep).

Symptoms are clinically attributable to periodic high-altitude breathing. PSG, if performed, reveals recurrent central apneas or hypopneas primarily during non-rapid eye movement (NREM) sleep at a frequency of ≥5/h.

The disorder is not better explained by another current sleep disorder, medical or neurological disorder, medication use (e.g., narcotics), or substance use disorder. The periodic breathing that occurs in high altitudes is likely due to hypobaric hypoxia stimulation of the peripheral chemoreceptors (White et al. 1987). The periodic breathing cycle length in high-altitude CBS is shorter than that in patients with heart failure (Wedewardt et al. 2010).

Pathophysiology of CSA: Pathophysiology varies with specific categories and the sleep stages of CSA. In particular, the transitional pathophysiological changes are significant between wake and sleep REM stages.

5.4 Breathing in Non-Rapid Eye Movement (NREM) Sleep

Skatrud and Dempsey studied the effect of sleep state on ventilatory rhythmicity following graded hypocapnia was determined in two normal subjects and one patient with a chronic tracheostomy. Passive positive-pressure hyperventilation (PHV) was performed for 3 min awake and during non-rapid eye movement (NREM) sleep with hyperoxia [fractional inspired O_2 concentration (FIO_2) = 0.50], normoxia, and hypoxia (FIO_2 = 0.12). They concluded that effective ventilatory rhythmogenesis in the absence of stimuli associated with wakefulness is critically dependent on chemoreceptor stimulation secondary to $PCO_2{}^{-}[H^+]$. This is also accompanied by increased upper airway resistance (Skatrud and Dempsey 1983). An average of 2 to 8 mmHg rise in $PaCO_2$ is seen during NREM sleep. During NREM sleep, ventilation is totally under metabolic control (Muza 2015). Douglas et al. measured the ventilatory response to hypercapnia in 12 sleeping adults using a modified rebreathing method. The HCVR (hypercapnic ventilatory response) was significantly reduced in all stages of sleep compared with that during wakefulness (1.60 ± 0.19 SEM L/min/mmHg CO_2), falling to less than half the waking HCVR during non-REM sleep (0.75 ± 0.08 L/min/mmHg CO_2), with a further significant drop during REM sleep when HCVR was less than a third of that during wakefulness (0.45 ± 0.10 L/min/mmHg CO_2) (Douglas et al. 1982). The decreased ventilatory responses to hypercapnia and hypoxia in REM sleep help explain rapid eye movement (REM)-related hypoxemic episodes (Sin et al. 1999; Douglas et al. 1982).

5.5 Breathing in REM Sleep

Irregular breathing is a common feature of REM sleep. Several studies have investigated the correlation between phasic eye movements and changes in respiratory patterns. Gould et al. showed that increasing eye-movement activity was associated with a reduction in tidal volume (VT) and a shortening of total breath time (TT), whereas inspiratory time (TI) was unchanged (Gould et al. 1988). This association of phasic events and changes in respiration causing the irregularity remained unchanged in the night, although the frequency of events increased from early to late REM sleep. Schafer and Schafke showed that the typical changes in respiration associated with rapid eye movements, for example, rapid, shallow breathing with predominant effects on rib cage excursions and TE, are by no means attenuated during increased respiratory drive during hypercapnia but are even more pronounced (Schafer and Schlafke 1998).

In general, central apnea is less frequent during REM sleep than NREM sleep (particularly the lighter stages of NREM sleep), which is thought to be related to the greater arousability during the lighter stages of NREM sleep (Xie et al. 1994), decreased hypercapnic and hypoxic ventilatory responses during REM sleep, and hence reduced loop gain probably due to inhibition of accessory chest wall muscles (Douglas et al. 1982; Badr et al. 2019).

5.6 Cheyne–Stokes Breathing Pattern

CSR is a breathing pattern characterized by cycles of crescendo-decrescendo changes in tidal volume followed by central sleep apneas. This condition leads to variations in hemodynamic, respiratory, and neurologic functions, as well as in acid-base balance. Both historical and current theories regarding its causation emphasize a disruption in central respiratory control as the primary factor in the development of periodic respiration. Tissue perfusion and metabolically determined changes in acid-base balance have dominant roles in the production of CSR (Dowell et al. 1971).

In patients with heart failure, a complete cycle of apnea, hypopnea, hyperpnea, and hypopnea leading to the next apnea usually takes about 45 seconds but longer. At least ten consecutive minutes of these cycles should be recorded before CSR can be diagnosed. The *ICSD-3* specifies that at least ten central apneas and hypopneas per hour of sleep should occur, accompanied by arousals and derangement of sleep structure (AASM 2014).

5.6.1 Predisposing Factors

Congestive heart failure

Stroke
Renal failure
Lower resting $PaCO_2$ than normal
Narcotics

It is challenging to distinguish heart failure symptoms from sleep-related breathing disorders, as both may have overlapping symptoms like poor sleep quality, daytime somnolence or insomnia, paroxysmal nocturnal dyspnea, and easy fatigability. Central apneas with CSB and heart failure characteristically have a long cycle, typically in the 40 seconds or more. Polysomnographic studies demonstrate, a typical crescendo-decrescendo pattern. Due to longer circulation time, there is a long delay in the nadir in the oxygen saturation tracing after event termination. The typical CSB cycle time is long at 60–90 s. It is associated with left ventricular systolic dysfunction but has also been seen with diastolic dysfunction. Cardiovascular diseases are by far the most relevant causes of central disturbances, especially HF and atrial fibrillation (Randerath and Javaheri 2016). Most of these episodes are explained by an increase in loop gain, with overshooting and undershooting of the ventilation resulting in the pattern of periodic breathing. Even the arousals may play a role in these events (Douglas et al. 1982; Xie et al. 1994; Badr et al. 2019).

5.7 Treatment-Emergent Central Sleep Apnea (Formerly Complex Sleep Apnea)

Treatment-emergent central sleep apnea (TECSA, formerly complex sleep apnea) describes the appearance of central sleep apnea (CSA) and/or hypopnea while undergoing treatment for obstructive sleep apnea (OSA) (AASM 2014). Some patients with OSA or mixed apneas developed central sleep apnea after the significant resolution of the obstructive events by treatment with a positive airway pressure (PAP) device without a backup rate therapeutic failure (BaHammam 2017). However, new-onset CSA has also been reported during other non-CPAP treatments for OSA, including endoscopic sinus and nasal surgery and the use of oral appliances (Goldstein and Kuzniar 2012). Additionally, TECSA has been reported following tracheostomy, mandibular-advancement device (MAD), and maxillomandibular advancement surgery (BaHammam 2017).

Mohan et al noted that a patient with moderate OSA who used MAD initially had CSA (Mohan et al. 2016). The spontaneous resolution of MAD-emergent CSA was observed in this patient 1 year after MAD treatment. There have been three reports on the occurrence of similar MAD-emergent CSA during treatment with MAD. Therefore, the definition of TECSA should include both transient and persistent CSA that may occur after not only CPAP treatment but also following any form of therapy for OSA (Avidan et al. 2006).

The reported prevalence rate of TECSA has been strikingly inconsistent; a systemic review that included nine studies reported prevalence rates of 5.0% to

20.3% (Nigam et al. 2016). The prevalence of TECSA during whole-night titration studies varies from 5.0% to 12.1%, whereas it was between 6.5% and 20.3% during split-night titration studies (Nigam et al. 2016). However, the cumulative prevalence of TECSA was 8% (Nigam et al. 2016). This broad difference in prevalence can be attributed to the characteristics of the study groups like patient demographics, body mass index (BMI), gender, age, comorbidities, and causes related to the diagnostic method such as split-night titration and titration in the supine position (Liu et al. 2017).

An extensive study that tracked patients with OSA for 90 days after the start of CPAP therapy found that older age, higher residual AHI and central apnea index (CAI), and higher leak are the main predictors of TECSA; henceforth, patients with these characteristics should always necessitate early and close monitoring (Liu et al. 2017). In general, patients with TECSA have reduced adherence to PAP therapy.

5.7.1 Pathophysiology of TECSA

To comprehend the pathophysiology of TESCA, we need to investigate the pathophysiology of central sleep apnea. During sleep, in the absence of a wakefulness drive to breathe, metabolic control is solely responsible for breathing. Additionally, the hypercapnic and hypoxic drives are reduced, and there is also increased upper airway resistance. The pathophysiological mechanism of this phenomenon revolves around the "loop gain," which has two components: plant gain and controller gain. Controller gain refers to the chemosensitivity and ventilatory responses to hypoxia and hypercapnia. On the other hand, plant gain refers to the changes in $PaCO_2$ levels resulting from a given change in ventilation. Central sleep apnea can result secondary to unstable breathing caused by high loop gain, or secondary to reduced output from the central neurons. Therefore, high loop gain results in hyperventilation and a lower $PaCO_2$ level below the apneic threshold. When the $PaCO_2$ level drops below the apneic threshold, ventilation will cease, and apnea will occur and last until the $PaCO_2$ level increases above the threshold (BaHammam 2017). This phenomenon is more profound and common during non-rapid eye movement sleep when the CO_2 reserve, which is the difference between eupneic $PaCO_2$ and the apneic threshold, is highly labile.

PAP therapy settings have a minor influence on the emergence of TECSA; nevertheless, high PAP pressure, quick titration, and the usage of BPAP therapy are allegedly connected with more mask leaks and pressure intolerance and, consequently, sleep disruption and consequent CSA. Several theories have been proposed, including that the upper airway resistance owing to OSA can reduce the ventilatory control system, thus dropping the efficiency of CO_2 washout; when the resistance improves by PAP therapy, $PaCO_2$ washout occurs, hence reducing the $PaCO_2$ tension to a value under the CO_2 apnea threshold resulting in CSA. Over several days to weeks, the CO_2 apnea threshold is known to adjust, causing the resolution of CSA. Another suggested theory is that, in the occurrence of a significant mask leak or mouth breathing, when a nasal mask is used, it may lead to a washout of CO_2 from the anatomical dead space, which causes in a fall in CO_2 below the apnea threshold. A third theory, though less probable, is that PAP therapy may deteriorate sleep quality, resulting in recurrent shifts from sleep to wake and wake to sleep conditions, causing repeated arousals with resultant stimulation of the ventilatory drive, which decreases $PaCO_2$ below the CO_2 apnea threshold and the appearance of CSA during consequent sleep.

5.8 High-Altitude Periodic Breathing

The term "high-altitude illness" is used to describe the cerebral and pulmonary syndromes that can develop in unacclimatized persons shortly after ascent to high altitude. Acute mountain sickness and high-altitude cerebral edema refer to cerebral abnormalities and high-altitude pulmonary edema to pulmonary abnormalities (Hackett and Roach 2001). Most people will gradually acclimatize to altitude if there is a gradual ascent. Polysomnographic studies demonstrate recurrent central apneas primarily during NREM sleep at a frequency of >5/h. The cycle length is shorter than central apneas in heart failure in the range of 12–34 s (Muza 2015).

5.9 CSA Due to Medications (Narcotics/Opiates)

Most used drugs include morphine, fentanyl, and methadone—both prescriptions use and illicit drug abuse. Some of these patients may exhibit a low respiratory rate, accompanied by prolonged episodes of both obstructive and central sleep apneas. They also can have variations in tidal volume and ataxic breathing. The most important risk factors for the severity of CSA are morphine equivalent daily dose (MEDD) >200 mg and low or normal body mass index (Correa et al. 2015).

5.10 Differential Diagnosis

Most of the patients with central sleep apnea manifest with excessive daytime drowsiness, daytime fatigue, and/or fragmented sleep; nevertheless, some may present with insomnia as central apneas may manifest during the transition from wake to sleep causing disruption of sleep. As usual, it is essential to take a detailed patient history, conduct a physical examination, and review all medications being taken by the patient, including prescribed, over-the-counter, and any substances of illicit use. If CSA is a consideration, they should undergo diagnostic polysomnography. Other differential diagnosis includes:

- Obstructive sleep apnea (OSA)
- Periodic limb movements of sleep (PLMS)
- Rotating shift workers
- Narcolepsy

5.11 Management of Central Sleep Apnea

Treatment of CSA should be individualized based on the patient's severity of symptoms and the presence or absence of known comorbidities and risk factors. When CSA is discovered incidentally, the treatment can also be watchful waiting and close follow-up. At least 15–20% of patients with CSA resolve with no significant intervention. This applies to TECSA, where most apnea events during initial PAP titration are self-limited and go away over time. A follow-up is necessary to make sure that the emergent central apneas disappeared. This can be accomplished by tracing the PAP stored data. Nonetheless, the therapist should search for problems causing TECSA and resolve them. If over-titration is suspected, reducing the PAP settings may be necessary to resolve the issue. When a significant mask leak is identified, the mask should be adapted or changed.

5.11.1 Continuous Positive Airway Pressure

Continuous positive airway pressure (CPAP) improves cardiac function in patients with congestive heart failure (CHF) who also have Cheyne–Stokes's respiration and central sleep apnea (CSR-CSA). Sin et al. conducted a randomized, controlled trial that showed that CPAP improves cardiac function in CHF patients with CSR-CSA but not in those without it. Stratified analysis of patients with and without CSR-CSA revealed that those with CSR-CSA experienced both a significant improvement in LVEF at 3 months and a relative risk reduction of 81% in the mortality-cardiac transplantation rate of those who used CPAP (Sin et al. 2000). However, the Canadian prospective continuous positive airway pressure (CANPAP) trial had different results. Bradley et al. conducted the study and tested the hypothesis that CPAP would improve the survival rate without heart transplantation of patients with central sleep apnea and heart failure (Bradley et al. 2005). Although CPAP attenuated central sleep apnea, improved nocturnal oxygenation, increased the ejection fraction, lowered norepinephrine levels, and increased the distance walked in 6 min, it did not affect survival. Thus, the data do not support the use of CPAP to extend life in patients who have central sleep apnea and heart failure (Bradley et al. 2005). Nevertheless, in the post hoc analysis of the study, patients with residual AHI < 15/h on CPAP had improvement in LVEF and heart transplant-free survival (Arzt et al. 2007). Thus, the authors advocated a 1-month trial of CPAP, to be maintained only if it significantly reduces AHI. Based on those findings, CPAP should not be consistently used to manage CSR in patients with heart failure and CSR.

The goals of PAP titration in CSA are extrapolated from the studies in patients with obstructive sleep apnea. The strict criteria for real success are AHI < 5/h, even though some clinicians do accept AHI < 15/h as a partial success criterion. The therapy needs to be individualized.

5.11.2 Bilevel Positive Airway Pressure

Bilevel positive airway pressure (BIPAP) is an effective therapy for treating patients with central sleep apnea. Cheyne–Stokes respiration with central sleep apnea (CSR-CSA) is not associated with a good prognosis in patients with CHF. However, some patients do not respond to CPAP, so other therapeutic modalities should be considered, such as bilevel positive airway pressure (BPAP).

Dohi et al. studied 20 patients with HF because of left ventricular systolic dysfunction. The AHI improved significantly in 11 (AHI <15), but only slightly in 9, in whom the AHI remained high (>or= 15). Bilevel PAP titration significantly improved the AHI in the latter group. Those who were unresponsive to CPAP had significantly lower $PaCO_2$, higher plasma brain natriuretic peptide (BNP), longer mean duration of CSR, and fewer obstructive episodes than CPAP responders. Bi-level PAP could be an effective alternative for patients with HF and pure CSR-CSA who are unresponsive to CPAP (Dohi et al. 2008).

5.11.3 Oxygen

Sasayama et al. studied the sustained efficacy of HOT (home oxygen therapy) in the long-term treatment in 51 patients with CSA and CHF [New York Heart Association (NYHA) functional classes II–III] assigned to receive either nocturnal oxygen or usual breathing for 52 weeks (Sasayama et al. 2009). In the HOT group, Sasayama et al. (2009) observed a more significant reduction in apnea and hypopnea events, along with a notable increase in nocturnal oxygen saturation (Sasayama et al. 2009). These changes correlated with a marked improvement in the Specific Activity Scale, NYHA functional class, and ejection fraction. Consequently, HOT may be considered a valuable non-pharmacological therapeutic addition for HF patients with CSA (Sasayama et al. 2009).

Campbell et al. compared two common treatments for CSA-CSR. Subjects with CSA-CSR and stable congestive heart failure were randomized to 8 weeks of treatment: oxygen 2 L/min through nasal prongs and concentrator or 8 weeks adaptive servo-ventilation (ASV) using a crossover design separated by a 3-week washout (Campbell et al. 2012). Ten subjects (age 64 ± 10 years, left ventricular ejection fraction (LVEF) 28 ± 10.5%, apnoea-hypopnoea index (AHI) 63 ± 30/h) were recruited. On therapy, an AHI of <10/h was achieved in two out of seven using oxygen (29%), six of seven using ASV (86%) and six of seven with (86%). Median AHI on oxygen therapy: 13.4/h, ASV, 1.4/h. Over 8 weeks, ASV reduced CSA-CSR more effectively than oxygen therapy, but it was not accepted for long-term use. Neither treatment improved prognostic indices of heart failure or symptoms in the short term. However, oxygen may be of value as an adjunct therapy (Campbell et al. 2012).

Oxygen is effective in patients with high-altitude periodic breathing by improving the sleep architecture. The severity of the CSA can be reduced by descent, supplemental oxygen therapy and oral or intravenous acetazolamide. Bilevel noninvasive ventilation can also be used in addition to oxygen. Sleep quality can also be improved independently of the presence of CSA using benzodiazepines (Burgess and Ainslie 2016).

5.11.4 Adaptive Servo-Ventilation (ASV)

ASV is used for treatment for CSA in selected patients. It is important to point out that there are many controversies regarding the use of ASV in the treatment of central sleep apnea. ASV provides both positive expiratory airway pressure (EPAP) and inspiratory pressure support (IPAP), which is servo-controlled based on the presence of CSA based on an algorithm. Adaptive servo-ventilation (ASV) provides a varying amount of inspiratory pressure superimposed on a low level of CPAP, with a backup respiratory rate based on the underlying algorithm. Phillippe et al. studied patients regarding the compliance and effectiveness of ASV versus CPAP in treating CSR in heart failure over 6 months (Philippe et al. 2006). Both ASV and CPAP decreased the AHI, but noticeably, only ASV completely corrected CSA-CSR, with AHI below 10/h (Philippe et al. 2006).

Cowie et al. reported the results of their trial on the use of ASV in patients with CSA (Cowie et al. 2016). They randomly assigned 1325 patients with a left ventricular ejection fraction of 45% or less, an apnea-hypopnea index (AHI) of 15 or more events (occurrences of apnea or hypopnea) per hour, and a predominance of central events to receive guideline-based medical treatment with ASV or guideline-based medical treatment alone (control) (Cowie et al. 2016). In the ASV group, the mean AHI at 12 months was 6.6 events per hour. The incidence of the primary endpoint did not differ significantly between the adaptive servo-ventilation group and the control group (54.1% and 50.8%, respectively; hazard ratio, 1.13; 95% confidence interval [CI], 0.97 to 1.31; $P = 0.10$). All-cause mortality and cardiovascular mortality were significantly higher in the ASV group than in the control group (hazard ratio for death from any cause, 1.28; 95% CI, 1.06 to 1.55; $P = 0.01$; and hazard ratio for cardiovascular death, 1.34; 95% CI, 1.09 to 1.65; $P = 0.006$) (Cowie et al. 2016). ASV had no significant effect on the primary endpoint in heart failure patients with reduced ejection fraction and predominantly central sleep apnea, but all-cause and cardiovascular mortality were both increased with this therapy (Cowie et al. 2016).

Based on the SERVE-HF trial, AASM recommendations for the use of ASV in central sleep apnea are as follows (Aurora et al. 2016):

Recommendation 1: ASV targeted to normalize the AHI should not be used for the treatment of CSA syndrome (CSAS) related to CHF in adults with an ejection fraction ≤45% and moderate or severe CSA predominant, sleep-disordered breathing.

Recommendation 2: ASV targeted to normalize the AHI can be used to treat CSAS related to CHF in adults with an ejection fraction >45% or mild CHF-related CSAS.

The ADVENT-HF trial is designed to test whether ASV improves cardiovascular outcomes in heart failure patients with sleep apnea and, so far, has better compliance than previous trials, with no safety concerns to date. More than 600 patients have been enrolled in ADVENT-HF to date, and the investigators hope to enroll more than 800. On average, patients with obstructive sleep apnea (OSA) used the device for 4.6 hours per night at 1 month and 4.1 hours at 12 months. In contrast, patients with CSA consistently used the device for 5.2 hours per night at both 1 month and 12 months. The results of this study may help us in defining the future of CSA Management (Burgess and Ainslie 2016; Philippe et al. 2006; Canadian Institutes of Health Research 2021).

5.11.5 Overdrive Atrial Pacing

Garrigue et al. observed that patients who had received a pacemaker with atrial overdrive pacing to reduce the incidence of atrial tachyarrhythmia had a reduction in breathing disorders after the implantation of the pacemaker (Garrigue et al. 2002). They studied 15 patients with central or obstructive sleep apnea who had received permanent atrial-synchronous ventricular pacemakers for symptomatic sinus bradycardia. All patients underwent three polysomnographic evaluations on consecutive nights, the first night for baseline evaluation and then, in random order, one night in spontaneous rhythm and one in dual chamber pacing mode with atrial overdrive. "In patients with sleep apnea syndrome, atrial overdrive pacing significantly reduces the number of episodes of central or obstructive sleep apnea without reducing the total sleep time." The number of patients who are eligible for this therapy may be small. But this may help individual patients (Garrigue et al. 2002). However, others are not able to reproduce the same result (Luthje et al. 2005).

5.12 Phrenic Nerve Stimulation/Implantable Device (*Rremedē* System)

Costanzo et al. studied patients with CSA associated with poor prognosis in CHF patients. Their study aimed to evaluate if using phrenic nerve stimulation to treat CSA in patients with CSA and HF was associated with changes in HF-specific metrics. In patients with HF, 53% (41/77) had a ≥50% reduction in AHI from baseline to 6 months and 57% (40/70) from baseline to 12 months. A reduction in AHI occurred in 61 of 70 (87%) patients at 12 months. In October 2017, the US Food and Drug Administration (FDA) approved an implantable device for the treatment of central sleep apnea. The *remedē* system, as described by Costanzo et al. (2016), consists of a battery pack implanted under the skin on the upper chest. It delivers electrical currents to the phrenic nerve to stimulate the diaphragm (Costanzo et al. 2016). The most common adverse effects in the trial were implant site infection, concomitant device interaction, local tissue damage or pocket erosion, and swelling (Sin et al. 2000; Costanzo et al. 2016). The follow-up study by Fox et al. showed that at least 60% of patients in the treatment group achieved at least 50% reduction in AHI through 24 months (Fox et al. 2019). There are no safety concerns over a period of 36 months (Fox et al. 2019).

5.13 Theophylline

Theophylline is a phosphodiesterase-5 inhibitor has been used to manage CSR that is associated with premature birth, the ascent to high altitudes, and heart failure (AlDabal and BaHammam 2010). The mechanism by which theophylline treats CSR is not yet clear. Patients with chronic heart failure have elevated levels of adenosine, leading to a depression of respiration, and theophylline may stimulate respiration by binding to adenosine receptors in the central nervous system (Funaya et al. 1997). On the other hand, theophylline may cause hyperventilation, thus inhibiting an increase in $PaCO_2$ and stopping an essential step in the induction of CSA. It may alter the central chemosensitivity reaction to $PaCO_2$, or it may reverse hypoxia-induced hypoventilation (Sanders et al. 1980). Irrespective of its mechanism, many studies have shown that taking oral theophylline to patients with CHF successfully treats CSR by dropping the number of episodes of central apnea and hypopnea per hour (AlDabal and BaHammam 2010). Unfortunately, all these studies were short-term and used small sample sizes. Nevertheless, there is insufficient data on the long-term safety profile of this potentially hazardous medication and the possibility of drug interactions. In particular, theophylline appears to be connected with arrhythmia, which could negatively affect a poorly functioning left ventricle (Huerta et al. 2005).

5.14 Acetazolamide (Diamox)

Acetazolamide is a weak diuretic that inhibits carbonic anhydrase in the brain, kidneys, vascular endothelium, red blood cells, and peripheral and central chemoreceptors with varying effects, depending on means and site of administration, duration, and dose (Sharp et al. 1985). Acetazolamide interrelates with the central and peripheral respiratory system in a compound way, and elements of its mechanism of action are not yet clearly understood. Nevertheless, it is recognized that acetazolamide inhibits carbonic anhydrase and hence produces mild metabolic acidosis. This stimulates the central respiratory system, causing hyperventilation, a reduction in $PaCO_2$, and a rise in PaO_2. Additionally, studies have demonstrated that acetazolamide results in a condition of peripheral and central chemoreceptor depression (Teppema and Dahan 1999). Nevertheless, acetazolamide is likely to lose its effects over time, and there are no long-term studies in patients with CSR and heart failure.

References

AASM (2014) American Academy of Sleep Medicine (AASM). International classification of sleep disorders (ICSD), 3rd edn. AASM, Darien

AlDabal L, BaHammam AS (2010) Cheyne-stokes respiration in patients with heart failure. Lung 188(1):5–14. https://doi.org/10.1007/s00408-009-9200-4

Arzt M, Floras JS, Logan AG, Kimoff RJ, Series F, Morrison D, Ferguson K, Belenkie I, Pfeifer M, Fleetham J, Hanly P, Smilovitch M, Ryan C, Tomlinson G, Bradley TD, CANPAP Investigators (2007) Suppression of central sleep apnea by continuous positive airway pressure and transplant-free survival in heart failure: a post hoc analysis of the Canadian Continuous Positive Airway Pressure for Patients with Central Sleep Apnea and Heart Failure Trial (CANPAP). Circulation 115(25):3173–3180. https://doi.org/10.1161/circulationaha.106.683482

Aurora RN, Bista SR, Casey KR, Chowdhuri S, Kristo DA, Mallea JM, Ramar K, Rowley JA, Zak RS, Heald JL (2016) Updated adaptive servo-ventilation recommendations for the 2012 AASM guideline: "the treatment of central sleep apnea syndromes in adults: practice parameters with an evidence-based literature review and meta-analyses". J Clin Sleep Med 12(5):757–761. https://doi.org/10.5664/jcsm.5812

Avidan AY, Guilleminault C, Robinson A (2006) The development of central sleep apnea with an oral appliance. Sleep Med 7(2):187–191. https://doi.org/10.1016/j.sleep.2005.06.013

Badr MS, Dingell JD, Javaheri S (2019) Central sleep apnea: a brief review. Curr Pulmonol Rep 8(1):14–21. https://doi.org/10.1007/s13665-019-0221-z

BaHammam AS (2017) Treatment-emergent central sleep apnea (complex sleep apnea). Sleep Vigilance 1:53–56. https://doi.org/10.1007/s41782-017-0015-8

Berry RB, Brooks R, Gamaldo C, Harding SM, Lloyd RM, Quan SF, Troester MT, Vaughn BV (2020) AASM Scoring Manual updates for 2020 (Version 2.6). J Clin Sleep Med 13(5): 665–666. https://doi.org/10.5664/jcsm.6576

Bradley TD, Logan AG, Kimoff RJ, Series F, Morrison D, Ferguson K, Belenkie I, Pfeifer M, Fleetham J, Hanly P, Smilovitch M, Tomlinson G, Floras JS, Investigators C (2005) Continuous positive airway pressure for central sleep apnea and heart failure. N Engl J Med 353(19): 2025–2033. https://doi.org/10.1056/NEJMoa051001

Burgess KR, Ainslie PN (2016) Central sleep apnea at high altitude. Adv Exp Med Biol 903:275–283. https://doi.org/10.1007/978-1-4899-7678-9_19

Campbell AJ, Ferrier K, Neill AM (2012) Effect of oxygen versus adaptive pressure support servo-ventilation in patients with central sleep apnoea-Cheyne Stokes respiration and congestive heart failure. Intern Med J 42(10):1130–1136. https://doi.org/10.1111/j.1445-5994.2011.02623.x

Canadian Institutes of Health Research (2021) Effect of adaptive servo ventilation (ASV) on survival and hospital admissions in heart failure (ADVENT-HF) (trial registration). ClinicalTrials.gov Identifier: NCT01128816. https://clinicaltrials.gov/ct2/show/results/NCT01128816?view=results

Carmona-Bernal C, Ruiz-Garcia A, Villa-Gil M, Sanchez-Armengol A, Quintana-Gallego E, Ortega-Ruiz F, Baron-Esquivias G, Capote F (2008) Quality of life in patients with congestive heart failure and central sleep apnea. Sleep Med 9(6):646–651. https://doi.org/10.1016/j.sleep.2007.11.008

Correa D, Farney RJ, Chung F, Prasad A, Lam D, Wong J (2015) Chronic opioid use and central sleep apnea: a review of the prevalence, mechanisms, and perioperative considerations. Anesth Analg 120(6):1273–1285. https://doi.org/10.1213/ANE.0000000000000672

Costanzo MR, Ponikowski P, Javaheri S, Augostini R, Goldberg L, Holcomb R, Kao A, Khayat RN, Oldenburg O, Stellbrink C, Abraham WT, Remede System Pivotal Trial Study Group (2016) Transvenous neurostimulation for central sleep apnoea: a randomised controlled trial. Lancet 388(10048):974–982. https://doi.org/10.1016/S0140-6736(16)30961-8

Cowie MR, Wegscheider K, Teschler H (2016) Adaptive servo-ventilation for central sleep apnea in heart failure. N Engl J Med 374(7):690–691. https://doi.org/10.1056/NEJMc1515007

Dohi T, Kasai T, Narui K, Ishiwata S, Ohno M, Yamaguchi T, Momomura S (2008) Bi-level positive airway pressure ventilation for treating heart failure with central sleep apnea that is unresponsive to continuous positive airway pressure. Circ J 72(7):1100–1105. https://doi.org/10.1253/circj.72.1100

Douglas NJ, White DP, Weil JV, Pickett CK, Zwillich CW (1982) Hypercapnic ventilatory response in sleeping adults. Am Rev Respir Dis 126(5):758–762. https://doi.org/10.1164/arrd.1982.126.5.758

Dowell AR, Buckley CE 3rd, Cohen R, Whalen RE, Sieker HO (1971) Cheyne-Stokes respiration. A review of clinical manifestations and critique of physiological mechanisms. Arch Intern Med 127(4):712–726. https://doi.org/10.1001/archinte.127.4.712

Fox H, Oldenburg O, Javaheri S, Ponikowski P, Augostini R, Goldberg LR, Stellbrink C, McKane S, Meyer TE, Abraham WT, Costanzo MR (2019) Long-term efficacy and safety of phrenic nerve stimulation for the treatment of central sleep apnea. Sleep 42(11):zsz158. https://doi.org/10.1093/sleep/zsz158

Funaya H, Kitakaze M, Node K, Minamino T, Komamura K, Hori M (1997) Plasma adenosine levels increase in patients with chronic heart failure. Circulation 95(6):1363–1365. https://doi.org/10.1161/01.cir.95.6.1363

Garrigue S, Bordier P, Jais P, Shah DC, Hocini M, Raherison C, Tunon De Lara M, Haissaguerre M, Clementy J (2002) Benefit of atrial pacing in sleep apnea syndrome. N Engl J Med 346(6):404–412. https://doi.org/10.1056/NEJMoa011919

Goldstein C, Kuzniar TJ (2012) The emergence of central sleep apnea after surgical relief of nasal obstruction in obstructive sleep apnea. J Clin Sleep Med 8(3):321–322. https://doi.org/10.5664/jcsm.1926

Gould GA, Gugger M, Molloy J, Tsara V, Shapiro CM, Douglas NJ (1988) Breathing pattern and eye movement density during REM sleep in humans. Am Rev Respir Dis 138(4):874–877. https://doi.org/10.1164/ajrccm/138.4.874

Hackett PH, Roach RC (2001) High-altitude illness. N Engl J Med 345(2):107–114. https://doi.org/10.1056/NEJM200107123450206

Huerta C, Lanes SF, Garcia Rodriguez LA (2005) Respiratory medications and the risk of cardiac arrhythmias. Epidemiology 16(3):360–366. https://doi.org/10.1097/01.ede.0000158743.90664.a7

Liu D, Armitstead J, Benjafield A, Shao S, Malhotra A, Cistulli PA, Pepin JL, Woehrle H (2017) Trajectories of emergent central sleep apnea during CPAP therapy. Chest 152(4):751–760. https://doi.org/10.1016/j.chest.2017.06.010

Luthje L, Unterberg-Buchwald C, Dajani D, Vollmann D, Hasenfuss G, Andreas S (2005) Atrial overdrive pacing in patients with sleep apnea with implanted pacemaker. Am J Respir Crit Care Med 172(1):118–122. https://doi.org/10.1164/rccm.200409-1258OC

Mohan A, Henderson J, Mador MJ (2016) Mandibular advancement device-emergent central sleep apnea can resolve spontaneously: a case report. J Clin Sleep Med 12(1):137–138. https://doi.org/10.5664/jcsm.5414

Muza RT (2015) Central sleep apnoea-a clinical review. J Thorac Dis 7(5):930–937. https://doi.org/10.3978/j.issn.2072-1439.2015.04.45

Nigam G, Pathak C, Riaz M (2016) A systematic review on prevalence and risk factors associated with treatment—emergent central sleep apnea. Ann Thorac Med 11(3):202–210. https://doi.org/10.4103/1817-1737.185761

Philippe C, Stoica-Herman M, Drouot X, Raffestin B, Escourrou P, Hittinger L, Michel PL, Rouault S, d'Ortho MP (2006) Compliance with and effectiveness of adaptive servoventilation versus continuous positive airway pressure in the treatment of Cheyne-Stokes respiration in heart failure over a six month period. Heart 92(3):337–342. https://doi.org/10.1136/hrt.2005.060038

Randerath W, Javaheri S (2016) Sleep-disordered breathing in patients with heart failure. Curr Sleep Medicine Rep 2:99–106. https://doi.org/10.1007/s40675-40016-40047-y

Sanders JS, Berman TM, Bartlett MM, Kronenberg RS (1980) Increased hypoxic ventilatory drive due to administration of aminophylline in normal men. Chest 78(2):279–282. https://doi.org/10.1378/chest.78.2.279

Sasayama S, Izumi T, Matsuzaki M, Matsumori A, Asanoi H, Momomura S, Seino Y, Ueshima K, CHF-HOT Study Group (2009) Improvement of quality of life with nocturnal oxygen therapy in heart failure patients with central sleep apnea. Circ J 73(7):1255–1262. https://doi.org/10.1253/circj.cj-08-1210

Schafer T, Schlafke ME (1998) Respiratory changes associated with rapid eye movements in normo- and hypercapnia during sleep. J Appl Physiol (1985) 85(6):2213–2219. https://doi.org/10.1152/jappl.1998.85.6.2213

Sharp JT, Druz WS, D'Souza V, Diamond E (1985) Effect of metabolic acidosis upon sleep apnea. Chest 87(5):619–624. https://doi.org/10.1378/chest.87.5.619

Sin DD, Fitzgerald F, Parker JD, Newton G, Floras JS, Bradley TD (1999) Risk factors for central and obstructive sleep apnea in 450 men and women with congestive heart failure. Am J Respir Crit Care Med 160(4):1101–1106. https://doi.org/10.1164/ajrccm.160.4.9903020

Sin DD, Logan AG, Fitzgerald FS, Liu PP, Bradley TD (2000) Effects of continuous positive airway pressure on cardiovascular outcomes in heart failure patients with and without Cheyne-Stokes respiration. Circulation 102(1):61–66. https://doi.org/10.1161/01.cir.102.1.61

Skatrud JB, Dempsey JA (1983) Interaction of sleep state and chemical stimuli in sustaining rhythmic ventilation. J Appl Physiol Respir Environ Exerc Physiol 55(3):813–822. https://doi.org/10.1152/jappl.1983.55.3.813

Teppema LJ, Dahan A (1999) Acetazolamide and breathing. Does a clinical dose alter peripheral and central CO(2) sensitivity? Am J Respir Crit Care Med 160(5 Pt 1):1592–1597. https://doi.org/10.1164/ajrccm.160.5.9903088

Wedewardt J, Bitter T, Prinz C, Faber L, Horstkotte D, Oldenburg O (2010) Cheyne-Stokes respiration in heart failure: cycle length is dependent on left ventricular ejection fraction. Sleep Med 11(2):137–142. https://doi.org/10.1016/j.sleep.2009.09.004

White DP, Gleeson K, Pickett CK, Rannels AM, Cymerman A, Weil JV (1987) Altitude acclimatization: influence on periodic breathing and chemoresponsiveness during sleep. J Appl Physiol (1985) 63(1):401–412. https://doi.org/10.1152/jappl.1987.63.1.401

Xie A, Wong B, Phillipson EA, Slutsky AS, Bradley TD (1994) Interaction of hyperventilation and arousal in the pathogenesis of idiopathic central sleep apnea. Am J Respir Crit Care Med 150(2): 489–495. https://doi.org/10.1164/ajrccm.150.2.8049835
Xie A, Skatrud JB, Puleo DS, Rahko PS, Dempsey JA (2002) Apnea-hypopnea threshold for CO_2 in patients with congestive heart failure. Am J Respir Crit Care Med 165(9):1245–1250. https://doi.org/10.1164/rccm.200110-022OC

Chapter 6
Current Clinical Perspective of Adult Obstructive Sleep Apnea: Symptomatology, Clinical Features, and Different Phenotypes

Ahmed S. BaHammam, Mahadevappa Hunasikatti, and Seithikurippu R. Pandi-Perumal

Abstract The incidence and prevalence of obstructive sleep apnea (OSA) continue to rise in the United States and the world. The prevalence of OSA is conservatively estimated to be 3% among women and 10% among men 30 to 49 years of age and 9% among women and 17% among men 50 to 70 years of age, including an estimated 24 million persons in the United States who have not received a diagnosis. Symptoms of OSA generally begin gradually and insidiously, and are often present for years before the patient is referred for treatment. Also, in general, OSA patients use healthcare resources at approximately twice the rate of controls as far back as 10 years before their diagnosis.

OSA is a complex and heterogeneous disorder, and the used severity criteria for OSA based on the apnea-hypopnea index (AHI) do not capture the diverse spectrum of the condition. Therefore, we may need to subcategorize the OSA into well-recognized phenotypes, which help us in the management and outcome of clinical studies. A "phenotype" is defined as an observable expression of an individual's characteristics that result from the interaction between the individual's genes (genotype) and the environment, without any implication of a mechanism. A phenotype is distinct from an "endotype," which is the subtype of a disease defined by a unique or distinctive functional or pathophysiologic mechanism.

A. S. BaHammam
The University Sleep Disorders Center, Department of Medicine, College of Medicine, King Saud University, Riyadh, Saudi Arabia

King Saud University Medical City, King Saud University, Riyadh, Saudi Arabia

M. Hunasikatti (✉)
Division of Respiratory, Sleep and Anesthesiology (DSRA), OHT1, Food and Drug Administration (FDA), Silver Spring, MD, USA

S. R. Pandi-Perumal
Somnogen Canada Inc., College Street, Toronto, ON, Canada

Saveetha Medical College and Hospitals, Saveetha Institute of Medical and Technical Sciences, Saveetha University, Chennai, Tamil Nadu, India

A. S. BaHammam, M. Hunasikatti (eds.), *Sleep Apnea Frontiers*, Progress in Sleep Research, https://doi.org/10.1007/978-981-99-7901-1_6

Some of the specific OSA phenotypes may have an impact on the management of OSA. REM-related SDB was more prevalent in women, younger individuals, and African Americans. Compared to non-stage-specific OSA, patients with REM-related OSA were less symptomatic with less sleepiness. REM sleep is associated with greater sympathetic activity and cardiovascular instability in healthy individuals than NREM sleep. REM-related OSA has been found to be associated with an increased risk of hypertension, metabolic complications such as insulin resistance, and metabolic syndrome. Patients with OSA have high sympathetic activity when awake, with further increases in blood pressure and sympathetic activity during sleep. These increases are attenuated by treatment with CPAP. This may have an impact on the treatment of these specific populations.

Keywords Obstructive sleep apnea (OSA) · Apnea-hypopnea index (AHI) · REM-related sleep-disordered breathing (SDB) · Phenotypes and endotypes · Continuous positive airway pressure (CPAP)

6.1 Introduction

The incidence and prevalence of obstructive sleep apnea (OSA) continue to rise in the United States as well as in the world. The prevalence of OSA is conservatively estimated to be 3% among women and 10% among men 30 to 49 years of age and 9% among women and 17% among men 50 to 70 years of age, including an estimated 24 million persons in the US who have not received a diagnosis (Peppard et al. 2013). This has resulted in major economic, social, and health consequences. Approximately one billion of the world's population of 7.3 billion people, between the ages of 30 and 69 years, are estimated to have the most common type of sleep-disordered breathing, OSA. OSA prevalence is rising and affects all countries. The increase in prevalence is driven by the global increase in obesity, the major risk factor for OSA, and improved diagnostic methods.

Obstructive sleep apnea is characterized by episodic sleep state-dependent collapse of the upper airway, resulting in periodic reductions or cessations in ventilation, with consequent hypoxia, hypercapnia, or arousals from sleep (Dempsey et al. 2010).

6.2 Epidemiology

6.2.1 Global Burden Versus United States Burden

Benjafield et al. estimated that nearly one billion adults aged 30–69 years worldwide could have OSA, and the number of people with moderate to severe obstructive sleep apnea, for which treatment is generally recommended, is estimated to be almost 425 million. This analysis represents the first set of global estimates of OSA prevalence (Benjafield et al. 2019). The estimated number of individuals with OSA was highest in China, followed by the United States, Brazil, and India. The worldwide obesity epidemic and the aging population demographic are likely to contribute to the rising global prevalence of obstructive sleep apnea. Additionally, the high prevalence in countries such as China is likely to be due to racial and genetic differences in common anatomical features that increase the likelihood of obstructive sleep apnea, such as a narrower airway (Lam et al. 2005). The high estimated global prevalence suggests that the development of cheaper alternative devices to diagnose and treat OSA is required. The new wearable devices with adequate sensitivity and specificity in diagnosing OSA may fill this need. Some fear that this large prevalence of this may overburden the healthcare system. However, the timely management of obstructive sleep apnea is not just cost-effective but potentially cost-saving because of the prevention of major complications like cardiovascular disease, stroke, and reduction in motor vehicle accidents with associated improved quality of health (Ayas et al. 2006). The cost of treatment devices might be an important barrier to treatment not only in low-income and middle-income countries, but also in industrialized countries like the United States where disparity of health care results in inaccessibility to the treatment of OSA (Frost, and Sullivan, American Academy of Sleep Medicine 2016).

Definition: Symptoms of OSA generally begin gradually and insidiously, and are often present for years before the patient is referred for treatment. Also, in general, OSA patients use healthcare resources at approximately twice the rate of controls as far back as 10 years before their diagnosis. Analysis of the two main components of medical costs, physician claims and hospital stay, demonstrates greater utilization of both in the patient group (BaHammam et al. 1999; Ronald et al. 1999). Even though most patients are referred to sleep clinics because of snoring, excessive daytime sleepiness, and witnessed apneas, a comprehensive list of presenting symptoms are as follows:

6.2.2 Major Nocturnal Symptoms

- Snoring, usually loud, habitual, and bothersome to others.
- Witnessed apneas, which often interrupt the snoring and end with a snort.
- *Gasping and choking sensations*: Gasping and choking sensations that arouse the patient from sleep, though in an extremely low proportion relative to the number of apneas they experience. Choking or gasping during sleep or awakening with a dry mouth are commonly reported, often in association with snoring. Myers et al. found that nocturnal choking or gasping was the most predictive symptom for identifying patients with OSA, with a summary likelihood ratio of 3.3, when the diagnosis was established by an AHI of 10 or more events per hour. Snoring is common in sleep apnea patients but is not useful for establishing the diagnosis (summary LR, 1.1). Patients with mild snoring and a body mass index lower than 26 are unlikely to have moderate or severe obstructive sleep apnea (LR, 0.07 at

threshold of AHI ≥15/h). Nocturnal gasping (choking) and witnessed apnea are the most reliable indicators of OSA, whereas snoring is not very specific (Myers et al. 2013).
- *Nocturia*: Nocturia is one of the commonly associated symptoms of OSA. When patients repeatedly awaken from apneic events, they experience the urge to urinate, which in turn disturbs their sleep in addition to sleep apnea (Margel et al. 2006).
- *Insomnia*: Sleep maintenance insomnia with repetitive awakenings should prompt evaluation of OSA, particularly in women (Alotair and BaHammam 2008). Approximately one-third of patients with OSA complain of insomnia rather than daytime sleepiness, which is more common in females (Cho et al. 2018; Subramanian et al. 2011; Alotair and BaHammam 2008).
- Restless sleep, with patients, often experiencing frequent arousals and tossing or turning during the night.

6.2.3 Daytime Symptoms

- Nonrestorative sleep (i.e., "waking up as tired as when they went to bed").
- *Morning headache*: Morning headaches are reported by 10% to 30% of patients with untreated OSA for some time. They are generally bifrontal and squeezing in quality. However, they may not have associated nausea, photophobia, or phonophobia. They typically occur daily or most days of the week and may last for several hours after awakening in the morning. Early morning headaches may not have a good correlation with the severity of the disease (Russell et al. 2014).
- *Excessive daytime sleepiness (EDS)*: Daytime sleepiness is among the most common feature of OSA. Sleepiness is the inability to remain fully awake or alert during the wakefulness portion of the sleep-wake cycle during work hours and driving situations. Many patients may use the term fatigue interchangeably with drowsiness during sleep, and some may also use "tiredness," "low energy," or "poor focus in day-to-day activities related to work." Many patients also may underestimate it because of its gradual and insidious onset of the clinical feature. Developing clinical skills to elicit their presenting symptoms is important in the management of patients with OSA. It is easier to question the spouse or the bed partner who may give better information about the patient's daily performance related to the disease. This is more noticeable in sedentary activities, passive, boring, and monotonous activities. For example, the wife may notice that her husband falls asleep while watching TV or attending church services.
- Daytime fatigue/tiredness.
- Cognitive deficits; memory and intellectual impairment (short-term memory, concentration).
- Decreased vigilance.
- Morning confusion.
- Personality and mood changes, including depression and anxiety.

- Sexual dysfunction, including impotence and decreased libido.
- Awakening with a sensation of choking, gasping, or smothering.
- Awakening with a dry mouth or sore throat.
- Restless, fitful sleep.
- Episodes of cessation of breathing.
- Periods of silence terminated by loud snoring.
- Moodiness.
- Lack of concentration/decreased memory.
- Awakening with angina pectoris (BaHammam and Almeneessier 2019).
- Nightmares in patients with REM-OSA (BaHammam and Almeneessier 2019).

History of hypertension, cardiovascular disease, cerebrovascular disease, renal disease, type 2 diabetes mellitus, or gastroesophageal reflux disease.

Physical examination: Most Common physical finding may be a normal physical examination. However, sometimes the following may be abnormal physical examination findings in patients with OSA:

Moderate/morbid obesity: All adult patients should be screened for overweight and obesity by measuring height, weight, and calculating body mass index (BMI) as part of the routine physical examination in any primary care medical settings. In addition, measuring waist circumference in those with a BMI between 25 and 35 kg/m^2 as abdominal adiposity (and associated risk) may not be captured in this BMI range, particularly for people of Asian descent. BMI provides a better estimate of total body fat compared with body weight alone.

Calculation: It is calculated from the weight and square of the height as follows:

$$\text{BMI} = \text{body weight (in kg)} \div \text{height (in m) squared}$$

Obesity is defined as a BMI ≥ 30, whereas a BMI ≥ 25.0 indicates the person is overweight. The major contributing factors to obesity include environment, eating behavior, and physical inactivity. Psychosocial circumstances like alcohol intake, other drug abuse, and genetics also play important roles in obesity. Obesity and low physical activity levels are associated with moderate to severe OSA. Exercise helps decrease weight, blood pressure, depression, anxiety, and fatigue. In the adult population, the prevalence of OSA is estimated to be ~25%, and as high as 45% in obese subjects. Obesity predisposes to and potentiates OSA. The prevalence of OSA and its consequences are likely to increase, considering the current obesity epidemic. Recent estimates suggest that 60% of the adult population in industrialized countries is overweight (BMI $\geq$ 25 kg/m^2), and at least 30% is obese (BMI $\geq$ 30 kg/m^2). It is imperative to address obesity in every clinical setting when possible. Weight loss, especially through bariatric surgery, has been shown to reduce the severity and symptoms of OSA and sometimes, though not always, result in its complete resolution (Fritscher et al. 2007; Romero-Corral et al. 2010).

Narrow/crowded oropharyngeal airway: It is customary practice to use the modified Mallampati classification to evaluate upper airway narrowing. The Mallampati classification and Friedman tongue position assessment techniques significantly predict obstructive sleep apnea severity (Friedman et al. 2013). Craniofacial conditions like retrognathia, micrognathia, lateral peritonsillar narrowing, macroglossia, tonsillar hypertrophy, an elongated or enlarged uvula, a high arched or narrow palate, nasal septal deviation, and nasal polyps can narrow the upper airway contributing to the OSA (Epstein et al. 2009).

Large neck and/or waist circumference: OSA is more strongly correlated with an increased neck size or waist circumference than general obesity. OSA is particularly prominent among men with a collar size greater than 17 in. and women with a neck size greater than 16 in. In addition, neck circumference can be used to predict the presence as well as the severity of OSA in snoring Asian patients. In another study, cutoff values for waist circumference and waist-to-height ratio for females with OSA were 95.5 cm and 0.595, respectively, whereas the values for males were 100.5 cm, and 0.575, respectively (Unal et al. 2019; Kim et al. 2015).

6.2.4 Classification of Severity

The American Academy of Sleep Medicine Task Force report on "Sleep-Related Breathing Disorders in Adults: Recommendations for Syndrome Definition and Measurement Techniques in Clinical Research" classified OSA into mild, moderate, or severe disease based on the AHI and symptoms. This classification is based on consensus, and depending on definitions of "hypopnea," there can be great discordance in the AHI.

6.2.5 Severity Criteria: AASM

The severity of the OSA has two components: severity of daytime sleepiness and overnight monitoring. A severity level should be specified for both components. The rating of severity for the syndrome should be based on the most severe component (Sakhamuri et al. 2020). The simplicity of severity criteria may be too simplistic at best. As Sakhamuri et al. argue "In addition, the current severity classification should be further updated with the use of either the Z-scores or ideally outcomes-based cut-off values, as being practised for other diseases. It is not only that a fixed threshold for defining clinically relevant OSA is inappropriate in the light of the age- and sex-dependent effects on AHI produced by Boulos et al. (Margel et al. 2006), but also that the current AHI threshold (i.e., 5) and the other parameters used by the AASM to define clinically relevant OSA are far too liberal. Overdiagnosis could be a greater problem than underdiagnosis. The current practice of simplifying the diagnosis of a complex disease can lead to imprecision" (Sakhamuri et al. 2020).

6.2.5.1 Sleepiness (Level of Evidence—3)

1. *Mild*: Unwanted sleepiness or involuntary sleep episodes occur during activities that require little attention. Examples include sleepiness that is likely to occur while watching television, reading, or traveling as a passenger. Symptoms produce only minor impairment of social or occupational function.
2. *Moderate*: Unwanted sleepiness or involuntary sleep episodes occur during activities that require some attention. Examples include uncontrollable sleepiness that is likely to occur while attending activities such as concerts, meetings, or presentations. Symptoms produce moderate impairment of social or occupational function.
3. *Severe*: Unwanted sleepiness or involuntary sleep episodes occur during activities that require more active attention. Examples include uncontrollable sleepiness while eating, during a conversation, walking, or driving. Symptoms produce marked impairment in social or occupational function.

6.2.5.2 Sleep-Related Obstructive Breathing Events (Level of Evidence—2)

1. Mild: 5–15 events/h
2. Moderate: 15–30 events/h
3. Severe: greater than 30 events/h

It is important to note that the sleep stages and slow-wave sleep are generally preserved in mild OSA. However, asymptomatic, mild OSA is associated with an increased risk of hypertension, and may have a stronger association at younger ages. There are conflicting data regarding the risk of cardiovascular events. Patients traditionally classified as having moderate OSA are those with an AHI/RDI/REI (Apnea-Hypopnea Index/Respiratory Disturbance Index/Respiratory Event Index) between 15 and 30 respiratory events per hour of sleep, can continue their daily activities, but at reduced levels, but may have an increased incidence of motor vehicle violations or accidents. In contrast, patients classified as having severe OSA are those with an AHI/RDI/REI greater than 30 respiratory events per hour of sleep, and more often have daytime sleepiness that interferes with normal daily activities. They are at risk for accidental injury from sleepiness. Patients with severe OSA are at increased risk for all-cause mortality and a variety of cardiovascular comorbidities, including hypertension, coronary artery disease, and arrhythmia. At times, we do see paradoxically, patients with severe OSA may have few clinical symptoms limiting their activities of daily living (Anon 1999).

6.2.6 *Phenotypes of Sleep Apnea*

OSA is a complex and heterogeneous disorder, and the AASM severity of OSA alone does not capture the diverse spectrum of the condition (BaHammam and Almeneessier 2019). We may need to subcategorize the OSA into well-recognized phenotypes, which will help us in the management and clinical outcome assessment studies. A "phenotype" is defined as an observable expression of an individual's characteristics that result from the interaction between the individual's genes (genotype) and the environment, without any implication of a mechanism. A phenotype is distinct from an "endotype," which is the subtype of a disease defined by a unique or distinctive functional or pathophysiologic mechanism. Pathogenic mechanisms of OSA based on craniofacial morphology, obesity, arousal functions, upper airway muscle activity, ventilatory control stability, and nocturnal rostral fluid shift constitute potential endotypes of OSA.

6.3 Subramani et al. Classifies the Phenotypes and Endotypes as Follows (Subramani et al. 2017)

6.3.1 *OSA Phenotypes and Corresponding Endotypes*

OSA phenotypes	Predominant endotypes		Mechanism
OSA in elderly patients	Low arousal threshold rostral fluid shift	Hyporesponsive genioglossus	Anatomical (abnormal fat distribution)
OSA in males	Anatomical (android obesity)	Rostral fluid shift (age > 40 years)	
OSA in fluid overloaded states	Rostral fluid shift		
OSA in menopausal women	Anatomical (abnormal fat distribution)		
Ethnic OSA phenotypes			
African Americans	Anatomical (obesity)		
Asians	Anatomical (abnormal craniofacial morphology)		
Caucasians	Anatomical (both obesity and abnormal craniofacial morphology)		
REM-related OSA phenotype	Hyporesponsive		
	Genioglossus		
Supine-related OSA phenotype	Anatomical	Hyporesponsive genioglossus	

6.3.1.1 Anatomical Endotypes

Obesity and craniofacial abnormalities account for two-thirds of the variation in OSA severity measured by AHI. Dempsey et al. showed that the single most important cephalometric variable in predicting AHI severity was the horizontal dimension of the maxilla (i.e., porion vertical to supradentale [PV-A] distance). When the PV-A distance was relatively narrow (< 97 mm), the probability of having mild (AHI, 15 to 30/h) to severe (AHI > 30/h) SDB increased fivefold to sevenfold in nonobese subjects and threefold in obese subjects. Thus, in nonobese subjects (average BMI, 25 ± 2) and subjects with narrow upper airway dimensions, four cephalometric dimensions were the dominant predictors of AHI, accounting for 50% of the variance. However, in subjects with a large anteroposterior facial dimension, BMI was the major predictor of AHI, and a BMI > 28 increased the probability of moderate-to-severe sleep apnea by approximately fivefold. Finally, the combination of cephalometric dimensions and BMI accounted for an increasing amount of AHI variance as the severity of AHI increased. Across the population spectrum of SDB, four cephalometric dimensions of the upper airway in combination with BMI accounted independently for up to two-thirds of the variation in AHI; and the relative contribution of these two sets of determinants of AHI varied depending on airway size, obesity, and the amount of SDB.

Obesity and craniofacial abnormalities contribute synergistically to increases in collapsibility of the passive pharyngeal airway in patients with SDB.

6.3.1.2 Physiologic Endotypes

OSA Endotype Related to Upper Airway Dilator Muscle

The genioglossus is the main upper airway dilator. In sleeping obstructive apnea patients: (1) Mechanoreceptor feedback is responsible for most of the genioglossus response to chemical drive; (2) Mechanoreceptor feedback is effective only above a threshold chemical drive, which varies greatly among patients. The contraction of the genioglossus directly dilates the upper airway by pulling the base of the tongue forward (Loewen et al. 2011).

Pharyngeal dilator muscle activation (GGEMG) during wakefulness is greater in patients with obstructive sleep apnea (OSA) than in healthy control subjects, representing a neuromuscular compensatory mechanism for a more collapsible airway, but it is reduced at sleep onset in both (Fogel et al. 2001; Worsnop et al. 1998).

Findley et al. studied nocturnal sleep studies of 12 patients with obstructive sleep apnea and a matched control group of 12 subjects without the sleep apnea syndrome were analyzed to compare arterial oxyhemoglobin saturation (SaO_2) during REM and non-REM sleep. The obstructive events increase in frequency and duration and are associated with more pronounced hypoxemia during REM versus NREM sleep (Findley et al. 1985).

OSA Due to Fluid Overload

In both healthy male subjects and men with heart failure, the severity of obstructive sleep apnea (OSA) is related to the amount of fluid displaced from their legs into the neck overnight. Elias et al. studied 26 patients with ESRD (14 men) on conventional hemodialysis. All subjects underwent polysomnography. LFV (left leg volume) was measured by bioelectric impedance at bedtime and repeated in the next morning on awakening. The nocturnal rostral fluid shift is associated with the severity of OSA in ESRD (Elias et al. 2012).

According to the "rostral fluid shift" hypothesis, during the day, fluid accumulates in the intravascular and interstitial spaces of the legs due to gravity, and upon lying down at night, redistributes rostrally, again owing to gravity. Some of this fluid may accumulate in the neck, increasing tissue pressure and causing the upper airway to narrow, thereby increasing its collapsibility, and predisposing to OSA. In heart failure patients, with increased rostral fluid shift, fluid may additionally accumulate in the lungs, provoking hyperventilation and hypocapnia, driving below the apnea threshold, leading to CSA (White and Bradley 2013).

6.3.2 Clinical Phenotypes of OSA

Sex Differences in OSA: Males are two to three times more likely to have OSA than females. The studies show that the severity of OSA was significantly higher in males (RDI 41.2 ± 27.9 in men vs. 30.0 ± 26.7 in women, despite a lower BMI and age in the men). The severity of OSA, as measured by RDI, varied significantly with changes in anthropometric measures such as neck circumference and waist and hip sizes in all genders. The waist-to-hip ratio was associated with OSA severity in men but not in women. Despite having a higher mean BMI and age, women have lower RDI than men. The waist-to-hip ratio is more predictive of the severity of OSA in men than in women (Gasa et al. 2019).

Obesity and increased neck circumference are risk factors for obstructive sleep apnea/hypopnea syndrome. OSA is more common in men than in women, despite women having higher rates of obesity and greater overall body fat. One factor in this apparently paradoxical sex distribution may be the differing fat deposition patterns adjacent to the upper airway in men and women. In a study, total neck soft tissue volume was greater in men [1295 (62)], but the volume of fat did not differ between the sexes [291 (29) vs. 273 (18) cm^3)] The only regions impinging on the pharynx, which showed a larger absolute volume of fat in men [3.2 (0.7) vs. 1.1 (0.3) cm^3] and a greater proportion of neck fat in men [1.3 (0.3)% vs. 0.4 (0.1)%] were the anterior segments inside the mandible at the palatal level. There are differences in neck fat deposition between the sexes, which, together with the greater overall soft tissue loading on the airway in men, may be factoring in the sex distribution of SAHS (Whittle et al. 1999).

OSA in the postmenopausal state: The menopausal transition is associated with an increased likelihood of SDB indicated by the AHI level after controlling for age, body habitus, and several lifestyle factors. Compared with premenopausal women, postmenopausal women were 2.6 times more likely to have SDB defined by an AHI of five or more and 3.5 times more likely with SDB defined by an AHI of 15 or more (Young et al. 2003). It is well-accepted now that menopause, pregnancy, and polycystic ovarian syndrome increase the risk for OSA in women (Young et al. 2003).

Bixler et al. conducted a study on this phenotype of patients. They evaluated the question of the difference of OSA in relation to gender. They used a two-phase random sample from the general population The results of their study indicated that, for clinically defined OSA (apnea/hypopnea index > or =10 and daytime symptoms), men had a prevalence of 3.9% and women 1.2%, resulting in an overall ratio of sleep apnea for men to women of 3.3:1. The prevalence of OSA was relatively low in premenopausal women (0.6%) as well as postmenopausal women with hormone replacement therapy (HRT). Further, in these women, the presence of OSA appeared to be associated exclusively with obesity (BMI > or =32.3 kg/m^2). Postmenopausal women without HRT had a prevalence of OSA that was significantly higher than the prevalence in premenopausal women with HRT (2.7% vs. 0.6%) and was more like the prevalence in men (3.9%), although it remained significantly less when controlling for age and BMI. Combined, these data indicate that menopause is a significant risk factor for sleep apnea in women and that hormone replacement appears to be associated with reduced risk (Bixler et al. 2001).

6.3.3 OSA in Ethnic Patients

Anatomical factors, such as obesity and craniofacial morphology, are key determinants of the predisposition to airway collapse; however, their relative importance for OSA risk likely varies between ethnicities. Asian OSA populations primarily display features of craniofacial skeletal restriction, African Americans display more obesity and enlarged upper airway soft tissues, while Caucasians show evidence of both bony and soft tissue abnormalities. Caucasians are more obese, and the Chinese have more skeletal restrictions. However, the ratio of obesity to craniofacial bony size (or anatomical balance, an important determinant of upper airway volume and OSA risk) was similar between Caucasians and Chinese OSA patients. Ethnicity appears to influence OSA craniofacial phenotype but furthermore, the relative contribution of the anatomical factors underlying OSA risk. The skeletal restriction craniofacial phenotype may be particularly vulnerable to increasing obesity rates (Sutherland et al. 2012).

Lee et al. studied to explore the differences in craniofacial structures and obesity as risk factors for OSA in Caucasian and Chinese patients. They found that craniofacial structures and obesity contributed differentially to OSA in these two ethnic groups. For the same degree of obesity, Chinese patients had more severe OSA and more craniofacial bony restriction. When OSA severity was similar, Caucasian patients were more overweight and had larger neck circumference, whereas the Chinese patients exhibited more craniofacial bony restriction (Lee et al. 2010).

African American individuals appear to be more predisposed to SDB than white persons. This increased predisposition varies according to age. The odds ratio is greater than three in children younger than 13 years and is 1.88 in persons younger than 25 years. In elderly African Americans, the risk is increased twofold. Examination of craniofacial morphology found that brachycephaly is associated with an increased AHI in whites but not in African Americans (Cakirer et al. 2001).

While African Americans had a similar prevalence of OSA as whites (32% and 30%, respectively) in a community-based survey of individuals 65 years of age and older, this group was 2.1 times more likely to have severe OSA (Ancoli-Israel et al. 1995).

REM-predominant OSA: A commonly used diagnostic criterion is based on the ratio of the apnea-hypopnea index (AHI) during REM sleep to the AHI during NREM sleep (AHI_{REM}/AHI_{NREM}), with a value of at least two indicating a predominance of disordered breathing during REM sleep (Pirzada and BaHammam 2021). Patients with REM-related OSA who are willing to assess whether nightly CPAP use improves their daytime sleepiness or fatigue may be denied therapy if their overall AHI does not meet a threshold for the diagnosis of OSA (e.g., ≥5 events/h) (Mokhlesi and Punjabi 2012).

REM-related OSA can be categorized as REM-predominant and REM-isolated or REM-only OSA, even though this is not well established. Studies do show that the prevalence of REM-related SDB ranged from 13.5% to 36.7%. There were no clinically significant differences between the strict definition and the traditional definition of REM-related OSA. REM-related OSA was more prevalent in women, younger individuals, and African Americans. Compared to non-stage-specific obstructive sleep apnea (OSA), patients with REM-related SDB were equally symptomatic and hypersomnolent (Conwell et al. 2012).

Current evidence indicates that REM-OSA is prevalent among younger age groups and women and is independently associated with cardiometabolic complications, particularly hypertension, metabolic complications such as insulin resistance and metabolic syndrome. However, currently, there is no consensus on the accepted diagnostic criteria for REM-OSA. Available data suggest that adherence to positive airway pressure (PAP) therapy in patients with REM-OSA is suboptimal. Moreover, the currently accepted criteria for good adherence to PAP therapy of 4 h/night, 70% of the days may not be suitable for REM-OSA, as it will not cover most of the REM sleep periods. A high index of suspicion is needed to diagnose this disorder, and close follow-up is required to enhance adherence to therapy (Pirzada and BaHammam 2021).

REM sleep is associated with greater sympathetic activity and cardiovascular instability in healthy individuals and OSA patients versus NREM sleep. REM-related OSA has been found to be associated with a risk of hypertension. Patients with obstructive sleep apnea have high sympathetic activity when awake, with further increases in blood pressure and sympathetic activity during sleep. These increases are attenuated by treatment with CPAP. This may have an impact on the treatment of these specific populations (Somers et al. 1995).

References

Alotair H, BaHammam A (2008) Gender differences in Saudi patients with obstructive sleep apnea. Sleep Breath 12(4):323–329. https://doi.org/10.1007/s11325-008-0184-8

Ancoli-Israel S, Klauber MR, Stepnowsky C, Estline E, Chinn A, Fell R (1995) Sleep-disordered breathing in African American elderly. Am J Respir Crit Care Med 152(6 Pt 1):1946–1949

Anon (1999) Sleep–related breathing disorders in adults: recommendations for syndrome definition and measurement techniques in clinical research. The Report of an American Academy of Sleep Medicine Task Force. Sleep 22(5):667–689

Ayas NT, FitzGerald JM, Fleetham JA et al (2006) Cost-effectiveness of continuous positive airway pressure therapy for moderate to severe obstructive sleep apnea/hypopnea. Arch Intern Med 166:977–984

BaHammam AS, Almeneessier AS (2019) Dreams and nightmares in patients with obstructive sleep apnea: a review. Front Neurol 10:1127. https://doi.org/10.3389/fneur.2019.01127

BaHammam A, Delaive K, Ronald J, Manfreda J, Roos L, Kryger MH (1999) Health care utilization in males with obstructive sleep apnea syndrome two years after diagnosis and treatment. Sleep 22(6):740–747. https://doi.org/10.1093/sleep/22.6.740

Benjafield AV, Ayas NT, Eastwood PR, Heinzer R, Ip MSM, Morrell MJ, Nunez CM, Patel SR, Penzel T, Pépin J-L, Peppard PE, Sinha S, Tufik S, Valentine K, Malhotra A (2019) Estimation of the global prevalence and burden of obstructive sleep apnea: a literature-based analysis. Lancet Respir Med 7(8):687–698. https://doi.org/10.1016/S2213-2600(19)30198-5

Bixler EO, Vgontzas AN, Lin HM et al (2001) Prevalence of sleep-disordered breathing in women: effects of gender. Am J Respir Crit Care Med 163:608–613

Cakirer B, Hans MG, Graham G, Aylor J, Tishler PV, Redline S (2001) The relationship between craniofacial morphology and obstructive sleep apnea in whites and in African Americans. Am J Respir Crit Care Med 163(4):947–950

Cho YW, Kim KT, Moon HJ, Korostyshevskiy VR, Motamedi GK, Yang KI (2018) Comorbid insomnia with obstructive sleep apnea: clinical characteristics and risk factors. J Clin Sleep Med 14(3):409

Conwell W, Patel B, Doeing D et al (2012) Prevalence, clinical features, and CPAP adherence in REM-related sleep-disordered breathing: a cross-sectional analysis of a large clinical population. Sleep Breath 16:519–526

Dempsey JA, Veasey SC, Morgan BJ, O'Donnell CP (2010) Pathophysiology of sleep apnea. Physiol Rev 90:47–112

Elias RM, Bradley TD, Kasai T, Motwani SS, Chan CT (2012) Rostral overnight fluid shift in end-stage renal disease: relationship with obstructive sleep apnea. Nephrol Dial Transplant 27(4):1569–1573. https://doi.org/10.1093/ndt/gfr605

Epstein LJ, Kristo D, Strollo PJ Jr, Friedman N, Malhotra A, Patil SP, Ramar K, Rogers R, Schwab RJ, Weaver EM, Weinstein MD, Adult Obstructive Sleep Apnea Task Force of the American Academy of Sleep Medicine (2009) Clinical guideline for the evaluation, management, and long-term care of obstructive sleep apnea in adults. J Clin Sleep Med 5(3):263

Findley LJ, Wilhoit SC, Suratt PM (1985) Apnea duration and hypoxemia during REM sleep in patients with obstructive sleep apnea. Chest 87:432–436

Fogel RB, Malhotra A, Pillar G et al (2001) Genioglossal activation in patients with obstructive sleep apnea versus control subjects. Mechanisms of muscle control. Am J Respir Crit Care Med 164:2025–2030

Friedman M, Hamilton C, Samuelson CG, Lundgren ME, Pott T (2013) Diagnostic value of the Friedman tongue position and mallampati classification for obstructive sleep apnea: a meta-analysis. Otolaryngol Head Neck Surg 148(4):540

Fritscher LG, Mottin CC, Canani S, Chatkin JM (2007) Obesity and obstructive sleep apnea-hypopnea syndrome: the impact of bariatric surgery. Obes Surg 17(1):95–99

Frost & Sullivan, American Academy of Sleep Medicine (2016) Hidden health crisis costing America billions: underdiagnosing and undertreating sleep apnea draining health care system. https://aasm.org/resources/pdf/sleep-apnea-economic-crisis.pdf

Gasa M, López-Padrós C, Monasterio C, Salord N, Mayos M, Vilarrasa N, Fernandez-Aranda F, Montserrat JM, Dorca J (2019) Anthropometrical phenotypes are important when explaining obstructive sleep apnea in female bariatric cohorts. J Sleep Res 28(5):e12830. https://doi.org/10.1111/jsr.12830

Kim SE, Park BS, Park SH et al (2015) Predictors for presence and severity of obstructive sleep apnea in snoring patients: significance of neck circumference. J Sleep Med 12(2):34–38

Lam B, Ip M, Tench E, Ryan C (2005) Craniofacial profile in Asian and white subjects with obstructive sleep apnea. Thorax 60:504–510

Lee RW, Vasudavan S, Hui DS et al (2010) Differences in craniofacial structures and obesity in Caucasian and Chinese patients with obstructive sleep apnea. Sleep 33:1075–1080

Loewen AH, Ostrowski M, Laprairie J, Maturino F, Hanly PJ, Younes M (2011) Response of genioglossus muscle to increasing chemical drive in sleeping obstructive apnea patients. Sleep 34(8):1061–1073. https://doi.org/10.5665/SLEEP.1162

Margel D, Shochat T, Getzler O et al (2006) Continuous positive airway pressure reduces nocturia in patients with obstructive sleep apnea. Urology 67:974

Mokhlesi B, Punjabi NM (2012) "REM-related" obstructive sleep apnea: an epiphenomenon or a clinically important entity? Sleep 35(1):5–7. https://doi.org/10.5665/sleep.1570

Myers KA, Mrkobrada M, Simel DL (2013) Does this patient have obstructive sleep apnea: the rational clinical examination systematic review? JAMA 310(7):731–741. https://doi.org/10.1001/jama.2013.276185

Peppard PE, Young T, Barnet JH, Palta M, Hagen EW, Hla KM (2013) Increased prevalence of sleep-disordered breathing in adults. Am J Epidemiol 177:1006–1014

Pirzada AR, BaHammam AS (2021) Rapid eye movement predominant obstructive sleep apnoea: prognostic relevance and clinical approach. Curr Opin Pulm Med 27(6):514–522. https://doi.org/10.1097/MCP.0000000000000817

Romero-Corral A, Caples SM, Lopez-Jimenez F, Somers VK (2010) Interactions between obesity and obstructive sleep apnea: implications for treatment. Chest 137(3):711–719. https://doi.org/10.1378/chest.09-0360

Ronald J, Delaive K, Roos L, Manfreda J, BaHammam A, Kryger MH (1999) Health care utilization in the 10 years prior to diagnosis in obstructive sleep apnea syndrome patients. Sleep 22(2):225–229

Russell MB, Kristiansen HA, Kværner KJ (2014) Headache in sleep apnea syndrome: epidemiology and pathophysiology. Cephalalgia 34(10):752–755

Sakhamuri S, Teelucksingh S, BaHammam AS (2020) Is it time to redefine obstructive sleep apnoea? Respirology 25:802–803. https://doi.org/10.1111/resp.13880

Somers VK, Dyken ME, Clary MP, Abboud FM (1995) Sympathetic neural mechanisms in obstructive sleep apnea. J Clin Invest 96(4):1897–1904. https://doi.org/10.1172/JCI1182

Subramani Y, Singh M, Wong J, Kushida CA, Malhotra A, Chung F (2017) Understanding phenotypes of obstructive sleep apnea: applications in anesthesia, surgery, and perioperative medicine. Anesth Analg 124(1):179–191. https://doi.org/10.1213/ANE.0000000000001546

Subramanian S, Guntupalli B, Murugan T, Bopparaju S, Chanamolu S, Casturi L, Surani S (2011) Gender and ethnic differences in prevalence of self-reported insomnia among patients with obstructive sleep apnea. Sleep Breath 15(4):711–715

Sutherland K, Lee RW, Cistulli PA (2012) Obesity, and craniofacial structure as risk factors for obstructive sleep apnoea: impact of ethnicity. Respirology 17(2):213–222. https://doi.org/10.1111/j.1440-1843.2011.02082.x

Unal Y, Ozturk DA, Tosun K, Kutlu G (2019) Association between obstructive sleep apnea syndrome and waist-to-height ratio. Sleep Breath 23:523

White LH, Bradley TD (2013) Role of nocturnal rostral fluid shift in the pathogenesis of obstructive and central sleep apnoea. J Physiol 591(5):1179–1193. https://doi.org/10.1113/jphysiol.2012.245159

Whittle AT, Marshall I, Mortimore IL, Wraith PK, Sellar RJ, Douglas NJ (1999) Neck soft tissue and fat distribution: comparison between normal men and women by magnetic resonance imaging. Thorax 54(4):323–328. https://doi.org/10.1136/thx.54.4.323

Worsnop C, Kay A, Pierce R, Kim Y, Trinder J (1998) Activity of respiratory pump and upper airway muscles during sleep onset. J Appl Physiol (1985) 85:908–920

Young T, Finn L, Austin D, Peterson A (2003) Menopausal status, and sleep-disordered breathing in the Wisconsin Sleep Cohort Study. Am J Respir Crit Care Med 167:1181–1185

Chapter 7
Is Screening of Obstructive Sleep Apnea Necessary? Current Perspectives on Screening/Diagnosis

Barry Krakow and Mahadevappa Hunasikatti

Abstract It is well known that the associations between untreated moderate or severe OSA with significant medical consequences are much clearer when compared to patients with mild sleep apnea. In a prospective, population-based cohort study of middle-aged women and men, followed up to 24 years, SDB was associated with incident CHD or heart failure after adjusting for age, sex, BMI, and smoking. The association was strongest in the most severe SDB category and remained significant after further controlling for sleep duration, total cholesterol, and HDL cholesterol.

We are not sure if the screening of obstructive sleep apnea is associated with any significant medical benefit to the population. In 2017, U.S. Preventive Services Task Force (USPTF) concluded that "The USPSTF concludes that the current evidence is insufficient to assess the balance of benefits and harms of screening for obstructive sleep apnea (OSA) in asymptomatic adults."

The position of USPTF was not well received by the sleep community. USPTF is revising its recommendations in 2021. The truth is somewhere in the middle. Both overdiagnosis and underdiagnosis are possible in some segment of the population if we screen all asymptomatic adults. It is in the best interest of the public that screening for obstructive sleep apnea is done in targeted, selective population:

1. Asymptomatic adults with congestive heart failure, COPD, Down syndrome, or patients with stroke.
2. Asymptomatic adults who work in sensitive occupations, which involve the safety of public like: bus drivers, pilots, air traffic controllers, heavy machine operators, etc.
3. Asymptomatic adults who are morbidly obese with BMI $> 35/\text{m}^2$ or who have neck circumference of $>$40 cm.

B. Krakow
Maimonides Sleep Arts and Sciences, Savannah, GA, USA

M. Hunasikatti (✉)
Division of Respiratory, Sleep and Anesthesiology (DSRA), OHT1, Food and Drug Administration (FDA), Silver Spring, MD, USA

A. S. BaHammam, M. Hunasikatti (eds.), *Sleep Apnea Frontiers*, Progress in Sleep Research, https://doi.org/10.1007/978-981-99-7901-1_7

Keywords Obstructive sleep apnea (OSA) · US preventive services task force (USPSTF) · Screening strategies · Asymptomatic adults · Medical consequences

7.1 Introduction

7.1.1 Why Is Diagnosis of Obstructive Sleep Apnea (OSA) Important?

7.1.1.1 Mild OSA

There is some evidence that mild OSA may have some adverse consequences on the health of patients eventually. It is a highly prevalent disorder in adults. However, whether mild OSA has significant neurocognitive and cardiovascular complications is uncertain (Laratta et al. 2017). Mild OSA is diagnosed with OSA severity (apnea–hypopnea index, respiratory disturbance index, or oxygen desaturation index) in the sleep study was at least 5/h and less than 15/h. Most randomized controlled trials with continuous positive airway pressure (CPAP) demonstrate that treatment of mild OSA improves subjective assessments of daytime sleepiness by a small amount and that this effect may only be present in patients with an elevated level of sleepiness at baseline. However, limited existing data show no impact of treatment on objective measures of sleepiness. We may need to evaluate and manage the patients with this limitation (Jonas DE, Amick HR, Feltner C et al. (2017). Nonetheless, mild OSA and its variant upper airway resistance syndrome (UARS) may be common among insomnia patients in the syndrome described as complex insomnia. Recent research suggests treatment of mild to moderate OSA will enhance sleep quality and decrease insomnia symptoms, more so with advanced PAP machines compared to standard CPAP.

7.1.1.2 Moderate and Severe OSA

Associations between untreated moderate or severe OSA with significant medical consequences are much clearer. In a prospective, population-based cohort study of middle-aged women and men, followed up to 24 years, SDB was associated with incident CHD or heart failure after adjusting for age, sex, BMI, and smoking. The association was strongest in the most severe SDB category and remained significant after further controlling for sleep duration, total cholesterol, and HDL cholesterol (Hla et al. 2015).

In another study, a total of 5422 participants without a history of stroke at the baseline examination and untreated for sleep apnea were followed for a median of 8.7 years. One hundred ninety-three ischemic strokes were observed. A significant positive association between ischemic stroke and OAHI was observed in men. Men in the highest OAHI quartile (>19) had an adjusted hazard ratio of 2.86. In the mild to moderate range (OAHI, 5–25), each one-unit increase in OAHI in men was estimated to increase stroke risk by 6%. In women, stroke was not significantly associated with OAHI quartiles, but increased risk was observed at an OAHI greater than 25 (Redline et al. 2010).

With changing times, the diagnosis of OSA has changed. Due to a large economic burden, the expensive polysomnogram may not be justified despite its greater accuracy and validity through its actual measurements of sleep staging and, thus, sleep fragmentation. AASM recommendations regarding diagnostic testing for indications for PSG and HST are the current practice of medicine. Due to the high prevalence of OSA, there is a significant cost associated with evaluating all patients suspected of having OSA with PSG (currently considered the gold standard diagnostic test). Further, there also may be limited access to laboratory testing in some areas. HSAT is an alternative to in-lab sleep studies for diagnosing OSA in adults. While it may be less costly and more efficient for some populations, it has notable limitations, including the absence of sleep scoring and inconsistencies in determining the severity of OSA/UARS. The AASM guidelines have adopted an evidence-based approach to address some of these issues. However, they have not fully addressed the challenges associated with UARS diagnosis.

7.1.1.3 AASM Recommendations

1. We recommend that clinical tools, questionnaires, and prediction algorithms not be used to diagnose OSA in adults, in the absence of polysomnography or home sleep apnea testing. (STRONG)
2. We recommend that polysomnography, or home sleep apnea testing with a technically adequate device, be used for the diagnosis of OSA in uncomplicated adult patients presenting with signs and symptoms that indicate an increased risk of moderate to severe OSA. (STRONG)
3. We recommend that if a single home sleep apnea test is negative, inconclusive, or technically inadequate, polysomnography be performed for the diagnosis of OSA. (STRONG)
4. We recommend that polysomnography, rather than home sleep apnea testing, be used for the diagnosis of OSA in patients with significant cardiorespiratory disease, potential respiratory muscle weakness due to neuromuscular condition, awake hypoventilation or suspicion of sleep related hypoventilation, chronic opioid medication use, history of stroke, or severe insomnia. Moreover, emerging research suggests mental health patients with sleep disorders may receive more reliable assessments with PSG. (STRONG)
5. We suggest that, if clinically appropriate, a split-night diagnostic protocol, rather than a full-night diagnostic protocol for polysomnography, be used to diagnose OSA. (WEAK)
6. We suggest that when the initial polysomnogram is negative and clinical suspicion for OSA remains, a second polysomnogram be considered for the diagnosis of OSA. (WEAK) Keywords: obstructive sleep apnea, diagnosis, polysomnography, home sleep testing (Kapur et al. 2017).

Diagnostic testing for OSA should be performed in conjunction with a comprehensive sleep evaluation and adequate follow-up.

Table 7.1 Indications for evaluation by a sleep medicine physician

Primary criteria	Secondary criteria
1. BMI > 40 kg/mm^2	1. Symptoms of OSA:
2. BMI > 33 kg/mm^2	– Loud, habitual snoring
With either	– Witnessed apnea
(a) Hypertension requiring two medications or more	– Sleepiness during the major wake period
(b) Type II diabetes	– Chronic insomnia
3. Sleepiness-related crash or accident, off-road deviation/ near miss	– Nocturia
4. Fatigue or sleepiness during duty period	– Morning headache
5. Reports of unexplained poor-quality sleep	– Chronic allergic or nonallergic rhinitis

Polysomnography is the standard diagnostic test for the diagnosis of OSA in adult patients in whom there is a concern for OSA based on a comprehensive sleep evaluation.

While PSG remains the gold standard for the diagnosis of OSA, it involves cumbersome sensors and devices that, if minimized and less obtrusive, could make PSG more tolerable for patients. So AASM has published the guidelines regarding HST (Kapur et al. 2017). Nevertheless, the deficits in HSAT, such as the absence of sleep scoring and the inconsistencies in scoring AHI or RDI for OSA and UARS, have not yet been addressed (Table 7.1).

7.1.2 Is Screening of Sleep Apnea Necessary?

The US Preventive Services Task Force (USPSTF) makes recommendations about the effectiveness of specific preventive care services for patients without obvious related signs or symptoms. It bases its recommendations on the evidence of both the benefits and harms of the service and an assessment of the balance. The USPSTF does not consider the costs of providing a service in this assessment.

"The USPSTF concludes that the current evidence is insufficient to assess the balance of benefits and harms of screening for OSA in asymptomatic adults" (Kapur et al. 2017).

The American Academy of Family Physicians' recommendation is consistent with that of the USPSTF and concludes that the current evidence is insufficient to assess the balance of benefits and harms of screening for OSA in asymptomatic adults.

The American College of Physicians recommends conducting a sleep study for patients with unexplained daytime sleepiness (grade: weak recommendation, low-quality evidence). It also recommends polysomnography for diagnostic testing in patients with suspected OSA. For patients without serious comorbid conditions, portable sleep monitors are recommended when polysomnography is not available (grade: weak recommendation, moderate-quality evidence).

The American Academy of Sleep Medicine recommends that routine health maintenance evaluations include questions about OSA and evaluation for risk factors (obesity, retrognathia, and treatment-refractory hypertension). Positive findings should trigger a comprehensive sleep evaluation.

7.1.3 Should the Screening Be Individualized?

In view of contradictory recommendations regarding screening of OSA, it may be prudent to consider Screening of asymptomatic people (Jonas DE, Amick HR, Feltner C et al. (2017), National Academies of Sciences, Engineering, and Medicine (2013).

1. Asymptomatic patients with patients with congestive heart failure, and cardiac arrythmias (atrial fibrillation)
2. Asymptomatic patients with pulmonary conditions like COPD (chronic obstructive pulmonary disease)
3. Obese patients (BMI $>$ 30 kg^2/m)
4. Refractory hypertension
5. Craniofacial deformities: micrognathia, retrognathia
6. Commercial drivers: truck drivers, taxi drivers, train drivers, airline pilots, air traffic controllers
7. Mental health patients, particularly those with posttraumatic stress disorder (PTSD) or depressive disorders

Even if all patients who screened positive were tested with a home-based portable monitoring device at a cost of $162 each, the increase in health care costs would be an estimated $4.4 billion. When we consider the cost of diagnosing patients with mild OSA, with no known long-term medical consequences without (compared to with) treatment, the balance tips heavily against screening in the general population (Frost and Sullivan 2016).

However, the overall cost to society can be reduced, need to be considered in terms of the following:

- Treatment of OSA can reduce the risk of crashes.
- Although commercial truck drivers are safer per vehicle-mile traveled compared to passenger car drivers.
- Large vehicle crashes are particularly deadly, disabling, and expensive.
- Commercial truck crashes are estimated to cost between $304,500 and $7 million, depending on the size of the truck and the severity of the incident (Federal Motor Carrier Safety Administration 2015).
- These costs are generated by property and vehicular damage, lost wages from absenteeism, and insurance and related medical expenses. Rail safety incidents can cost much more.

"Management of Obstructive Sleep Apnea in Commercial Motor Vehicle Operators: Recommendations of the AASM Sleep and Transportation Safety Awareness Task Force" by Gurubhagavathula et al. are important for the safety of the public. (Gurubhagavatula et al. 2017) Routine screening should be conducted as part of the preexisting required fitness-for-duty medical evaluation. If this screening suggests substantial risk for OSA, we believe a comprehensive sleep evaluation should be performed by a board-certified sleep medicine physician.

7.2 How Can We Control the Total Cost Related to OSA?

If OSA-related health costs remain unchecked, the United States will continue to incur billions of dollars in unnecessary expenses. Reducing these costs requires a multidisciplinary approach that involves many constituency groups, including patients, families, physicians, payors, employers, and government agencies (Gurubhagavatula et al. 2017).

References

Federal Motor Carrier Safety Administration (2015) 2015 Pocket guide to large truck and bus statistics. U.S. Department of Transportation, Washington, DC

Frost and Sullivan (2016) Hidden health crisis costing America Billions. https://aasm.org/resources/pdf/sleep-apnea-economic-crisis.pdf

Gurubhagavatula I, Sullivan S, Meoli A et al (2017) Management of obstructive sleep apnea in commercial motor vehicle operators: recommendations of the AASM sleep and transportation safety awareness task force. J Clin Sleep Med 13(5):745–758. https://doi.org/10.5664/jcsm.6598

Hla KM, Young T, Hagen EW, Stein JH, Finn LA, Nieto FJ, Peppard PE (2015) Coronary heart disease incidence in sleep disordered breathing: the Wisconsin Sleep Cohort Study. Sleep 38(5): 677–684. https://doi.org/10.5665/sleep.4654

Jonas DE, Amick HR, Feltner C et al (2017) Screening for obstructive sleep apnea in adults: an evidence review for the U.S. Preventive Services Task orce. Evidence synthesis no. 146. AHRQ publication no. 14-05216-EF-1. Rockville, Agency for Healthcare Research and Quality

Kapur VK, Auckley DH, Chowdhuri S, Kuhlmann DC, Mehra R, Ramar K, Harrod CG (2017) Clinical practice guideline for diagnostic testing for adult obstructive sleep apnea: an American Academy of Sleep Medicine clinical practice guideline. J Clin Sleep Med 13(3):479–504

Laratta CR, Ayas NT, Povitz M, Pendharkar SR (2017) Diagnosis and treatment of obstructive sleep apnea in adults. CMAJ 189(48):E1481–E1488. https://doi.org/10.1503/cmaj.170296

National Academies of Sciences, Engineering, and Medicine (2013) National Cooperative Highway Research Program report 755: comprehensive costs of highway-rail grade crossing crashes. Transportation Research Board. http://www.trb.org/Main/Blurbs/169061.aspx

Redline S, Yenokyan G, Gottlieb DJ, Shahar E, O'Connor GT, Resnick HE, Diener-West M, Sanders MH, Wolf PA, Geraghty EM, Ali T, Lebowitz M, Punjabi NM (2010) Obstructive sleep apnea-hypopnea and incident stroke: the sleep heart health study. Am J Respir Crit Care Med 182(2):269–277. https://doi.org/10.1164/rccm.200911-1746OC

Chapter 8
Diagnosis and Management of OSA Comorbid with Insomnia (COMISA)

Ravi Gupta and Lokesh Kumar Saini

Abstract Research addressing comorbid insomnia with obstructive sleep apnea (OSA) (COMISA) has been conducted in two different ways—firstly, looking for the insomnia among subjects with OSA and secondly, assessment of OSA among subjects with chronic insomnia. This research has shown that prevalence of insomnia among subjects with OSA is higher than the prevalence of OSA among subjects with insomnia. While investigated through polysomnography, subjects with pure insomnia and COMISA show shorter total sleep time, greater percentage of wake after sleep onset and, poorer sleep efficiency compared to subjects having OSA and normal controls. Patients with COMISA suffer from two disorders and this combination is likely to produce greater disability and poorer quality of life among them compared to patients with either disorders alone. Pathophysiological models based on the scientific evidences favor that (i) OSA may lead to insomnia, (ii) insomnia may also lead to OSA and both can work in feed forward manner to develop COMISA. Insomnia is difficult to be diagnosed in the presence of other sleep disorders including OSA. It needs to be remembered that overlap of symptoms between OSA and insomnia may have a positive influence the diagnosis of either disorders before and, even after the treatment of both the disorders making the diagnosis difficult. A number of other challenges also make diagnosis of COMISA difficult that include physician's knowledge of all versus selected sleep disorders and, methods of assessment of sleep disorders. In view of limited literature, no consensus exists on the management of cases having COMISA. However, available evidences suggest that concurrent therapy (CBTi with PAP) could be better than sequential therapies where management of one disorder is followed by other.

R. Gupta (✉)
Department of Psychiatry and Division of Sleep Medicine, All India Institute of Medical Sciences, Rishikesh, Uttarakhand, India

L. K. Saini
Department of Pulmonary Medicine and Division of Sleep Medicine, All India Institute of Medical Sciences, Rishikesh, Uttarakhand, India

A. S. BaHammam, M. Hunasikatti (eds.), *Sleep Apnea Frontiers*, Progress in Sleep Research, https://doi.org/10.1007/978-981-99-7901-1_8

Keywords Sleep-apnea · Insomnia · COMISA · Sleep · PAP · Cognitive behavior therapy for insomnia

8.1 History of Insomnia and OSA Comorbidity Research

Comorbidity of obstructive sleep apnea (OSA) and insomnia (COMISA) was first recognized by Guilleminault et al. (1973) in two cases and later in a study of 56 patients with chronic insomnia (Sweetman et al. 2019a). However, it took another three decades before the research picked up in this area, perhaps due to the fact that two disorders were earlier considered to be affecting two different populations and, which rarely co-occur (Fig. 8.1) (Sweetman et al. 2019a). While OSA was considered to be disorders affecting primarily middle-aged obese males and characterized by daytime sleepiness, insomnia was thought to affect middle-aged females presenting with worries, anxiety, and depression (Sweetman et al. 2019a). However, researchers at that time failed to recognize that disorders have a number of symptoms that overlap with each other (Fig. 8.2).

Patients with COMISA differ across a number of demographic and clinical variables from the patients having either of the disorders alone. Moreover, presence of these two disorders together influence the outcome of the therapy for either of the disorders, as discussed in this chapter.

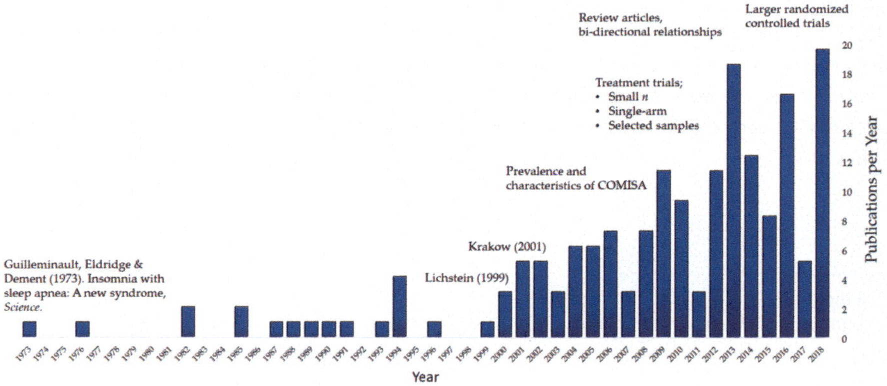

Fig. 8.1 Evolution of research on coexisting insomnia and sleep apnea, beginning with the 1973 article by Guilleminault and colleagues, with notable gaps in extensive research until the emergence of pivotal articles by Lichstein et al. (1999) and Krakov et al. (2001)

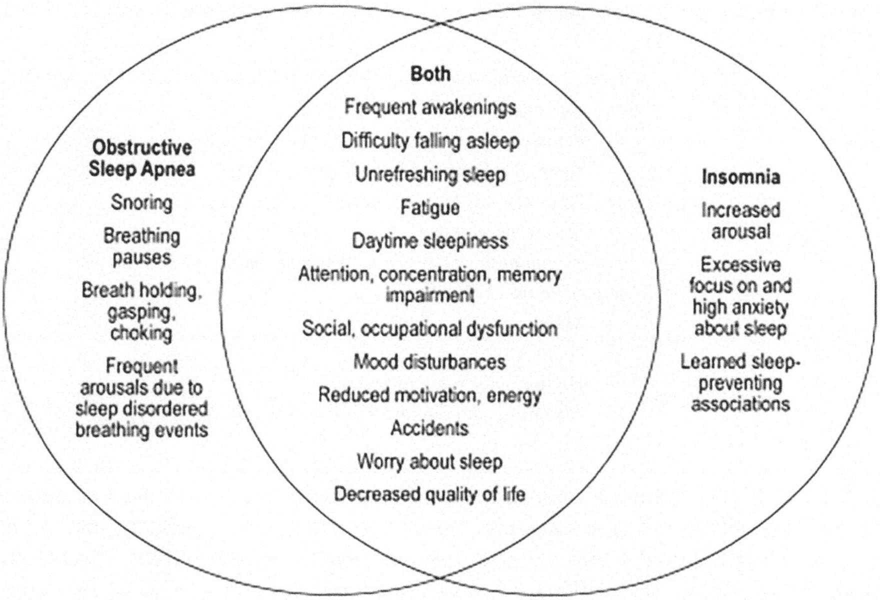

Fig. 8.2 The connection between symptoms and clinical elements linked to obstructive sleep apnea and insomnia. With permission from: Faith and colleagues. https://doi.org/10.5664/jcsm.27772

8.2 Prevalence of COMISA

Research addressing COMISA has been conducted in two different ways—firstly, looking for the insomnia among subjects with OSA and secondly, assessment of OSA among subjects with chronic insomnia. This research has shown that prevalence of insomnia among subjects with OSA is higher than the prevalence of OSA among subjects with insomnia (Zhang et al. 2019). Moreover, there is paucity of studies that have assessed OSA among subjects with chronic insomnia. This could be related to the fact that insomnia is still not considered as an indication for polysomnography, unless patients show some prominent symptoms suggestive of comorbid sleep disorders (Schutte-Rodin et al. 2008; Riemann et al. 2017).

Results of the research in this area have been confounded by varying definitions of insomnia and difference in methods to define OSA (Zhang et al. 2019). Studies have used different clinical criteria (Diagnostic and Statistical Manual versus International Classification of Sleep Disorders), validated questionnaires and scales, viz, insomnia severity index to define insomnia (Sweetman et al. 2019a; Zhang et al. 2019; Ong and Crawford 2013). In some of the studies, diagnostic criteria for insomnia were relaxed to presence of at least one nighttime symptom while in others, daytime impact of nighttime symptoms was considered mandatory to diagnose insomnia (Ong and Crawford 2013). Studies following former criteria for the diagnosis of insomnia showed greater prevalence of insomnia compared to those following latter criteria (Ong and Crawford 2013).

Table 8.1 Differences in patient characteristics based upon primary complaint (Ong and Crawford 2013)

1.	*Primary complaint insomnia having COMISA compared to pure OSA*
	Subjects more likely to:
	• Female
	• Have dysfunctional beliefs about sleep
	• Have restless legs syndrome
	• Lesser alcohol use
2.	*Primary complaint OSA having COMISA compared to No-OSA*
	Subjects more likely to:
	• Male
	• Older
	• Higher body mass index
	• Nighttime and daytime symptoms of OSA

Similar to the definition of insomnia, different definitions were used to define OSA across literature, for example, apnea-hypopnea index greater than 5 or 15 which were scored using either polysomnography or polygraphy (Sweetman et al. 2019a; Zhang et al. 2019). Most of the studies are cross-sectional in nature and thus can underestimate the true prevalence as insomnia may follow a relapsing remitting pattern (Ong and Crawford 2013). Many of the studies have also excluded diagnosed cases of OSA from the study, further reducing the prevalence (Ong and Crawford 2013). These factors have resulted in heterogeneity in the results of various studies. Studies also show different characteristics based on primary complaint (Table 8.1).

8.2.1 Prevalence of Insomnia Among Subjects with OSA

Around one third of patients with OSA also suffer from comorbid insomnia; however, it varies between 6% and 84% across different studies (Zhang et al. 2019; Ong and Crawford 2013). A recent study used the international classification of sleep disorders 3rd edition (ICSD-3) criteria for the diagnosis of insomnia and reported that 29% subjects with OSA report comorbid insomnia, predominantly difficulty in maintenance of sleep and falling asleep in half of them (Stelzer et al. 2021). Insomnia in patients with OSA can be predicted by presence of female gender, age greater than 60 years, tobacco use, presence of daytime sleepiness, symptoms of depression, and anxiety (Stelzer et al. 2021; Cho et al. 2018). These factors further highlight the role of neurodegeneration (including areas regulating sleep-wake cycle), hypothalamo-pituitary-adrenal (HPA) axis and cortical arousal for the occurrence of insomnia among OSA patients (Stelzer et al. 2021).

Most of the patients report difficulty in maintaining the sleep while initial and terminal insomnia are less prevalent (Zhang et al. 2019; Lichstein et al. 2013). It could be related to frequent awakenings (micro-arousals as well as complete arousals) associated with episodes of apnea, sympathetic stimulation, and activation of HPA axis, latter two being implicated in pathogenesis of insomnia *(discussed in Sect. 8.5 in detail)*.

Interestingly, prevalence of insomnia among patients with OSA differs across geographical regions. A recent review and metanalysis reported that pooled prevalence of insomnia among patients with OSA was higher in European and American regions compared to the Western Pacific region (Fig. 8.3) (Zhang et al. 2019). Authors explained that these differences could be related to difference in the population included in the studies. Studies from Western Pacific region included younger participants, had fewer women and with lower body mass index compared to other two regions. However, in this metanalysis—age, body mass index, as well as severity of OSA as measured by apnea-hypopnea index (AHI) were not associated with insomnia (Zhang et al. 2019). Insomnia disorder and early and middle phenotypes of insomnia are more common among females compared to males having OSA (Zhang et al. 2019). This meta-analysis further reported that criteria used to diagnose insomnia varied across studies and could have been a source of heterogeneity.

8.2.1.1 Risk Factors for Insomnia Among OSA

Older age (60 years or more) and use of tobacco are risk factors for insomnia among subjects having OSA (Stelzer et al. 2021). Most of these patients report initial and sleep-maintenance insomnia (Stelzer et al. 2021). Severity of OSA was lower in this group, but there was no difference observed between genders with regards to prevalence of insomnia (Stelzer et al. 2021).

8.2.2 Prevalence of OSA Among Subjects with Insomnia

Occult OSA is common among veterans. Among the veterans (mean age 72 years), nearly 50% subjects with insomnia were found to have comorbid OSA (defined as AHI $\geq$ 15); however, the prevalence ranges between 7% and 69% across studies (Ong and Crawford 2013; Kinugawa et al, 2012; Hein, et al, 2017; Fung et al. 2013). None of the nighttime symptoms of OSA, but only history of excessive daytime sleepiness was found to correlate with the diagnosis of OSA (Fung et al. 2013). Patients having comorbid OSA had higher body mass index and shorter mean sleep onset latency on multiple sleep latency test and are usually older (Hein et al, 2017; Kinugawa et al. 2012). However, body mass index did not fall in "obese" range among these patients (Kinugawa et al. 2012). Thus, elderly persons suffering from insomnia and daytime sleepiness should be screened for presence of comorbid OSA, even when they are not obese.

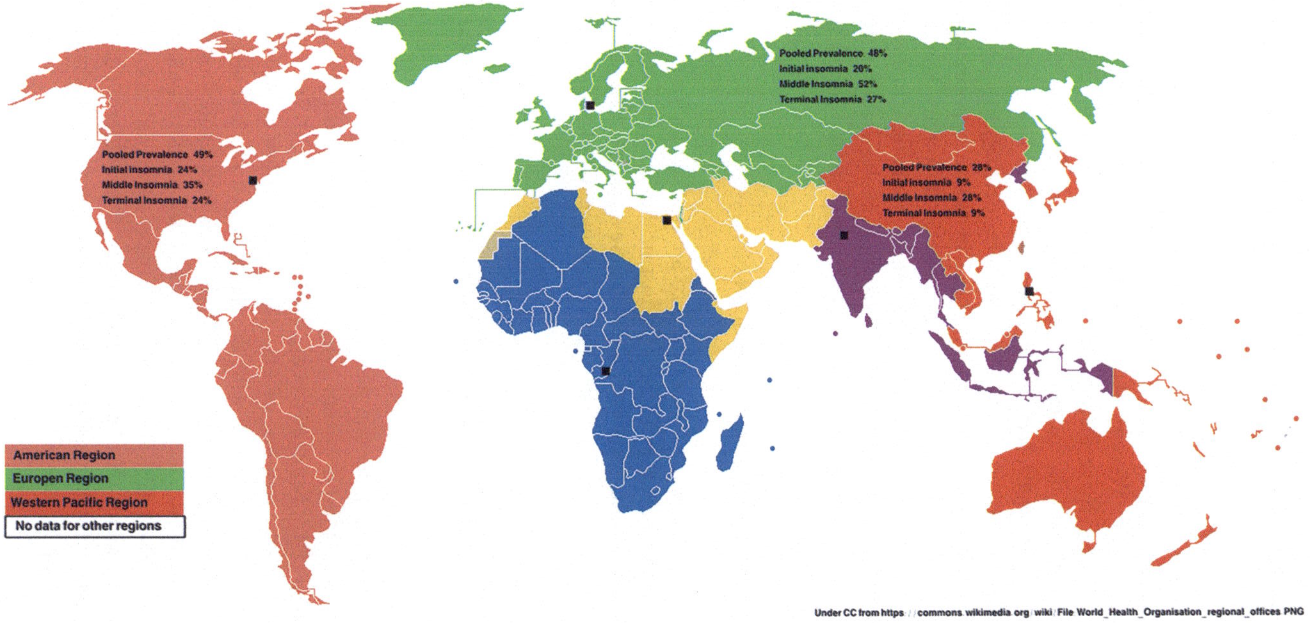

Fig. 8.3 The prevalence of insomnia among patients with OSA was higher in European and American regions compared to the Western Pacific region

8.2.2.1 Risk Factors for OSA Among Patients with Insomnia

To find out the OSA among patients with insomnia, it is important to find out the risk factors associated with it. In general, regular, loud snoring and witnessed pauses in breath are considered as signs of OSA. However, among veterans having primary diagnosis of insomnia, these signs appear inadequate to raise the suspicion. For example, Fung et al. reported that these signs could not identify patients with OSA, while excessive daytime sleepiness appeared to be a reliable indicator (Fung et al. 2013). On the other hand, during middle age, subjects having OSA were more likely to be male, older than 50 years, obese, had history of snoring, had metabolic syndrome, raised C-reactive protein, reported difficulty in maintenance of sleep and excessive daytime sleepiness (Hein et al. 2017). There are two possibilities for these findings. First, patients with obvious nighttime symptoms of OSA attracts attention early and hence, do not remain untreated (Fung et al. 2013). On the other hand, excessive daytime sleepiness may remain unnoticed for a long time and hence it indicated occult OSA in both middle-aged and veterans (Fung et al. 2013; Hein et al. 2017). Secondly, symptoms profile varies according to the age. Lower prevalence of nighttime symptoms among elderly could be explained by shorter and fragmented sleep that does not allow nighttime symptoms of OSA to get noticed (Fung et al. 2013).

8.3 Objective Measures of Sleep in COMISA

Subjects with pure insomnia and COMISA show shorter total sleep time, greater percentage of wake after sleep onset, poorer sleep efficiency compared to subjects having OSA and normal controls (Choi et al. 2016, 2018). These parameters do not appear unexpected considering the pathophysiology of insomnia. However, percentage of light sleep (N1) and arousal index was significantly greater in OSA group compared to pure insomnia and COMISA (Choi et al. 2016). Interestingly, while respiratory events related arousal dominated in OSA group, most of arousals in other two groups were spontaneous (Choi et al. 2016). Thus, despite having greater time spent in lighter sleep and higher arousal index, patients with OSA do not report insomnia. Other studies did not find these differences and reported that sleep-architecture, arousal index, periodic limb movement index, and oxygen saturation were comparable between OSA subjects with and without insomnia (Stelzer et al. 2021; Cho et al. 2018; Lichstein et al. 2013). These findings suggest that insomnia is an independent disorder and its presence leads to misperception of sleep as wakefulness despite having comparable sleep macro-architecture (Choi et al. 2016).

Misperception of sleep also follows a gradient and it is greatest among patients with insomnia compared to patients having OSA or COMISA (Choi et al. 2016; Bianchi et al. 2013). In other words, presence of OSA reduces the difference between subjective and objective sleep (Bianchi et al. 2013). Interestingly, mismatch between objective and perceived sleep is not related to the proportion of N1 sleep or number of nocturnal awakenings (Bianchi et al. 2013).

Insomnia in patients with OSA is not a byproduct of OSA for several reasons. Proportion of patients with initial and sleep maintenance insomnia are similar across groups. Time spent is wake after sleep onset is also comparable across groups. Lastly, severity of respiratory parameters, namely, respiratory disturbance index, minimum oxygen desaturation, respiratory event-related arousal index are also comparable between OSA and COMISA (Lichstein et al. 2013).

Taken together, evidence suggests that patients with COMISA form a different subgroup where two disorders are comorbid and warrants independent treatment.

8.4 Need for Identification of COMISA

Patients with COMISA suffer from two disorders and this combination is likely to produce greater disability and poorer quality of life among them compared to patients with either disorders. This has been confirmed in clinical investigations where subjects with COMISA were found to have poorer scores on physical and social domains compared to subjects with pure OSA (Cho et al. 2018; Tasbakan et al. 2018). On the other hand, compared to pure OSA and insomnia, presence of both the disorders together among patients with COMISA do not have addictive effects on cardiovascular risk, which is contrary to the theoretic possibility, as reported by Luyster et al. (2014). They also reported that prevalence of diabetes, hypertension, and dyslipidemia was also comparable between OSA and COMISA (Luyster et al. 2014). Diameter of brachial artery was also comparable between these two groups (OSA vs COMISA), though it was greater than pure insomnia and control groups (Luyster et al. 2014). These findings suggest that OSA is main culprit for the cardiovascular and metabolic changes in these patients with little, if at all, contribution from insomnia. However, further research is required whereby a number of factors related to insomnia, for example, chronicity of insomnia, duration of sleep in a day among patients with insomnia (insomnia with objective short sleep assessed longitudinally) are required before refuting absence of additive effect (Luyster et al. 2014).

Two studies have assessed the effect of gender on quality of life among patients with COMISA. One study showed that perception of quality of life was dependent upon gender as women with COMISA report poorer quality of life in all domains than men (Tasbakan et al. 2018). However, another study reported opposite findings with men having COMISA had greater fatigue and poorer quality of life compared to opposite gender (Lee et al. 2014). Thus, the available evidence indicates towards poorer quality of life among patients with COMISA, however, further research is required in this area.

Besides affecting quality of life, longitudinal data suggest that comorbid insomnia adversely affects adherence to positive airway pressure (PAP) therapy resulting lesser daily use and lesser nights with PAP use greater than 4 h (Ong and Crawford

Table 8.2 Why should COMISA be addressed? (Ong and Crawford 2013; Cho et al. 2018; Ong and Crisostomo 2013; Luyster et al. 2010; Jeon et al. 2021)

Patients with COMISA are at increased risk for
• Reduced vigilance
• Depression
• Sleepiness
• Poor quality of life
• Functional impairment
• Chronic pain
• Hypertension
• Coronary artery disease
• Congestive heart failure
• Diabetes mellitus
• Metabolic syndrome
• Stroke
• Work absenteeism
• Poor adherence to positive airway pressure therapy
• Accidents: Motor vehicle and industrial

2013; Wallace et al. 2018). However, contradictory results have been shown in another study which reported that presence of insomnia does not affect adherence to PAP (Cho et al. 2018). Thus, this area is open for further research. Concomitant occurrence of insomnia and OSA increase risk for psychiatric and other medical disorders as well (Table 8.2).

Pathophysiologically, O occurrence of metabolic, cardiovascular, and cerebrovascular morbidities could be ascribed to pathophysiology of insomnia as well OSA that includes recurrent hypoxemia, formation of reactive oxygen species and other free radicals, sympathetic surge, and hypercortisolemia due to activation of HPA axis (Gupta and Knapp 2014).

Presence of OSA among subjects with insomnia does not affect effectiveness of cognitive behavior therapy for insomnia (CBTi) (Ong and Crawford 2013; Sweetman et al. 2017). In other words, magnitude of improvement in symptoms of insomnia is comparable between patients with COMISA and insomnia following CBTi (Sweetman et al. 2017). Thus, CBTi before institution of PAP therapy among patients with COMISA can not only improve insomnia, but also improve adherence to PAP therapy.

8.5 Pathogenesis of COMISA

Though exact mechanism leading to COMISA is not clear, however, certain theoretical models have been proposed (Ong and Crawford 2013). These models are based on cross-sectional as well and longitudinal studies. Longitudinal studies, nonetheless, challenge the models proposed based on cross-sectional studies (Ong

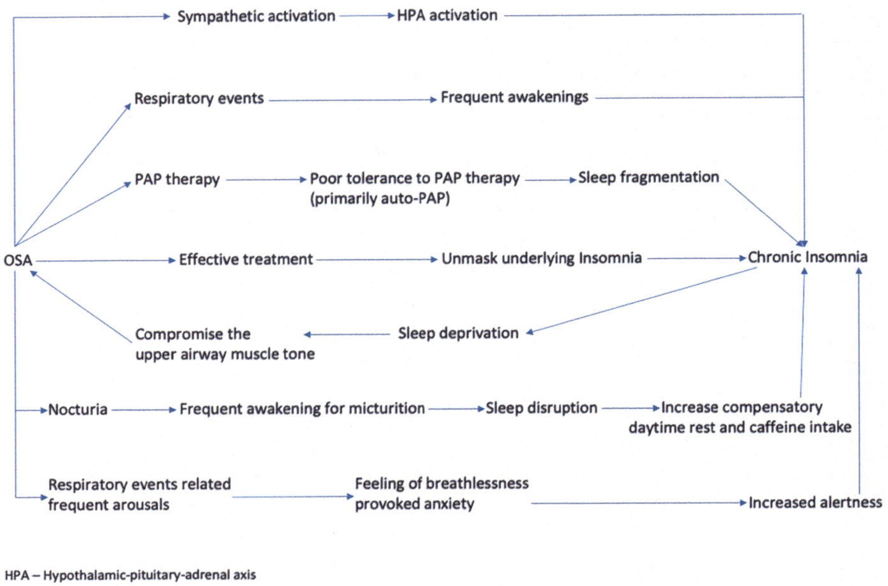

Fig. 8.4 Pathophysiological mechanism involved in OSA as well as insomnia may precipitate each other and can work in feed forward manner to develop COMISA

and Crawford 2013). For the sake of understanding, these models are discussed here (Fig. 8.4).

8.5.1 OSA Leading to Insomnia

A number of factors are considered to be instrumental in occurrence of insomnia among patients having OSA. First, it is a well-known fact that respiratory events occurring during OSA may lead to micro-arousals. Sleep deprivation occurring because of these micro-arousals actives stress pathway, which in turn, may lead to insomnia (Ong and Crawford 2013). Respiratory events also cause sympathetic surge at the termination, which can facilitate hyperarousal leading to insomnia (Ong and Crawford 2013) (Fig. 8.5). Thirdly, a number of patients are not able to tolerate either mask or the pressure of PAP and thus remains awake or experience disturbed sleep. When sleep disturbances continue for a long time, through the kindling mechanisms insomnia becomes chronic and severe. Fourth, it is possible that patients having untreated OSA remains in the state of sleep-debt and thus do not experience insomnia. When sleep-debt improves after treatment of OSA, insomnia gets unmasked (Ong and Crawford 2013). Lastly, OSA is usually associated with nocturnal polyuria which compels the person to wake up repeatedly. Repeated awakenings may lead to behavioral changes that include extension of bedtime, use

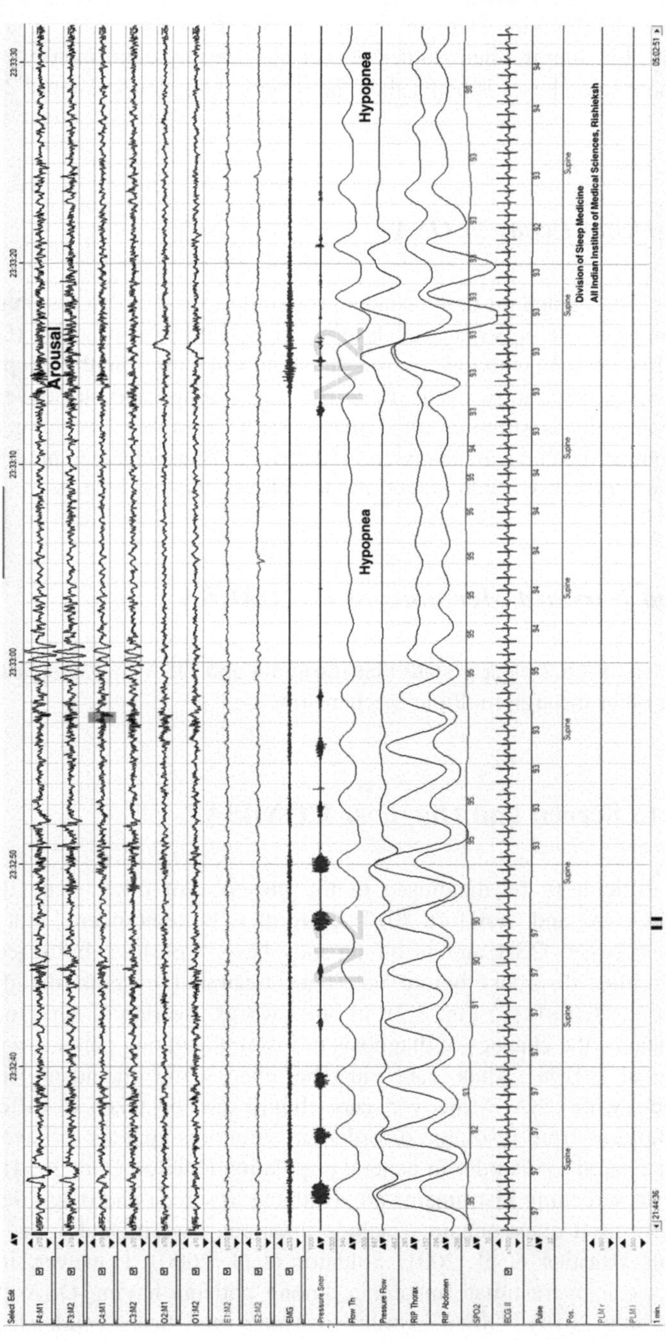

Fig. 8.5 An epoch showing that respiratory events can also lead to a sympathetic surge at the termination, facilitating hyperarousal and leading to insomnia

of stimulants like caffeine during the day, which in trun facilitate insomnia through psychophysiological model (Janssen et al. 2019). However, data from cross-sectional as well as longitudinal studies do not provide any confirmatory evidence that any of these hypotheses is better than the others. (Ong and Crawford 2013).

.

8.5.2 Insomnia Leads to OSA

Though never investigated properly, some of the studies suggest that insomnia alters the tone of muscles of upper airway leading to precipitation of OSA (Ong and Crawford 2013). Persons with low arousal threshold wake up from the sleep is close association with a respiratory event. Recurrent awakenings and subsequent sleep reduce the wakefulness stimulus and, predispose to unstable breathing through ventilatory control mechanisms (Janssen et al. 2019). These changes predispose to OSA. However, further research is required to validate this hypothesis.

8.5.3 Feed Forward Mechanisms to COMISA

Sections 8.5.1 and 8.5.2 clearly state that insomnia and OSA may increase occurrence of each other through multiple mechanisms.

8.6 How to Screen and Diagnose COMISA?

Insomnia is difficult to be diagnosed in the presence of other sleep disorders including OSA (Ong and Crawford 2013). It needs to be remembered that overlap of symptoms between OSA and insomnia may have a positive influence in the diagnosis of either disorders before and after treatment of both the disorders (Sweetman et al. 2019a) (see Fig. 8.2). In other words, overlap of symptoms may spuriously enhance the chances of diagnosis of insomnia among patients with OSA (Sweetman et al. 2019a). Thus, care must be taken while diagnosing insomnia among subjects with OSA. Moreover, care should also be taken to differentiate insomnia symptoms from insomnia disorder as symptoms appears to have greater prevalence compared to disorder in general population (Mallon et al. 2014).

A number of screening instruments are available to screen and diagnose insomnia. Commonly used questionnaires include insomnia severity index and Athens insomnia scale (Bastien et al. 2001; Soldatos et al. 2000). However, insomnia severity index can overestimate insomnia among patients having OSA (Wallace and Wohlgemuth 2019). Other instruments have not been validated in this population.

On the other hand, diagnosis of OSA requires specific nighttime symptoms, for example, snoring, witnessed pauses in breath which are less likely to be reported in subjects with pure insomnia. This could explain the lower prevalence of OSA among subjects with insomnia found in the epidemiological studies. Risk factors for OSA as mentioned in Sect. 8.2.2, except for the difficulty in maintaining the sleep, are known to increase risk for OSA and are included in various screening instruments, for example, Berlin Questionnaire and STOP-Bang Questionnaire (Chiu et al. 2017). These questionnaires may be applied to screen all subjects with chronic insomnia for OSA as sensitivity of Berlin Questionnaire and STOP-Bang Questionnaire varies between 76–84% and 88–93% for mild to severe OSA, respectively (Chiu et al. 2017). However, their use is limited by poor specificity that ranges between 38% and 59% for Berlin Questionnaire and 35–42% for STOP-Bang Questionnaire (Chiu et al. 2017). Further, specificity decreased as severity of OSA increased (Chiu et al. 2017). Authors could find only one study that has assessed the validity of OSA screening questionnaires among patients with COMISA. No-Apnea, NoSAS and STOP-Bang have been found to optimal psychometric properties to be used as screening methods for OSA, but not Epworth Sleepiness Scale (Duarte et al. 2019).

This must be noted that these instruments have been validated among subjects with either insomnia or OSA. Since, COMISA have overlapping symptoms, validity of these questionnaires is questionable in this population unless more data is available.

Luyster et al. (2010) suggested that in treatment refractory cases of insomnia should undergo diagnostic polysomnography to rule out other sleep disorders that mimic insomnia, particularly OSA.

8.7 Challenges to the Diagnosis of COMISA

Besides the issues discussed in previous section, a number of other factors may also interfere with the diagnosis of COMISA. While diagnosis is easier for isolated obstructive sleep apnea and insomnia, it is at times difficult for COMISA (Ong et al. 2021). When patients go to pulmonologists, primary focus may be limited to OSA, on the other hand, if the primary contact is with a mental health professional, primary focus remains on insomnia or purely on psychiatric disorders (depression) which are common among patients having COMISA (Ong et al. 2021; Jeon et al. 2021).

Chances that insomnia would remain undetected increase if the physician prefers to choose home sleep apnea testing (HSAT) for the diagnosis of OSA instead of attended in-laboratory polysomnography (Ong et al. 2021).

8.8 Management of COMISA

Though the guidelines are available for diagnosis and management of OSA and insomnia, yet patients with COMISA form a special group where these guidelines may not be applicable as a whole (Riemann et al. 2017; Patil et al. 2019). So far, guidelines for the management of COMISA are not available (Janssen et al. 2019). In this section, only the literature related to management (even if partial) of COMISA shall be reviewed.

Present guidelines recommend PAP therapy as the first line option for moderate to severe OSA (Patil et al. 2019). However, a number of factors influence adherence to PAP therapy that include tolerance to pressure and mask, continuity of sleep, and improvement on the night of titration (Ong and Crisostomo 2013). Thus, presence of insomnia may negatively affect adherence to PAP therapy. On the other hand, management of insomnia includes cognitive behavior therapy for insomnia (CBT-i) and use of hypnotic medications (Riemann et al. 2017). While hypnotic medication may provide immediate relief in symptoms, CBT-i shows long lasting effect, thereby reducing the risk of relapse of insomnia.

It has been argued that hypnotics may worsen OSA by reducing tone of upper airway muscles (Ong and Crisostomo 2013). However, available evidence from a study involving 52,504 women does not confirm this notion even after controlling for body mass index, age, waist circumference, physical activity, and alcohol consumption (Lin et al. 2017). If the treating physician is not confident in using hypnotics for insomnia, non-hypnotic medications (sedating antidepressants, dual orexin receptor antagonists) may be used and use of non-hypnotic medications among patients with OSA has been found to improve sleep quality without affecting severity of OSA and oxygen saturation (Zhang et al. 2014).

8.8.1 What Should Be Managed First?

A number of options are available for management of COMISA (Table 8.3). A recent randomized controlled trial assessed the effect of three different trajectories (Concurrent CBTi with PAP; CBTi followed by PAP and PAP alone) on adherence to PAP therapy and improvement in insomnia and sleep quality (Ong et al. 2020). Results suggested that use of PAP alone versus concurrent therapy did not influence adherence to PAP therapy. However, concurrent therapy improved severity of insomnia and sleep quality compared to PAP alone. Sequential therapy (CBTi followed by PAP) was not different from concurrent therapy on any of the measures. Thus, concurrent therapy appears superior to other approaches (Ong et al. 2020). Among patients having OSA or COMISA, who are not able to tolerate PAP, hypoglossal nerve stimulation improved not only measures of OSA and sleepiness in both groups, but also improved insomnia in COMISA group (Dhanda Patil et al. 2021).

Since enough experimental data is not available regarding sequencing of treatment, available literature suggest that decision should be left to the patient after explaining the risks and benefits of each of the possible management trajectories (Ong and Crisostomo 2013).

Table 8.3 Possible management options for patients with COMISA (Ong and Crisostomo 2013)

S. N.	Management technique	Advantages	Disadvantages
1.	Treat insomnia followed by OSA	Improves yield at diagnostic study for OSA	Sleep restriction therapy may worsen the excessive daytime sleepiness and fatigue among patients, though it appears temporary (Sweetman et al. 2020)
		Improves adherence to PAP during titration study	
		Improves long-term adherence to PAP therapy	
2.	Concurrent management of insomnia and OSA	May provide faster relief to both disorders	CBTi may take time to show effect, during this period, patient may experience issues related to poor tolerance. Can be overcome by prescribing hypnotics along with CBTi
		Reduces length of treatment	
3.	Manage OSA followed by insomnia	Reduces time to provide benefits of treatment of OSA to patients	May interfere with adherence to PAP
4.	PAP therapy only	May improve insomnia symptoms in some patients	Sleep quality and insomnia may not improve, especially if severe

8.8.2 Available Evidence for the Treatment Planning Among Patient Having OSA

8.8.2.1 Treat Insomnia First

Insomnia is a condition of hyperarousal that can be aggravated further by issues related to PAP pressure or mask intolerance (Luyster et al. 2010). This may produce or enhance anxiety and hence, insomnia should be treated before prescribing PAP therapy (Luyster et al. 2010).

Sweetman et al. (2019b) reported that compared to treatment as usual (PAP therapy without CBTi) among subjects having COMISA, CBTi earlier to PAP improves not only insomnia but also the adherence to PAP therapy.

A 6 -month follow-up data from veterans reported that insomnia negatively affects the adherence to PAP therapy with reduced duration of PAP use as well as lower proportion of patients using PAP for four or more hours per night in the COMISA group compared to pure OSA (Wallace et al. 2018). A long-term PAP adherence data of 7 years suggested that early and late insomnia, especially among lean patients is associated with poor treatment adherence among patients having OSA (Eysteinsdottir et al. 2017). It is recommended all patients having OSA who are prescribed PAP therapy should be screened for initial and terminal insomnia, and it should be treated to improve compliance to PAP therapy (Eysteinsdottir et al. 2017).

8.8.2.2 Treat OSA First

As discussed in the section of pathophysiology, if the symptoms of insomnia are less troublesome, it may be a good approach to manage OSA first. In many patients, resolution of OSA through PAP therapy improves sleep quality and symptoms of insomnia (Krakow et al. 2019). If the residual symptoms of insomnia persist, they may be addressed using routine guidelines for the management of insomnia (Krakow et al. 2019). A case series showed that maxillomandibular advancement in patients with mild OSA remits chronic insomnia, as well (Proothi et al. 2019).

8.8.2.3 Treat Both Conditions Concurrently

Concurrent treatment improves sleep quality and reduce insomnia severity but does not affect adherence to PAP therapy (Ong et al. 2020). However, a recent study reported that concurrent treatment improves both insomnia-related measures as well as adherence to PAP therapy (Alessi et al. 2021).

8.8.2.4 Treat OSA, Insomnia Improves

It has been reported that use of PAP therapy can improve middle insomnia, but not the initial insomnia (Björnsdóttir et al. 2013). Late insomnia improves among patients non-adherent to PAP therapy, perhaps related to accumulating sleep-debt (Björnsdóttir et al. 2013). Another study reported that use of auto-PAP can improve insomnia over the period of 24 months (Nguyên et al. 2013).

8.8.2.5 What Is to Be Done?

However, still the question remains-what should be treated first? One systematic review reported that evidence is insufficient to show that (i) PAP therapy improves insomnia (ii) to suggest that combined therapy is effective, and to conclude that (iii) treatment of insomnia improves compliance to PAP therapy (Bahr et al. 2018). In view of these findings, at present, it is recommended that clinician should consider a number of factors while planning the sequence of therapy (Table 8.4).

8.9 Multidisciplinary Approach

Considering the issues discussed above, it becomes important that such cases are seen by a trained sleep physician. Another option would be to have a multidisciplinary team consisting of a pulmonologist, psychiatrist, and neurologist,

Table 8.4 Points to be considered while planning management of COMISA

1.	Patient's concern regarding the severity of symptoms for either of the disorders	More severe disorder or for which patient is more concerned should be addressed first or more aggressively
2.	Improvement observed during the course of the treatment for either disorders	If adherence is an issue either of the disorders, assess if other disorder is responsible for same
		If one is improving slowly (as usually happens with insomnia during CBTi, use pharmacotherapy for quick relief)
3.	Adherence to the therapy for either disorder	Assess other reasons for noncompliance including beliefs and do appropriate intervention
4.	Use of adjunctive treatment methods for OSA, at least temporarily to reduce severity, for example, positional therapy	These methods may provide temporary relief till insomnia improves since insomnia may influence adherence to PAP therapy

having some background knowledge of sleep disorders is made so that not only COMISA, but also the disorders that mimic insomnia (e.g., restless legs syndrome, anxiety, circadian rhythm sleep disorders) can be diagnosed and addressed (Ong et al. 2021). This would improve not only the diagnostic process but also compliance to the treatment for each of the disorders.

8.10 Conclusion

Literature reviewed so far suggests that COMISA is a common problem and mostly remains undiagnosed. Questionnaires used for the screening for insomnia and OSA may either overestimate of underestimate the problem. Hence, best approach would be to screen the patients for all possible comorbidities during assessment. Use of standardized instruments, viz, sleep diary, actigraphy can help to uncover insomnia and insomnia mimics. On the other hand, patients complaining of "depression along with disturbed sleep" must be screened for not only insomnia mimics, but also for OSA. Management of these cases should include a trained sleep physician or a multidisciplinary team. Since the condition has received attention of scientific community, more research is required for drafting treatment guidelines. However, available literature shows that concurrent therapy using PAP and CBTi/Hypnotics may improve compliance to either of them.

References

Alessi CA, Fung CH, Dzierzewski JM, Fiorentino L, Stepnowsky C, Rodriguez Tapia JC et al (2021) Randomized controlled trial of an integrated approach to treating insomnia and

improving the use of positive airway pressure therapy in veterans with comorbid insomnia disorder and obstructive sleep apnea. Sleep 44(4):zsaa235
Bahr K, Camara RJA, Gouveris H, Tuin I (2018) Current treatment of comorbid insomnia and obstructive sleep apnea with CBTI and PAP-therapy: a systematic review. Front Neurol 9:804
Bastien C, Vallières A, Morin C (2001) Validation of the Insomnia Severity Index as an outcome measure for insomnia research. Sleep Med 2(4):297–307
Bianchi MT, Williams KL, McKinney S, Ellenbogen JM (2013) The subjective-objective mismatch in sleep perception among those with insomnia and sleep apnea. J Sleep Res 22(5):557–568
Björnsdóttir E, Janson C, Sigurdsson JF, Gehrman P, Perlis M, Juliusson S et al (2013) Symptoms of insomnia among patients with obstructive sleep apnea before and after two years of positive airway pressure treatment. Sleep 36(12):1901–1909
Chiu HY, Chen PY, Chuang LP, Chen NH, Tu YK, Hsieh YJ et al (2017) Diagnostic accuracy of the Berlin questionnaire, STOP-BANG, STOP, and Epworth sleepiness scale in detecting obstructive sleep apnea: a bivariate meta-analysis. Sleep Med Rev 36:57–70
Cho YW, Kim KT, Moon H-J, Korostyshevskiy VR, Motamedi GK, Yang KI (2018) Comorbid insomnia with obstructive sleep apnea: clinical characteristics and risk factors. J Clin Sleep Med 14(03):409–417
Choi SJ, Suh S, Ong J, Joo EY (2016) Sleep misperception in chronic insomnia patients with obstructive sleep apnea syndrome: implications for clinical assessment. J Clin Sleep Med 12(11):1517–1525
Dhanda Patil R, Hong MP, Ishman SL (2021) Hypoglossal nerve stimulation in veterans with comorbid insomnia and sleep apnea. Otolaryngol Head Neck Surg 164(6):1345–1353
Duarte RLM, Magalhaes-da-Silveira FJ, Oliveira ESTS, Rabahi MF, Mello FCQ, Gozal D (2019) Predicting obstructive sleep apnea in patients with insomnia: a comparative study with four screening instruments. Lung 197(4):451–458
Eysteinsdottir B, Gislason T, Pack AI, Benediktsdottir B, Arnardottir ES, Kuna ST et al (2017) Insomnia complaints in lean patients with obstructive sleep apnea negatively affect positive airway pressure treatment adherence. J Sleep Res 26(2):159–165
Fung CH, Martin JL, Dzierzewski JM, Jouldjian S, Josephson K, Park M et al (2013) Prevalence and symptoms of occult sleep disordered breathing among older veterans with insomnia. J Clin Sleep Med 9(11):1173–1178
Guilleminault C, Eldridge F.L, Dement W.C. Insomnia with sleep apnea: A new syndrome. Science 1973; 181: 856–858
Gupta MA, Knapp K (2014) Cardiovascular and psychiatric morbidity in obstructive sleep apnea (OSA) with insomnia (sleep apnea plus) versus obstructive sleep apnea without insomnia: a case-control study from a Nationally Representative US sample. PLoS One 9(3):e90021
Hein M, Lanquart JP, Loas G, Hubain P, Linkowski P (2017) Prevalence and risk factors of moderate to severe obstructive sleep apnea syndrome in insomnia sufferers: a study on 1311 subjects. Respir Res 18(1):135
Janssen HCJP, Venekamp LN, Peeters GAM, Pijpers A, Pevernagie DAA (2019) Management of insomnia in sleep disordered breathing. Eur Respir Rev 28(153):190080
Jeon B, Luyster FS, Callan JA, Chasens ER (2021) Depressive symptoms in comorbid obstructive sleep apnea and insomnia: an integrative review. West J Nurs Res 2021:193945921989656
Kinugawa K, Doulazmi M, Sebban C, Schumm S, Mariani J, Nguyen-Michel VH (2012) Sleep apnea in elderly adults with chronic insomnia. J Am Geriatr Soc 60(12):2366–2368
Krakow BJ, McIver ND, Obando JJ, Ulibarri VA (2019) Changes in insomnia severity with advanced PAP therapy in patients with posttraumatic stress symptoms and comorbid sleep apnea: a retrospective, nonrandomized controlled study. Mil Med Res 6(1):15
Krakov B, Melendrez D, Ferreira E, Clark J, Warner TD, Sisley B, Sklar D. Prevalence of insomnia symptoms in patients with sleep-disordered breathing. Chest 2001, 120, 1923–1929
Lee MH, Lee SA, Lee GH, Ryu HS, Chung S, Chung YS et al (2014) Gender differences in the effect of comorbid insomnia symptom on depression, anxiety, fatigue, and daytime sleepiness in patients with obstructive sleep apnea. Sleep Breath 18(1):111–117

Lichstein KL, Justin Thomas S, Woosley JA, Geyer JD (2013) Co-occurring insomnia and obstructive sleep apnea. Sleep Med 14(9):824–829

Lichstein, KL. Riedel, BW, Lester, KW, Aguillard, RN. Occult sleep apnea in a recruited sample of older adults with insomnia. J. Consult. Clin. Psychol. 1999, 67, 405–410.

Lin BM, Hu FB, Curhan GC (2017) Association between benzodiazepine receptor agonists and snoring among women in the nurses' health study. JAMA Otolaryngol Head Neck Surg 143(2): 162

Luyster FS, Buysse DJ, Strollo PJ Jr (2010) Comorbid insomnia and obstructive sleep apnea: challenges for clinical practice and research. J Clin Sleep Med 6(2):196–204

Luyster FS, Kip KE, Buysse DJ, Aiyer AN, Reis SE, Strollo PJ Jr (2014) Traditional and nontraditional cardiovascular risk factors in comorbid insomnia and sleep apnea. Sleep 37(3): 593–600

Mallon L, Broman J-E, Akerstedt T, Hetta J (2014) Insomnia in Sweden: a population-based survey. Sleep Disord 2014:843126

Nguyên X-L, Rakotonanahary D, Chaskalovic J, Fleury B (2013) Insomnia related to sleep apnoea: effect of long-term auto-adjusting positive airway pressure treatment. Eur Respir J 41(3): 593–600

Ong JC, Crawford MR (2013) Insomnia and obstructive sleep apnea. Sleep Med Clin 8(3):389–398

Ong JC, Crisostomo MI (2013) The more the merrier? Working towards multidisciplinary management of obstructive sleep apnea and comorbid insomnia. J Clin Psychol 69(10):1066–1077

Ong JC, Crawford MR, Dawson SC, Fogg LF, Turner AD, Wyatt JK et al (2020) A randomized controlled trial of CBT-I and PAP for obstructive sleep apnea and comorbid insomnia: main outcomes from the MATRICS study. Sleep 43(9):zsaa041

Ong JC, Crawford MR, Wallace DM (2021) Sleep apnea and insomnia: emerging evidence for effective clinical management. Chest 159(5):2020–2028

Patil SP, Ayappa IA, Caples SM, Kimoff RJ, Patel SR, Harrod CG (2019) Treatment of adult obstructive sleep apnea with positive airway pressure: an American Academy of sleep medicine systematic review, meta-analysis, and GRADE assessment. J Clin Sleep Med 15(02):301–334

Proothi M, Grazina VJR, Gold AR (2019) Chronic insomnia remitting after maxillomandibular advancement for mild obstructive sleep apnea: a case series. J Med Case Rep 13(1):252

Riemann D, Baglioni C, Bassetti C, Bjorvatn B, Dolenc Groselj L, Ellis JG et al (2017) European guideline for the diagnosis and treatment of insomnia. J Sleep Res 26(6):675–700

Schutte-Rodin S, Lauren B, Buysse D, Dorsey C, Sateia M (2008) Clinical guideline for the evaluation and management of chronic insomnia in adults. J Clin Sleep Med 5(4):487–504

Soldatos C, Dikeos D, Paparrigopoulos T (2000) Athens Insomnia Scale: validation of an instrument based on ICD-10 criteria. J Psychosom Res 48(6):555–560

Stelzer FG, Garcia E, Schorr F, Barea LM, Barros HT (2021) Prevalence of chronic insomnia in patients with obstructive sleep apnea. Braz J Psychiatry 43(4):370–375

Sweetman A, Lack L, Lambert S, Gradisar M, Harris J (2017) Does comorbid obstructive sleep apnea impair the effectiveness of cognitive and behavioral therapy for insomnia? Sleep Med 39: 38–46

Sweetman A, Lack L, Bastien C (2019a) Co-morbid insomnia and sleep apnea (COMISA): prevalence, consequences, methodological considerations, and recent randomized controlled trials. Brain Sci 9(12):371

Sweetman A, Lack L, Catcheside PG, Antic NA, Smith S, Chai-Coetzer CL et al (2019b) Cognitive and behavioral therapy for insomnia increases the use of continuous positive airway pressure therapy in obstructive sleep apnea participants with comorbid insomnia: a randomized clinical trial. Sleep 42(12):zsz178

Sweetman A, McEvoy RD, Smith S, Catcheside PG, Antic NA, Chai-Coetzer CL et al (2020) The effect of cognitive and behavioral therapy for insomnia on week-to-week changes in sleepiness and sleep parameters in patients with comorbid insomnia and sleep apnea: a randomized controlled trial. Sleep 43(7):zsaa002

Tasbakan MS, Gunduz C, Pirildar S, Basoglu OK (2018) Quality of life in obstructive sleep apnea is related to female gender and comorbid insomnia. Sleep Breath 22(4):1013–1020

Wallace DM, Wohlgemuth WK (2019) Predictors of Insomnia Severity Index profiles in United States veterans with obstructive sleep apnea. J Clin Sleep Med 15(12):1827–1837

Wallace DM, Sawyer AM, Shafazand S (2018) Comorbid insomnia symptoms predict lower 6-month adherence to CPAP in US veterans with obstructive sleep apnea. Sleep Breath 22(1): 5–15

Zhang XJ, Li QY, Wang Y, Xu HJ, Lin YN (2014) The effect of non-benzodiazepine hypnotics on sleep quality and severity in patients with OSA: a meta-analysis. Sleep Breath 18(4):781–789

Zhang Y, Ren R, Lei F, Zhou J, Zhang J, Wing Y et al (2019) Worldwide and regional prevalence rates of co-occurrence of insomnia and insomnia symptoms with obstructive sleep apnea: a systematic review and meta-analysis. Sleep Med Rev 45:1–17

Chapter 9
Sleep Telemedicine During COVID Pandemic and Post-Pandemic Period: Current Perspectives from East and West

Mahadevappa Hunasikatti, Ravi Gupta, and Vijay Krishnan

Abstract This chapter explores the significant growth and potential of sleep telemedicine during the COVID-19 pandemic and beyond. It discusses the rapid expansion of telemedicine, particularly in sleep medicine, driven by the pandemic and the subsequent changes in healthcare delivery. The chapter highlights the benefits of telemedicine in sleep medicine, such as increased accessibility and efficiency, and the potential for telemedicine to fill the gap in sleep disorder treatment due to a shortage of sleep medicine physicians. It also discusses the American Academy of Sleep Medicine's position on telemedicine and its recommendations for its use. The chapter further examines the impact of telemedicine on the diagnosis and management of sleep disorders, the role of digital health technology, and the future of sleep telemedicine. It concludes by discussing the challenges and potential solutions for telemedicine implementation, particularly in developing countries like India.

Keywords Sleep telemedicine · COVID-19 pandemic · Sleep disorder management · Digital health technology · Telemedicine challenges and solutions

9.1 Introduction

Sleep telemedicine was growing even before the COVID-19 pandemic worldwide. Now, with the COVID pandemic, the growth is phenomenal. The telemedicine market is expected to hit US$53.1 billion by 2026 according to a recent report from Acumen Research and Consulting (Smith 2020). Commencing on March 6, 2020, and throughout the COVID-19 Public Health Emergency, Medicare will provide reimbursement for telehealth services rendered to patients under expanded

M. Hunasikatti (✉)
Division of Respiratory, Sleep and Anesthesiology (DSRA), OHT1, Food and Drug Administration (FDA), Silver Spring, MD, USA

R. Gupta · V. Krishnan
Department of Psychiatry and Division of Sleep Medicine, All India Institute of Medical Sciences, Rishikesh, Uttarakhand, India

A. S. BaHammam, M. Hunasikatti (eds.), *Sleep Apnea Frontiers*, Progress in Sleep Research, https://doi.org/10.1007/978-981-99-7901-1_9

circumstances. These telehealth visits will be treated on par with in-person visits and will be reimbursed at the same rate. Although traditionally patients were required to be physically present at specific originating sites like a physician's office, skilled nursing facility, or hospital, Medicare will now cover telehealth services provided to beneficiaries in any healthcare facility or their own residence during the pandemic period (Centers for Medicare and Medicaid Services 2020). Out of all branches of medicine, sleep medicine is the most suitable for telemedical practices. It is a surprise why telemedicine for sleep practices has evolved so slowly. This is a dramatic change in the delivery of medical services in sleep medicine, as the last large-scale practice disruption of home sleep apnea testing was met by years of intense resistance (Tuohuti et al. 2023).

Long before the COVID-19 pandemic, the American Academy of Sleep Medicine (AASM) issued a Position Paper on the "Utilization of Telemedicine in the Diagnosis and Management of Sleep Disorders, to seek more efficient and accessible ways to provide services beyond the traditional office model" (Singh et al. 2015) (Table 9.1).

The American Telemedicine Association cites four key models of telemedicine services: synchronous and asynchronous (store-and-forward) telemedicine patient encounters, remote patient monitoring (RPM), and mobile health (mHealth) smartphone applications (American Telemedicine Association 2020). These modes apply very well to sleep telemedicine.

9.2 Shortage of Sleep Medicine Physicians and Its Impact on Tele Sleep Medicine

As of 2018, there are 6035 sleep medicine physicians board certified by the American Board of Medical Specialties, which is insufficient for the needs of the population, with almost 30 million patients with obstructive sleep apnea alone. Health Insurers (Payers) are placing a greater value on sleep. In addition, employers, regulators, and legislators recognize the importance of diagnosing and treating sleep disorders (Kiley et al. 2019). Needless to say that sleep telemedicine can fill the needs of patients with sleep disorders even after the current COVID pandemic.

In 2006, the United States National Academy of Sciences published a report titled Sleep Disorders and Sleep Deprivation: An Unmet Public Health Problem. The report stated that 50 to 70 million Americans had a chronic disorder of sleep and wakefulness, a population that "greatly outstrips" our healthcare system's ability to provide adequate sleep services for these patients. Since 2006, the number of people who experience sleep disorders has increased and US healthcare systems remain overwhelmed by the demand for sleep care services (Ford et al. 2015). It is clear there is a large unmet patient need for sleep specialists.

Between 2013 and 2018, the use of VA sleep care services increased by an average of 16% per year, while the average wait time for an appointment in a VHA sleep clinic remains greater than 40 days (VA Corporate Data Warehouse). The number of appointments for sleep care in VA facilities showed significant increases between 2013–2018 and exceeded 650,000 in 2018. There is an inherent advantage

Table 9.1 AASM task force recommendations (Singh et al. 2015)

• Identifying crucial medical conditions where it is useful: Clinical care standards for telemedicine services should mirror those of live office visits, including all aspects of diagnosis and treatment decisions as would be reasonably expected in traditional office-based encounters
• Identifying crucial medical conditions where it is useful: Clinical judgment should be exercised when determining the scope and extent of telemedicine applications in the diagnosis and treatment of specific patients and sleep disorders
• Identifying crucial medical conditions where it is useful: Live interactive telemedicine for sleep disorders, if employed in accordance with the principles delineated in this paper, telemedicine can be employed in a way that aligns with the guidelines and recommendations presented herein, should be recognized and reimbursed in a manner competitive or comparable with traditional in-person visits
• Identifying crucial medical conditions where it is useful: Roles, expectations, and responsibilities of providers involved in the delivery of sleep telemedicine should be defined, including those at originating sites and distant sites
• Identifying crucial medical conditions where it is useful: The practice of telemedicine should aim to promote a care model in which sleep specialists, patients, primary care providers, and other members of the healthcare team aim to improve the value of healthcare delivery in a coordinated fashion
• Identifying crucial medical conditions where it is useful: Appropriate technical standards should be upheld throughout the telemedicine care delivery process, at both the originating and distant sites, and specifically meet the regulations and requirements established by the Health Insurance Portability and Accountability Act (HIPAA)
• Identifying crucial medical conditions where it is useful: Methods that aim to improve the utility of telemedicine exist and should be explored, including the utilization of patient presenters, local resources and providers, adjunct testing, and add-on technologies
• Identifying crucial medical conditions where it is useful: Quality assurance processes should be in place for telemedicine care delivery models that aim to capture process measures, patient outcomes, and patient/provider experiences with the model(s) employed
• Identifying crucial medical conditions where it is useful: Time for data management, quality processes, and other aspects of care delivery related to telemedicine encounters should be recognized in value-based care delivery models
• Identifying crucial medical conditions where it is useful: The use of telemedicine services and equipment should adhere to strict professional and ethical standards so as not to violate the intent of the telemedicine interaction while aiming to improve overall patient access, quality, and/or value of care
• Identifying crucial medical conditions where it is useful: When billing for telemedicine services, it is recommended that patients, providers, and others rendering services understand payor reimbursements and that there be financial transparency throughout the process
• Identifying crucial medical conditions where it is useful: The use of telemedicine in the field of sleep medicine is expected to rapidly expand, as are broader telehealth applications in general; further research into the impact and outcomes of these are needed (Singh et al. 2015)

of closed healthcare systems like VA VISN (Veterans Affairs) and Kaiser Heath care systems. Synchronous telehealth encompasses all video teleconferencing between patients and care providers, regardless of patient or provider location. Video teleconferencing at VA facilities (medical centers and community-based outpatient clinics) is used for initial consultation and follow-up visits. In-home video teleconferencing (or video-on-demand) is a secure, encrypted means of video chatting with patients anywhere with their smartphone for reception. This technology also enables bringing other providers or family members into the virtual medical room for warm handoffs or family meetings. For sleep-specific visits, in-home video-on-demand enables viewing of bedroom environments (presence of a television, windows, lighting, etc.), positioning of PAP or other treatment devices within the bedroom, and other home conditions relevant to improving patients' sleep. The VA uses an internal system for patient data communication, but device manufacturers provide positive airway pressure (PAP) data, and compliance with the PAP usage is necessary. This comprehensive Tele Sleep program provides strategies and methods that have already been a model for other healthcare systems to improve access to quality and cost-efficiency of sleep care for all veterans (Sarmiento et al. 2019).

In other closed systems like Kaiser, a study demonstrated that implementing CPAP telemonitoring with automated feedback messaging led to enhanced adherence to CPAP therapy over a 90-day period among individuals diagnosed with obstructive sleep apnea (OSA). On the other hand, although telemedicine-based education did not have a significant impact on CPAP adherence, it did result in increased attendance for OSA evaluation at clinics. Notably, these improvements were achieved without the need for additional intervention from healthcare providers (Hwang et al. 2018).

One of the major barriers to the adoption of telesleep medicine is reimbursement from the health insurance industry. Now that CMS has adopted the policy of reimbursing telemedicine services in general, most other healthcare insurance industries will follow the lead of CMS. This has surfaced because diagnostic and treatment algorithms of several common disorders like obstructive sleep apnea are often data-driven, and sleep clinicians generally rely less on discrete physical examination findings than examinations in many other specialties. Even examining the oral cavity to check for oropharyngeal space may be possible with remote monitoring devices on video conferencing (Singh and Keer 2020).

9.3 Digital Health Technology

Digital healthcare innovations have the potential to empower individuals by equipping them with valuable information to enhance their health-related choices. These technologies offer alternative avenues for promoting preventive measures, timely identification of severe ailments, and effective management of chronic conditions beyond the confines of conventional healthcare environments. By leveraging software and technological solutions that facilitate diagnosis, treatment evaluations, health record storage and sharing, as well as workflow management, healthcare providers can streamline their practices, resulting in improved efficiency. Furthermore, through the integration of advanced analytics, medical software can play a crucial role in addressing pressing public health emergencies, such as the devastating opioid crisis in numerous American communities. It achieves this by promptly delivering information on nearby treatment options and emergency assistance, thus aiding in mitigating the impact of the crisis. FDA has encouraged digital innovations for medical devices (US Food and Drug Administration 2018).

Sleep telemedicine needs the support of both software and hardware digital applications. As of January 1, 2020, the combined market cap of the nine digital health companies that entered the public market in 2019 is a staggering US$17 billion. In 2019, the telemedicine industry witnessed significant revenue generation from various sectors. Specifically, companies involved in telemedicine software, remote patient monitoring (RPM) devices, infrastructure, and connectivity provision, as well as telemedicine services, collectively contributed to the revenue growth. The revenue generated by telemedicine software companies, including those offering RPM software and health IT infrastructure, amounted to US$7.07 billion. Additionally, manufacturers of remote patient monitoring devices accounted for US$5.94 billion in revenue. Infrastructure and connectivity providers followed closely with US$5.26 billion in revenue, while telemedicine service providers generated US$1.78 billion.

Major players such as Masimo, Medtronic, and GE Healthcare held significant market shares within the hardware segment, representing 15.75%, 14.04%, and 13.46% of the overall market, respectively. Microsoft emerged as the dominant player among software providers, capturing 33.50% of the global telemedicine market. IBM and Allscripts also held substantial market shares, accounting for 25.45% and 22.34%, respectively.

From a service provider perspective, Ping A Good Doctor emerged as the leader in the telemedicine services market, commanding a 40.10% share of the overall market in 2019. While Teladoc and Amwell held market shares of 31.00% and 11.60% in the global telemedicine services market, respectively (Wood 2020; Day and Gambon 2020).

9.3.1 Reimagined Sleep Diagnostics

Many have modified their approach to their clinical practice to suit their needs. Many device manufacturers have tried to reengineer their devices for this pandemic. This includes utilizing a disposable home sleep apnea test (HSAT) device with wireless data transfer (WatchPAT ONE, Itamar Medical), while HSAT and PAP (positive airway pressure) setups are supported by information sheets, online videos (YouTube), and automated interactive platforms.

9.3.2 Telephone-Based Care

The CMS and other health insurance have started paying for the reimbursements delivered by telephone-based care. A large trial ($N = 250$) randomized patients new to PAP to either an interactive, motivation theory-driven voice response system or to one that disseminated general health information over the telephone (control). Not only was PAP adherence greater in the intervention group at the end of 1 year (2.98 vs. 0.99 h/night; $p = 0.004$), but there was also greater improvement in sleep symptoms, depressive symptoms, and functional status. However, no comparison was made with INP care regarding adherence, symptomatic improvement, or cost (Sparrow et al. 2010).

9.3.3 Direct to Consumer

The current health insurance climate of health care insurance control is being slowly replaced by a direct-to-consumer telemedicine pathway. This may be a blessing in disguise, as the middleman making a profit at the cost of patients as well as healthcare professionals may not be sustainable. So, the new direct-to-consumer (DTC) telemedicine model has become increasingly popular. Patients select their clinical issue on a DTC company's website or cell phone application and submit a medical intake form online. A clinician reviews this information and then may or may not reach out to the patient for additional information. The healthcare provider will forward a prescription to a pharmacy or mail the medication to the patient's home if deemed appropriate.

Over the past 2 years, DTC telemedicine companies have provided more than one million care visits using this model and have experienced further growth during the coronavirus disease 2019 pandemic. Despite the rapid adoption of this telemedicine option, little is known about the patients who use these companies, visits provided, and how they compare with the US population and visits to primary care physicians (PCPs). The studies have found that the patients who use DTC telemedicine services were younger, tended to live in wealthier urban communities, and typically accessed care outside normal business hours. Among the services offered, only three issues (urinary tract infection, erectile dysfunction, contraception) accounted for 87.1% of visits compared with 2.3% of PCP visits. Direct-to-consumer telemedicine companies advertise their potential to improve healthcare access. Access has many dimensions: accommodation, acceptability, availability, and affordability. Younger, more technologically savvy patients may consider online care as simply more convenient. As many younger patients may not have many other comorbidities other than sleep disorders, the sleep telemedicine model of direct to consumer may be attractive (Jain and Mehrotra 2020; Jain et al. 2019).

9.3.4 Digital Health Equity/Digital Health Literacy

Sleep telemedicine cannot be delivered in isolation. The issuance of the Office of the National Coordinator for Health Information Technology's (ONC) Cures Act final regulations in March 2020 brings attention to the accessibility of patients' health information. These rules aim to enhance the computability of health data and empower patients to have greater control over their medical records. Ideally, patients will have improved interaction with their data through patient portals and mobile health applications, allowing them to conveniently compare costs, gain insights into potential treatments, and anticipate health outcomes while seeking healthcare services (Rodriguez et al. 2020). Digital health literacy refers to the degree a patient can obtain, process, and understand digital services and information. The combination of low health and digital literacy may limit a patient's ability to benefit from this expanded access. In digital health, one area of concern is security and privacy literacy. The new rules make it easier for patients to connect directly to their data using third-party apps. Even in a pandemic, the staff working with the sleep clinic can interact with the patients to troubleshoot their issues, with the access, and input of their clinical data into the system, and their ability to follow the recommendations of sleep medicine physicians. This would also include the details about the compliance of CPAP therapy in their portal.

9.4 The Future Changes That May Need to Help in Sleep Telemedicine

Federal and State Regulatory Agencies

CMS: Regulations regarding Interoperability as a mandatory requirement

Rural: Facilitates use in rural populations with grants from federal healthcare agencies

Access: Increase broadband access to minority and rural population

Manufacturers/Vendors

Diversity: Target the diverse population use, which is relevant.

Sleep Clinics/Academic Institutions

Patient Portal: Patients-friendly development to the use of patients in terms of the input of their concerns and ability of them be able to see the usage of their CPAP data.

Patients

All the efforts should stem from the end-user patient.

The development and modification should be directed from the experience of the patients.

9.5 Key Points in Telemedicine

Dehumanization: Dehumanization and depersonalization of care can be poorly perceived by patients (e.g., telephone-based programmers and automatic e-mails). Nurses and physicians should learn specific communication skills that apply to the use of TM technologies. We should use the best of the two worlds: preserving humanism in medicine and using new technologies to improve medical care. This skill may be learned with training (Miller 2010).

Intrusiveness: The data regarding the perception of the patients are mixed. In one study, patients were not worried about being remotely observed through telemonitoring systems, but in another study of patient acceptance of telemonitoring, even though the majority (78%) of the 160 studied patients expressed a favorable attitude toward telemonitoring, 40% consider this device intrusive. However, this perception may change with time. In addition, patients who continue to use telemedicine may not feel this is intrusive. This perception also may have some impact on the outcome of their primary disease process (Turino et al. 2017).

Cost-effectiveness: In the first study, 139 CPAP patients were randomized to TM follow-up (telemonitoring and teleconsultation) or usual care for 6 months. TM was more cost-effective: despite more visits, costs were lower, and no significant differences in effectiveness were observed. In the second study, in 100 CPAP patients randomized to telemonitoring or usual care for 3 months, the authors observed that telemonitoring was less expensive than standard management (€124 vs. €171) and was cost-effective. In these days of cost-containment, it may be of value to consider telemedicine in appropriate situations (Pelletier-Fleury et al. 2001).

CPAP usage: In a retrospective study (data from Philips) comparing CPAP use in an "old fashioned" OSA cohort versus a "new feature" OSA cohort, including remote monitoring, device display, and a patient support phone app. In the 817 included patients, the implementation of multiple telemedicine supporting tools increased daily CPAP use of 1.1 h night^{-1} during the first 12 weeks of CPAP therapy. The adoption of telemedicine may have a real value in increasing the outcome of the patients on CPAP therapy. In another retrospective study, the investigators assessed the impact of using a telemonitoring tool (proactive care, AirView; Resmed Inc., San Diego, CA, USA) versus telemonitoring plus a patient engagement tool (AirView+myAir; ResMed) on termination rates. Use was $\geq$4 h night^{-1} in 77% of the AirView group versus 63% in the proactive care group ($p < 0.001$). These results were confirmed in a much larger cohort (128, 037 patients) where adherence was 5.9 h night^{-1} in the AirView+myAir group versus 4.9 h night^{-1} in the matched usual care group ($p < 0.0001$) (Woehrle et al. 2018; Malhotra et al. 2018).

CPAP education: In a study of CPAP education, 40 OSA-naive patients and scheduled for CPAP training either by video conference (20 patients) or usual face-to-face practice (20 patients). After the training, patients answered questionnaires anonymously about what they had learned. The results showed that both knowledge and practical skills (e.g., mask and headgear placement and leak avoidance) related to OSA and CPAP therapy were equivalent between the groups (94% correct answers in the video conference group vs. 92%). This is already being practiced in the pandemic times to replace mask-fit sessions usually done at sleep clinics with equivalent success (Isetta et al. 2014).

9.6 Insomnia Treatment Over Sleep Telemedicine

As there is a significant shortage of behavioral therapists certified in delivering CBT, sleep telemedicine may be an option where there is significant efficacy demonstrated in many clinical studies.

Vedaa et al. did a parallel-group superiority randomized controlled trial comparing dCBT-I with online patient education about sleep. The interventions were available through a free-to-access website, publicized throughout Norway, which incorporated automated screening, informed consent, and randomization procedures, as well as outcome assessments. Adults (age $\geq$ 18 years) who had regular Internet access and scored 12 or higher on the insomnia severity index (ISI) were eligible for inclusion and were allocated (1:1) to receive dCBT-I (consisting of six core interactive sessions to be completed over 9 weeks) or patient education (control group). Participants were masked to group assignment and had no contact with researchers during the intervention period. The primary outcome was the change in ISI score from baseline to 9-week follow-up, assessed in the intention-to-treat population. Between February 26, 2016, and July 1, 2018, 5349 individuals commenced the online screening process, of which 1497 were ineligible or declined to participate, 2131 discontinued the screening process, and 1721 were randomly allocated (868 to receive dCBT-I and 853 to receive patient education). At the 9-week follow-up, 584 (67%) participants in the dCBT-I group and 534 (63%) in the patient education group completed the ISI assessment. The latent growth model showed that participants in the dCBT-I group had a significantly greater reduction in ISI scores from baseline (mean score 19.2 [SD 3.9]) to 9-week follow-up (10.4 [6.2]) than those in the patient education group (from 19.6 [4.0] to 15.2 [5.3]; estimated mean difference −4.7 (95% CI −5.4 to −4.1, Cohen's d −1.21; $p < 0.001$). Compared with patient education, the number needed to treat with dCBT-I was 2.7 (95% CI 2.4 to 3.2) for treatment response (ISI score reduction ≥8) and 3.2 (2.8 to 3.8) for insomnia remission (ISI score < 8). No adverse events were reported to the trial team. dCBT-I is effective in reducing the severity of symptoms associated with insomnia disorder. These findings support the widespread dissemination of dCBT-I. In days of the pandemic telemedicine using dCBT may be of great value in the care of patients with insomnia (Vedaa et al. 2020).

9.7 Glossary of Telemedicine Terminology

Asynchronous telemedicine	Non-real-time provision of care, including HSAT interpretation and PAP machine data download review via cloud-based platforms. Data transmission typically occurs in one direction from the originating site to the distant site
CVT	Clinical video telehealth
Distant site	The location of the provider
HSAT	Home sleep apnea testing
Originating site	The location of the patient (clinic, home, hospital)
Patient presenter	A medical professional at an originating site who ensures the technical integrity of the CVT session and may assist with physical examination components
Sleep telemedicine	Utilizing electronic communications to exchange sleep-related medical information between different locations with the aim of enhancing a patient's well-being
Synchronous telemedicine	Real-time provider-patient interaction via the internet, telephone, or CVT. Data flow is bidirectional, from and to the originating and distant sites
Telehealth	Leveraging electronic information and communication technologies to deliver and facilitate healthcare services when geographical distance separates the participants
Telemedicine	A narrower term than "telehealth"; it specifies direct patient-provider interactions, whether synchronous or asynchronous

dCBT digital cognitive behavior therapy

9.8 Telemedicine Services in the Practice of Sleep Medicine in India

Telemedicine (TM) refers to the use of information technology applications for treatment and providing medical care (Bali 2018). On the other hand, telehealth (TH) encompasses a broader area that includes its use for the advancement of healthcare professionals through education and research besides prevention and provision of medical care (Bali 2018).

TM services can be of great help for a country like India where most of the population lives in the villages. In most of these areas, medical facilities are barely available or elementary (Bagchi 2006). Indian Space Research Organization (ISRO) tried to fill the void by starting telemedicine services through its pilot project in 2001 to address this unmet need in the country (Bagchi 2006). As of 2021, the telemedicine network established by ISRO is providing superspecialty services to far-off places in the areas of ophthalmology, cardiology, radiology, diabetology, mammography, general medicine, and women and child healthcare (Chellaiyan et al. 2019). This approach is important in developing countries like India, where most of the patients bear the burden of healthcare from their pocket and these services can reduce their expenses (Bagchi 2006; Chandwani and Dwivedi 2015). However, telemedicine as a medium of patient care grew slowly in India compared to its use for education (Chandwani and Dwivedi 2015).

9.9 Challenges to Telemedicine in the Developing World

In a paper published in 2015, Chandwani and Dwivedi (2015) identified various challenges to the use of telemedicine that are listed in Table 9.2. Many patients do not feel comfortable with TM-based services, perhaps owing to the absence of human touch (Kustwar and Ray 2020).

Besides the challenges identified in available literature (Bali 2018; Chandwani and Dwivedi 2015), other challenges also remain for the use of TM. First, can patients be diagnosed in the absence of three essential pillars of clinical decision-making inspection, palpation, and auscultation? A recent study of substance use disorder services identified that nearly 30% of patients were called for a clinical examination (Ghosh et al. 2021). Though this may be handled by having a paramedic or primary care physician available at the periphery, still the margin of error remains (Kustwar and Ray 2020). Another report suggested that the history provided by the patients and relatives may be inaccurate at times; however, clinical examination may help the physician to arrive at the correct diagnosis in such cases (Lahiri and Mitra 2020).

In such a scenario, who should be held responsible if there is a gap in the diagnosis or if any medicolegal issue arises (Bali 2018)? Second, do all diagnostic services available at the lowest level where TM is made available? This may be overcome by asking the patients to get their tests done wherever available. In this manner, only selected patients may require traveling and for others, it can save money spent on treatment. Third, what can be prescribed through TM, and what cannot be prescribed? While on one hand, prescribed medication may not be available in the peripheral areas, free availability of other medications is also a problem (Kustwar and Ray 2020). Unlike in developed countries, in developing countries, many medications that have potential for the addiction are either freely available, especially in peripheral areas or patients may get their prescription refilled by consulting a local physician or another physician using TM (Bali 2018). Fourth, in most circumstances, previous medical history is essential to make a correct clinical decision. For this, developing countries need to develop a centralized database for each patient where every physician has access to previous records of a given patient. Developing such a system would incur a huge cost in terms of time, money, and infrastructure (Bali 2018). Lastly, even if such a system is developed, considering the burden of patients on physicians in developing countries like India, its feasibility and utility is also questionable.

Table 9.2 Challenges for Telemedicine™ in India (Bali 2018; Chandwani and Dwivedi 2015; Kustwar and Ray 2020)

S. N.	Challenges	Issues to be addressed
1.	Policy level:	
	Integrating it into the healthcare system	• Defining minimal requirements for infrastructure
		• Ensuring responsibility for treatment in case of multiple consultations
		• Payment of fees and expenditure of system
	Developing policies	• Developing uniform policies across states and nations
		• Clarifying roles and responsibilities of parties, for example, patients, healthcare workers, and doctors
		• Protection of privacy of data
		• Standardization of policy
	Accreditation of TM facilities by appropriate bodies	• Identifying areas of TM practice that need regulation
2.	Resources	
	Human and technological	• Availability of high-speed internet and electricity
		• Cost-effectiveness
		• Affordability
		• Diversifying the use of available technology to enlarge its application
		• Already overburdened doctors due to a poor doctor: Population ratio
		• Identifying crucial medical conditions where it is useful
		• Knowledge of internet and TM application use in population
	Paucity of efficient platforms	• Paucity of agencies efficient in TM
		• Paucity of information technology support efficiency in TM
		• Paucity of physicians and other healthcare workers trained to deliver services through TM
3.	Sociocultural	
	Application of TM in traditional healing systems	• Using telemedicine for primary care
		• Using the bottom approach after assessing the need of the community
		• Adapting the TM to the beliefs, expectations, and healthcare practices of the population
		• Language barrier
4.	Legal considerations	• Malpractice suites
		• Informed consent before a consultation
5.	Financial	• Cost of infrastructure
		• The cost incurred towards training
		• Reimbursement for TM consultation to patients
6.	Resistance to change	• Patients
		• Doctors and other healthcare workers

9.10 Current Status of Telemedicine in India

At present, telemedicine services are working in 12 states in India, which are funded by National Health Mission (Agarwal et al. 2020). Telemedicine services in India work through the "spoke and hub" model, where the hub is usually a tertiary care center like a medical college or an institute and spokes are centers located in the periphery ranging from the village (primary health care center) to major hospital (district hospital) (Agarwal et al. 2020).

Based on funding and operation, TM services are provided using two different models—first where these are funded as well as run by the government, and second, where the government is providing the funds but services are provided by private agencies (Public-Private-Partnership) (Agarwal et al. 2020).

Depending upon the interaction, TM services can be divided into two categories—real-time or synchronous mode where patients are given medical opinions in real-time using video consultation. In this model, an expert sitting at the hub provides medical opinion directly to the patient, to a paramedic accompanying the patient, or to a physician present at the spoke (Agarwal et al. 2020). In this mode, the expert is expected to take an immediate decision. In the store and transfer model of interaction also known as asynchronous mode, information (audio or visual or both) is stored at the peripheral center and then transferred to the hub. An expert sitting at the hub can have a look at the information and make decisions at his convenience (Agarwal et al. 2020).

9.11 Ground Reality

Though TM services were available in India for almost 20 years, as discussed, they became popular during the COVID-19 pandemic. During the pandemic TM services were used across specialties. Moreover, many of them complained of a lack of privacy during TM compared to face-to-face consultation as family members kept hovering around (Kustwar and Ray 2020; Ghosh et al. 2021). These factors also led to background noise that interfered with the consultation process (Ghosh et al. 2021). Though the Government of India developed web-based medical consultation interfaces like e-Sanjeevani during the COVID-19 pandemic, many patients were computer illiterate and were not able to navigate through the digital platforms (Ghosh et al. 2021). Many of these patients preferred mobile-based teleconsultations using various familiar applications like WhatsApp or Facetime (Ghosh et al. 2020). However, 90% of patients did not have access to smartphones or internet connectivity or lacked minimal technical expertise required for the video consultation (Ghosh et al. 2020, 2021). Physicians and healthcare workers reported that patients could not be consulted or consultation was interrupted because of having wrong contact numbers, non-response, fluctuating bandwidth, and absence of familiarity of patients with devices (Ghosh et al. 2020, 2021). Hearing and vision problems of patients were reported as other obstacles to successful teleconsultation (Ghosh et al. 2020).

9.12 Utility of Telemedicine Services in Sleep Medicine Practice

Diagnosis of sleep disorders, as common as insomnia, which is primarily considered behavioral, requires several other medical disorders, for example, neurological, cardiovascular, gastrointestinal, endocrinal, musculoskeletal, and rheumatological need to be ruled out (Schutte-Rodin et al. 2008). Similarly, diagnosis of other sleep disorders also requires laboratory and radiological investigations in addition to physical examination (Shelgikar and Chervin 2013). In such a scenario, the application of TM for the first visit appears difficult. However, for follow-up, TM can be effectively used. Indian Society for Sleep Research published guidelines for the practice of sleep medicine based on available scientific literature and prevailing guidelines (Table 9.3) (Gupta et al. 2020). In addition to teleconsultation, these guidelines also included indications, precautionary measures, and procedures for Home Sleep Apnea Testing, attended polysomnography (diagnostic as well as titration study), and PAP therapy (Gupta et al. 2020).

In a recent survey, 84% of sleep physicians were continuing services through TM services, mostly through synchronous mode (Kanchan et al. 2021). They served both old and new patients (Table 9.3), but only half of them found TM consultation useful (Kanchan et al. 2021). Only half of the sleep physicians who were doing cognitive behavior therapy for insomnia (CBT-I) continued it using TM. Only half were aware that PAP compliance data could be obtained using cloud-based technology (Kanchan et al. 2021).

Table 9.3 Teleconsultation guidelines developed by Indian Society for Sleep Research

1.	Defining patient consultation	
	First consultation	Follow-up consultation
	• Patient consulting for the first time	• Consultations other than those described in "first consultation"
	• Not consulted the same physician within past 6 months	
	• The consultation sought for a reason different from the initial consultation	
	• Physician not able to recall	
2.	Selecting patients:	
	• Prefer consultation with family members	
	• Prefer in the presence of a primary care physician given physical examination	
	• The physician is not compelled to provide consultation to patients who require laboratory or radiological work-up	
	• Patients requiring behavioral management should be managed using TM services	
3.	Responsibility of interface used for TM	
	• Displaying the professional credentials of a physician	
	• Addressing grievances of customers	
4.	Reliability of patient information	
	• Physicians should ensure the identity of patients	
	• For subjective or qualitative information, the presence of the caregiver is preferred	
	• Objective and quantitative data (actigraph, PAP compliance) may be obtained wherever applicable	
	• Wherever applicable and possible, objective data should be collected	
5.	Modes for TM	
	• Prefer synchronous mode during the first consultation	
	• For follow-up synchronous or asynchronous may be used	
6.	Documentation	
	• Similar to face to face consultation	

References

Agarwal D, Roy N, Panwar V, Basil A, Agarwal PM (2020) Bringing health care closer to people—a review of various telemedicine models under the National Health Mission in India. Indian J Community Med 45(3):274–277. https://doi.org/10.4103/ijcm.IJCM_334_19

American Telemedicine Association (2020) Telehealth: defining 21st century care. American Telemedicine Association. https://www.americantelemed.org/resource/why-telemedicine/. Accessed 1 Mar 2022

Bagchi S (2006) Telemedicine in rural India. PLoS Med 3(3):e82. https://doi.org/10.1371/journal.pmed.0030082

Bali S (2018) Barriers to development of telemedicine in developing countries. In: Thomas FH (ed) Telehealth. IntechOpen, Rijeka, pp 1–14. https://doi.org/10.5772/intechopen.81723

Centers for Medicare & Medicaid Services (2020) Medicare Telemedicine Health Care Provider fact sheet: medicare coverage and payment of virtual services. Centers for Medicare & Medicaid Services. https://www.cms.gov/newsroom/fact-sheets/medicare-telemedicine-health-care-provider-fact-sheet

Chandwani RK, Dwivedi YK (2015) Telemedicine in India: current state, challenges and opportunities. Transform Gov People Process Policy 9(4):393–400. https://doi.org/10.1108/TG-07-2015-0029
Chellaiyan VG, Nirupama AY, Taneja N (2019) Telemedicine in India: where do we stand? J Family Med Prim Care 8(6):1872–1876. https://doi.org/10.4103/jfmpc.jfmpc_264_19
Day S, Gambon E (2020) In 2019, digital health celebrated six IPOs as venture investment edged off record highs. Rock Health. https://rockhealth.com/insights/in-2019-digital-health-celebrated-six-ipos-as-venture-investment-edged-off-record-highs/. Accessed 27 Feb 2022
Ford ES, Cunningham TJ, Giles WH, Croft JB (2015) Trends in insomnia and excessive daytime sleepiness among U.S. adults from 2002 to 2012. Sleep Med 16(3):372–378. https://doi.org/10.1016/j.sleep.2014.12.008
Ghosh A, Dutta K, Tyagi K, Gupta R, Misra A (2020) Roadblock in application of telemedicine for diabetes management in India during COVID19 pandemic. Diabetes Metab Syndr 14(4):577–578. https://doi.org/10.1016/j.dsx.2020.05.010
Ghosh A, Mahintamani T, Subodh BN, Pillai RR, Mattoo SK, Basu D (2021) Telemedicine-assisted stepwise approach of service delivery for substance use disorders in India. Asian J Psychiatr 58:102582. https://doi.org/10.1016/j.ajp.2021.102582
Gupta R, Kumar VM, Tripathi M, Datta K, Narayana M, Ranjan Sarmah K, Bhatia M, Devnani P, Das S, Shrivastava D, Gourineni RD, Singh TD, Jindal A, Mallick HN (2020) Guidelines of the Indian Society for Sleep Research (ISSR) for practice of sleep medicine during COVID-19. Sleep Vigil 4(2):61–72. https://doi.org/10.1007/s41782-020-00097-2
Hwang D, Chang JW, Benjafield AV, Crocker ME, Kelly C, Becker KA, Kim JB, Woodrum RR, Liang J, Derose SF (2018) Effect of telemedicine education and telemonitoring on continuous positive airway pressure adherence. The Tele-OSA randomized trial. Am J Respir Crit Care Med 197(1):117–126. https://doi.org/10.1164/rccm.201703-0582OC
Isetta V, Leon C, Torres M, Embid C, Roca J, Navajas D, Farre R, Montserrat JM (2014) Telemedicine-based approach for obstructive sleep apnea management: building evidence. Interact J Med Res 3(1):e6. https://doi.org/10.2196/ijmr.3060
Jain T, Mehrotra A (2020) Comparison of direct-to-consumer telemedicine visits with primary care visits. JAMA Netw Open 3(12):e2028392. https://doi.org/10.1001/jamanetworkopen.2020.28392
Jain T, Lu RJ, Mehrotra A (2019) Prescriptions on demand: the growth of direct-to-consumer telemedicine companies. JAMA 322(10):925–926. https://doi.org/10.1001/jama.2019.9889
Kanchan S, Saini LK, Daga R, Arora P, Gupta R (2021) Status of the practice of sleep medicine in India during the COVID-19 pandemic. J Clin Sleep Med 17(6):1229–1235. https://doi.org/10.5664/jcsm.9172
Kiley JP, Twery MJ, Gibbons GH (2019) The National Center on Sleep Disorders Research-progress and promise. Sleep 42(6):zsz105. https://doi.org/10.1093/sleep/zsz105
Kustwar RK, Ray S (2020) Telemedicine in India: an overview on the health care need of the people. J Multidiscip Res Healthcare 6(2):2025–2036
Lahiri D, Mitra S (2020) COVID-19 is accelerating the acceptance of telemedicine in India. J Family Med Prim Care 9(7):3785–3786. https://doi.org/10.4103/jfmpc.jfmpc_580_20
Malhotra A, Crocker ME, Willes L, Kelly C, Lynch S, Benjafield AV (2018) Patient engagement using new technology to improve adherence to positive airway pressure therapy: a retrospective analysis. Chest 153(4):843–850. https://doi.org/10.1016/j.chest.2017.11.005
Miller AE (2010) Telemedicine and the provider-patient relationship: what we know so far. https://www.nuffieldbioethics.org/wp-content/uploads/Miller-E-2010-Evidence-review-Telemedicine-and-the-Provider-Patient-Relationship-what-we-know-so-far.pdf. Accessed 7 Mar 2022
Pelletier-Fleury N, Gagnadoux F, Philippe C, Rakotonanahary D, Lanoe JL, Fleury B (2001) A cost-minimization study of telemedicine. The case of telemonitored polysomnography to diagnose obstructive sleep apnea syndrome. Int J Technol Assess Health Care 17(4):604–611
Rodriguez JA, Clark CR, Bates DW (2020) Digital health equity as a necessity in the 21st century cures act era. JAMA 323(23):2381–2382. https://doi.org/10.1001/jama.2020.7858

Sarmiento KF, Folmer RL, Stepnowsky CJ, Whooley MA, Boudreau EA, Kuna ST, Atwood CW, Smith CJ, Yarbrough WC (2019) National expansion of sleep telemedicine for veterans: the TeleSleep program. J Clin Sleep Med 15(9):1355–1364. https://doi.org/10.5664/jcsm.7934

Schutte-Rodin S, Broch L, Buysse D, Dorsey C, Sateia M (2008) Clinical guideline for the evaluation and management of chronic insomnia in adults. J Clin Sleep Med 4(5):487–504

Shelgikar AV, Chervin R (2013) Approach to and evaluation of sleep disorders. Continuum (Minneap Minn) 19(1 Sleep Disorders):32–49. https://doi.org/10.1212/01.CON.0000427214.00092.0f

Singh J, Keer N (2020) Overview of telemedicine and sleep disorders. Sleep Med Clin 15(3): 341–346. https://doi.org/10.1016/j.jsmc.2020.05.005

Singh J, Badr MS, Diebert W, Epstein L, Hwang D, Karres V, Khosla S, Mims KN, Shamim-Uzzaman A, Kirsch D, Heald JL, McCann K (2015) American Academy of Sleep Medicine (AASM) position paper for the use of telemedicine for the diagnosis and treatment of sleep disorders. J Clin Sleep Med 11(10):1187–1198. https://doi.org/10.5664/jcsm.5098

Smith R (2020) Telehealth market to hit $53.1 billion by 2026—report. Insurance Business. https://www.insurancebusinessmag.com/us/news/breaking-news/telehealth-market-to-hit-53-1-billion-by-2026%2D%2Dreport-213866.aspx

Sparrow D, Aloia M, Demolles DA, Gottlieb DJ (2010) A telemedicine intervention to improve adherence to continuous positive airway pressure: a randomised controlled trial. Thorax 65(12): 1061–1066. https://doi.org/10.1136/thx.2009.133215

Tuohuti A, Lin Z, Cai J, Chen X (2023) Can portable sleep monitors replace polysomnography for diagnosis of pediatric OSA: a systematic review and meta-analysis. Eur Arch Otorhinolaryngol 280(10):4351–4359. https://doi.org/10.1007/s00405-023-08095-6

Turino C, de Batlle J, Woehrle H, Mayoral A, Castro-Grattoni AL, Gomez S, Dalmases M, Sanchez-de-la-Torre M, Barbe F (2017) Management of continuous positive airway pressure treatment compliance using telemonitoring in obstructive sleep apnoea. Eur Respir J 49(2): 1601128. https://doi.org/10.1183/13993003.01128-2016

US Food & Drung Administration (2018) Digital health innovation action plan. https://www.fda.gov/media/106331/download. Accessed 2 Mar 2022

Vedaa O, Kallestad H, Scott J, Smith ORF, Pallesen S, Morken G, Langsrud K, Gehrman P, Thorndike FP, Ritterband LM, Harvey AG, Stiles T, Sivertsen B (2020) Effects of digital cognitive behavioural therapy for insomnia on insomnia severity: a large-scale randomised controlled trial. Lancet Digit Health 2(8):e397–e406. https://doi.org/10.1016/S2589-7500(20)30135-7

Woehrle H, Arzt M, Graml A, Fietze I, Young P, Teschler H, Ficker JH (2018) Effect of a patient engagement tool on positive airway pressure adherence: analysis of a German healthcare provider database. Sleep Med 41:20–26. https://doi.org/10.1016/j.sleep.2017.07.026

Wood L (2020) Worldwide telemedicine industry to 2030—major market drivers, challenges, and opportunities. CISION PR Newswire. https://www.prnewswire.com/news-releases/worldwide-telemedicine-industry-to-2030%2D%2D-major-market-drivers-challenges-and-opportunities-301127826.html. Accessed 3 Mar 2022

Chapter 10
Wearable Technologies/Consumer Sleep Technologies in Relation to Sleep Disorders Developments in the Last Decade

Ahmed S. BaHammam, Seithikurippu R. Pandi-Perumal, and Mahadevappa Hunasikatti

Abstract Wearable devices [consumer sleep technologies (CSTs)] have disrupted sleep disorders management in the world today. In the USA, sleep disorders are managed in sleep disorders centers, usually accredited by AASM. The delivery of sleep health care is usually done by clinical evaluation by a sleep medicine physician after a referral from a primary care physician. The physician would decide if a home sleep test or in-lab polysomnogram is needed after a thorough patient evaluation. In-lab attended polysomnography (PSG) is the most reliable and valid measure of sleep and is the current gold standard for quantifying sleep, but it is inconvenient and expensive. To make this more convenient, many wearable devices are in the market with varying performance in addition to Type III sleep study devices.

Consumer sleep technologies (CSTs) are computer-based systems available to the public to improve or self-monitor sleep. Most of them do not require a physician's prescription. In addition, many smartphone technologies are being brought to the public for "sleep management." When the patients do come to see a sleep physician, minimal guidelines are available, which are based on evidence-based medicine. The authors have tried to dissect various technologies available in the market and their clinical utility despite the manufacturers' claims (Stones et al. 2020). This is purely a clinical analysis and should not be construed as any opinion regarding regulations.

A. S. BaHammam
The University Sleep Disorders Center, Department of Medicine, College of Medicine, King Saud University, Riyadh, Saudi Arabia

King Saud University Medical City, King saud University, Riyadh, Saudi Arabia

S. R. Pandi-Perumal
Somnogen Canada Inc., College Street, Toronto, ON, Canada

Saveetha Medical College and Hospitals, Saveetha Institute of Medical and Technical Sciences, Saveetha University, Chennai, Tamil Nadu, India

M. Hunasikatti (✉)
Division of Respiratory, Sleep and Anesthesiology (DSRA), OHT1, Food and Drug Administration (FDA), Silver Spring, MD, USA

A. S. BaHammam, M. Hunasikatti (eds.), *Sleep Apnea Frontiers*, Progress in Sleep Research, https://doi.org/10.1007/978-981-99-7901-1_10

The integration of wearable devices and consumer technologies into mainstream practice may be accelerated by the execution of extensive clinical trials that are rigorously compared with the gold-standard polysomnograms. Despite the variations in their performance, these technologies can serve as valuable tools for preliminary testing, prescreening, and providing notifications or early warnings. Their utility may extend from the initial stages of screening and diagnosis to long-term monitoring. It is essential that future research initiatives focus on large-scale clinical studies to derive robust conclusions. Furthermore, it would be beneficial for professional societies to conduct and publish an annual, impartial assessment of the latest technologies and their application in clinical settings.

Keywords Wearable technologies · Consumer sleep devices · Sleep disorders · Monitoring · Data analysis · Sleep apnea

10.1 Introduction

Publicly available wearable devices [consumer sleep technologies (CSTs)] have disrupted sleep disorders management in the world today. In the USA, sleep disorders are managed in sleep disorders centers, usually accredited by AASM. Sleep health care is typically administered following a referral from a primary care physician to a sleep medicine specialist. After a comprehensive evaluation of the patient, the specialist determines the necessity of conducting either a home sleep test or an in-lab polysomnography to further assess the patient's condition. In-lab attended polysomnography (PSG) is the most reliable and valid measure of sleep and is the current gold standard for quantifying sleep. PSG is most often performed in sleep laboratories and involves the application of electroencephalography (EEG) electrodes to the scalp, electromyography (EMG) electrodes to the facial muscles and limbs, electrooculography (EOG), electrocardiography (EKG), pulse oximetry, thoracic and abdominal belts, and nasal (or oronasal) flow sensors to measure breathing patterns. Sleep staging is determined from brain EEG, EOG, and chin EMG.

10.2 Consumer Sleep Technology

Consumer sleep technologies (CSTs) are computer-based systems available to the public to improve or self-monitor sleep. Most of them do not require a physician's prescription. Most of the manufacturers of these gadgets claim in their promotional labeling: self-guided sleep assessment, entertainment, social connection, information sharing, and sleep education. Unfortunately, there is a lack of well-designed clinical studies using PSG as the gold standard diagnostic test with these new devices. Some have compared their device with actigraphy and MSLT (multiple sleep latency testing). Many of these devices may find a place in the marketplace because of their convenience and affordability. Instead of resisting the inevitable usage of these devices, it is time for us to evaluate each of these devices with the gold standard PSG regarding their accuracy of the reported sleep parameters. Sleep medicine physicians can use this as adjunctive information in the assessment and management of sleep disorders. We are already seeing patients coming to see us with data from over-the-counter (OTC) products for their sleep-related medical problems.

Patients love to share OTC device-generated health data with their sleep clinicians to better understand their sleep. Even when not prescribed by any healthcare professional, patients expect the treating team to use the data to manage their disease. Nevertheless, how much data from these devices are valid and accurate enough to be included in disease management?

Companies that develop and manufacture CSTs are disconnected from the sleep research community and appear ambivalent about designing devices that appeal to sleep medicine professionals (Devine et al. 2021). Opening a line of communication between sleep researchers and CST developers is important for innovation and best achieved through an area of mutual interest 26.

10.2.1 Definitions

Consumer sleep technology	Nonprescription devices directly marketed to consumers that may assert sleep monitoring, tracking or other sleep-related interventions.
App	An application downloaded onto mobile devices. This may also function independently with/without a wearable device.
Patient-generated health data	Data generated directly by patients, not healthcare practitioners, during or between clinical visits. Examples include past medical history, biometric data, self-reported observational symptoms data, and lifestyle monitoring.
Wearable technology	A device that a consumer can wear for entertainment and communications that may track several physiological parameters during sleep.

AASM (2018): It is the position of the AASM that CST must be FDA-cleared and rigorously evaluated against current gold standards if it is intended to render a diagnosis and/or treatment. It is time for us to assess each CST independently if they can be used in the evaluation of sleep disorders (Khosla et al. 2018).

The AASM, in reference to CSTs (which includes sleep wearables devices), clearly stated that "*CSTs cannot be utilized for the diagnosis and/or treatment of sleep disorders at this time (2018).*".

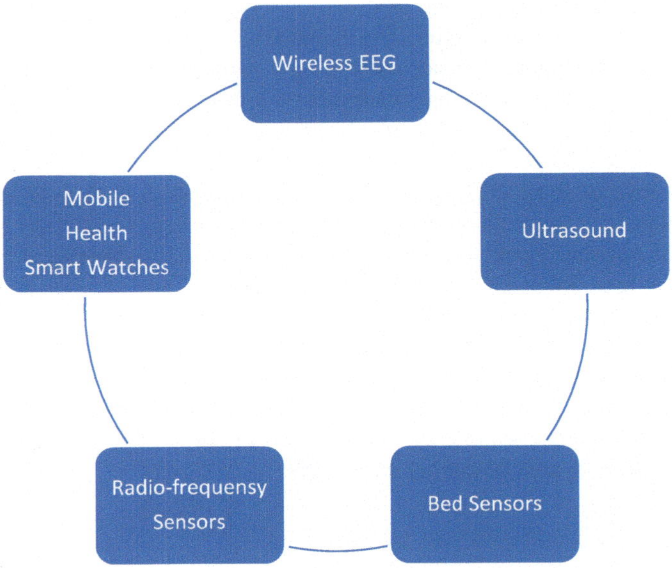

10.3 Types of Emerging Sleep Technologies

10.3.1 Sleep Monitoring Outside the Laboratory: Why and How?

What is the best way to assess sleep? Even though the polysomnogram has multiple channels, the patient is sleeping in an unnatural environment. Ideally, we should get information with sensors in natural home surroundings. Future low-cost, wearable sleep parameters detection systems are promising. This can be done at the population level by conducting large epidemiological studies.

10.3.2 Actigraphy and Accelerometry

Actigraphy and accelerometry are non-invasive methods to monitor physical activity in large population studies. The AASM does recognize the validity of actigraphy. The latest progress in artificial intelligence and machine learning have made the task much more useful. However, we may still need More Validation studies from consumer-grade to medical-grade devices, standardization for distinct types of human activity and motion, and validation studies to differentiate daytime activity from nighttime motion/activity. In practice, many of the devices that are coming to market usually have some combination of the following sensors: pulse oximetry, ECG (single lead), blood pressure monitors, gyroscopes, smartphone sensors, and underbed sensors (Perez-Pozuelo et al. 2020).

10.3.3 Bed Sensors/Smart Watch/Ultrasound Devices

Bed sensors may be defined as any sensor that sits on the bed and may be used for monitoring physiological processes—sleep-related parameters. In 2005, the investigators presented a mathematical model with their systems in sleep analysis. The cardiorespiratory signal is a fundamental vital sign used for the assessment of a patient's status. Additionally, the cardiorespiratory signal provides valuable information to healthcare providers wishing to monitor the health status of individuals. The air mattress sensor system allows the measurement of respiration and heartbeat movements without using a harness or sensor on the subject's body, eliminating the difficulties these pose for long-term measurements (Chee et al. 2005).

AAL (ambient-assisted living) platforms: Sadek et al. report that ambient-assisted living (AAL) platforms, which is a specific type of ZET (zero-effort technologies), target improve the quality of life—of both the monitored persons and their caregivers. Such a platform aims to empower people at risk without assistance, especially the elderly. It contributes to users' autonomy in their own living space rather than leaving them entirely dependent on others (e.g., a nursing home) (Sadek et al. 2020). They presented an unobtrusive novel method for sleep monitoring using fiber optics in an ambient-assisted living platform, featuring a concise user interface. Their system was evaluated for 30 days in a real deployment in three flats in Singapore. Personal interviews confirmed the results of their postprocessing.

Ultrasonic contactless sensor: Arlotto et al. described an ultrasonic contactless sensor for breathing monitoring. The researchers explored the feasibility of a noncontact ultrasonic device in quantifying breathing activity without the subject's awareness. The device utilizes a low-power ultrasonic active source and transducer to measure the frequency shift resulting from the difference in velocity between exhaled airflow and the surrounding environment, known as the Doppler effect. Through signal processing techniques, the effects of breath and subject movements are separated from the Doppler signal. With a distance of approximately 50 cm between the source and the sensor, and the use of ultrasound frequencies beyond the range of human hearing, the device enables monitoring of breathing activity without the subject's perception, thus preserving the quality of sleep and making it suitable for diagnosing sleep disorders. This may help in studying sleep in its natural state (Arlotto et al. 2014).

Beddit sleep tracker (BST): Tuominen et al. studied the sleep parameters accuracy of Beddit sleep tracker (BST), an unobtrusive and non-wearable sleep monitoring device based on ballistocardiography. This was evaluated by comparing it with polysomnography (PSG). The researchers assessed various sleep parameters, including the total amount of time spent asleep (total sleep time), the duration it takes to fall asleep (sleep onset latency), the duration of wakefulness during sleep (wake after sleep onset), and the effectiveness of sleep in terms of time spent asleep compared to time spent in bed (sleep efficiency). Additionally, they examined whether BST could differentiate sleep stages. During two nonconsecutive nights in a sleep laboratory, they performed sleep studies simultaneously with PSG and BST in ten healthy young adults (five females and five males). BST was able to distinguish SOL with some accuracy. However, it underestimated WASO and thus overestimated TST and SE. Also, it failed to discriminate between non-rapid eye movement sleep stages and did not detect the rapid eye movement sleep stage. Despite these limitations, the device may be of some value in assessing SOL (sleep onset latency) in patients in their natural sleep state at home (Tuominen et al. 2019).

This may be of value in comprehensive sleep evaluation as an adjunct device.

Withings: Yang et al. showed that a non-intrusive pneumatic sensor could be used to measure respiratory rate, heart rate, and variability during sleep. They evaluated the ability of the first commercially available mass-market pneumatic sleep sensor, the Withings AURA, to accurately quantify the respiratory and heart rates and their variability against PSG. The pneumatic sensor was included in a PSG study involving 42 participants in a sleep laboratory. The agreement between the pneumatic sensor and the PSG for respiratory rate, heart rate, and variability was quantified by Bland–Altman analysis. These results show that a non-intrusive pneumatic sensor can accurately estimate cardiorespiratory metrics overnight. Long-term follow-up with a non-intrusive sensor is still not widely used, especially during sleep; it could be particularly useful, especially for preventive care. The night heart rate, for example, has been proven as a marker of cardiovascular risk and may be useful in long-term clinical studies of patients with sleep apnea patient's response to the interventions (Yang et al. 2019).

Campo et al. evaluated patients with the diagnosis of obstructive sleep apnea syndrome (OSAS) with the WITHINGS sleep apnea detector (SAD), a non-intrusive pressure and sound sensor placed under the mattress. One hundred eighteen patients (67 F, 49 years, BMI 33 kg/m^2) suspected of OSAS had an in-laboratory PSG together with a sleep apnea detector. Sleep apnea detector derives respiratory and cardiac signals and movements from the pressure signal. From the microphone, snoring and snorting are detected. A convolutional neural network used these features to detect sleep periods with a random forest classifier and apnea and hypopnea events. Total sleep time (TST) and apnea-hypopnea index (AHI) deduced (TST_{sad}, AHI_{sad}) were compared with the PSG results scored according to AASM rules (TST_{psg}, AHI_{psg}). AHI and TST were compared using bias and mean absolute error (MAE). Sensitivity, specificity, likelihood ratios (LR), and AUROC were calculated for AHI thresholds of 15 and 30/h. The average (SD) TST_{psg} was 367 (61) min. The sleep apnea detector overestimated TST by 25 min, 7.0% of the average duration in the sample. The average AHI_{psg} was 32.5 (30.1) and AHI_{sad} 32.8 (29.9). The bias was 0.3 MAE = 10.3. The sensitivity (Se_{15}) and specificity (Sp_{15}) and their 95% confidence intervals were $Se_{15} = 88.0\%$ [79.0, 94.1] and $Sp_{15} = 88.6\%$. Positive and negative LR were, respectively, $LR^{+}{}_{15} = 7.70$ and $LR^{-}{}_{15} = 0.136$.

$AUROC_{15} = 0.926$. At the 30 thresholds, $Se_{30} = 86.0\%$ and $Sp_{30} = 91.2\%$. Positive and negative LR were $LR^{+}{}_{30} = 9.75$ and $LR^{-}{}_{30} = 0.153$. $AUROC_{30} = 0.954$. Thus, it can be used as an unattended OSAS screening device in patients likely to suffer from OSAS. If the study can be replicated in other clinical settings, this may have potential value in assessments of sleep apnea (Edouard et al. 2020).

WHOOP: Berryhill et al. study aimed to determine whether a wearable sleep tracker improves perceived sleep quality in healthy participants and test whether wearables reliably measure sleep quantity and quality compared to PSG. A wearable device (WHOOP, Inc.) was used to provide feedback regarding the sleep information of participants for one week and maintain sleep logs versus one week of maintained sleep logs alone. Self-reported daily sleep behaviors were documented in sleep logs. PSG was performed on one night when wearing the wearable. The sleep disturbance sleep scale of the Patient-Reported Outcomes Measurement Information System was evaluated at the initial assessment, day 7, and day 14 of study participation. In 32 participants, wearables improved nighttime sleep quality (Patient-Reported Outcomes Measurement Information System sleep disturbance) after adjusting for age, sex, baseline, and order effect. When wearing the device, there was a slight increase in self-reported daytime naps, but total daily sleep time remained unchanged. The wearable had low bias (13.8 min) and precision (17.8 min) errors for measuring sleep duration and measured dream sleep and slow-wave sleep accurately. In addition, bias and precision errors for heart rate and respiratory rate were incredibly low compared with those measured by electrocardiogram and inductance plethysmography during polysomnography. Thus, wearables can improve sleep quality and accurately measure sleep and cardiorespiratory variables in healthy people. While this study may be preliminary, improving the performance may result in the device being useful in patients with sleep disorders (Berryhill et al. 2020). Accurate cloud-based remote monitoring of sleep and cardiorespiratory variables is feasible and could facilitate sleep and cardiorespiratory research if this study can be replicated in other clinical centers.

Miller et al. did a comparative study with the WHOOP strap—a wearable device that estimates sleep based on movement and heart rate measures derived from actigraphy and photoplethysmography, respectively. Twelve healthy adults (six females, and six males, aged 22.9 ± 3.4 years) participated in a 10-day, laboratory-based protocol. A total of 86 periods of sleep were independently assessed in 30-s epochs using polysomnography and WHOOP. For WHOOP, researchers entered bedtimes and periods of sleep; WHOOP overestimated total sleep time by 8.2 ± 32.9 min compared to PSG, but this difference was non-significant. WHOOP was compared to PSG for two-stage (i.e., wake, sleep) and four-stage categorization (i.e., wake, light sleep [N1 or N2], slow-wave sleep [N3], and REM) of sleep periods. For the two-stage categorization, the agreement, sensitivity to sleep, specificity for a wake, and Cohen's kappa were 89%, 95%, 51%, and 0.49, respectively. For the four-stage categorization, the agreement, sensitivity to light sleep, SWS, REM, and wake, and Cohen's kappa were 64%, 62%, 68%, 70%, 51%, and 0.47, respectively. In situations where polysomnography is impractical (e.g., field settings), WHOOP is a reasonable method for estimating sleep, particularly for two-stage categorization, if accurate bedtimes are manually entered. As the subjects involved were healthy persons, we may have to wait for a study with a larger sample of patients with sleep disorders and epoch-by-epoch comparison of sleep stages with polysomnograms (Miller et al. 2020).

10.3.4 Smartphone Tools

Smartphones are widely in use. Even in developing countries, mobile phones are more commonly used than landlines. There is a significant overlap between consumer sleep technology and medical devices. There are too many variations in the performance of the devices. Some also can run on Microsoft/Google tablet platforms. Most of them may or may not require additional accessories. Accessories, when available, can use low-cost, low-performance pulse oximetry devices. It is to be noted that most of them do not have regulatory approval at this time and may need further clinical validation studies.

Currently, some of these devices claim sleep tracking, optimum sleep stage to wake up for sleep alarm, modification, and facilitating sleep onset.

Some of these phone applications may have additional sensors/signals like visual graphics, relaxing music, sleep-facilitating natural sounds, and white noise.

Advantages include the convenience of being able to use it at home and app accessibility from various devices. Disadvantages include the fact that the performance of the devices may vary, which may delay the diagnosis and therapy of sleep disorders.

Best-effort sleep (BES) model: Chen and colleagues assessed sleep duration using a unique model called the best-effort sleep (BES) model. BES utilizes smartphones to estimate sleep in a non-intrusive manner, where the user remains uninvolved in the monitoring process and carries out regular phone usage without any additional interaction (Chen et al. 2013). They performed quantitative and qualitative comparisons between two smartphone-only approaches that they developed (i.e., the BES model and a sleep-with-the-phone approach) and two popular commercial wearable systems (i.e., the Zeo headband and Jawbone wristband). Results from a 1-week eight-person study showed that the BES model could accurately infer sleep duration ($\pm$42 min) employing a fully "non-intrusive" method capable of adapting to the inherent fluctuations in individuals' sleep patterns and surroundings. If this can be improved, it may have a place in the day-to-day evaluation of patients with sleep disorders (Chen et al. 2013). The study involved eight healthy volunteers. If the findings can be replicated in patients with sleep disorders using the gold standard PSG and yield acceptable results, these methods may find a role in patient management.

The utilization of a phone oximeter as a screening tool to detect sleep-disordered breathing in children/SpO_2/PRV (pulse rate variability): Garde et al. combined the blood oxygen saturation (SpO_2) characterization and cardiac modulation, quantified by pulse rate variability (PRV), to detect sleep-disordered breathing (SDB) in children, the phone oximeter, a device that combines a smartphone with a pulse oximeter, is employed for identification purposes. They studied 146 children by analyzing the SpO_2 pattern and PRV as an estimate of heart rate variability calculated from the PPG. SpO_2 variability and SpO_2 spectral power at low frequencies were significantly higher in children with SDB due to the modulation provoked by airway obstruction during sleep. PRV analysis reflected a significant augmentation of sympathetic activity provoked by intermittent hypoxia in SDB children. A linear classifier was trained with the most discriminating features to identify children with SDB. "The classifier was validated with internal and external cross-validation, providing a high negative predictive value (92.6%) and a good balance between sensitivity (88.4%) and specificity (83.6%). Combining SpO_2 and PRV analysis improved the classification performance, providing an area under the receiver operating characteristic curve of 88%, beyond the 82% achieved using SpO_2 analysis alone. Their results show that this may be useful as a home sleep apnea test on multiple nights with natural sleep architecture at home" (Garde et al. 2014). While home sleep testing is not recommended for children to diagnose pediatric sleep apnea, this process may be used as a screening tool if this can be replicated in other clinical studies. The advantages are that the device may be used to monitor patients over multiple nights.

SpO_2/microphone/accelerometer algorithm: Al Mardini et al. algorithm design takes advantage of a smartphone's built-in sensors, pervasiveness, computational capabilities, and user-friendly interface to screen OSA. They extracted three main physiological signals to diagnose OSA, which are (1) oxygen saturation using an external oximeter, (2) respiratory effort using the smartphone's built-in microphone, and (3) body movement using the smartphone's built-in accelerometer. The signals are analyzed on the smartphone to screen the OSA. They examined the system's ability to screen the disease compared to the golden standard by evaluating it on 15 subjects. Results showed that 100% of the sick subjects were correctly classified as having OSA, and 85.7% of the healthy subjects were correctly classified as not having OSA. Even though the sample size is small and may need to be replicated with a larger, multicenter sample to validate this (Al-Mardini et al. 2013).

Sleep cycle: "Sleep Cycle is an intelligent alarm clock app that analyzes users' sleep, records findings, and wakes users during their lightest sleep phase, so they feel rested and refreshed. The app generates nightly sleep reports, tracks long-term sleep trends, and logs how daily activities impact sleep quality. The built-in microphone is used to analyze your movements as you sleep. You must place your phone correctly. The accelerometer is employed to analyze and interpret the movements occurring during sleep. The accurate location of the phone is an important concern." However, well-designed clinical studies to substantiate their claims are lacking. If the clinical performance of the device is acceptable in future clinical studies using PSG as the gold standard diagnostic test; in that case, this may have a place in the initial management of sleep disorders. At present, this device cannot be recommended for the clinical care of patients (Sleep Cycle 2014).

Sleep as Android: This app has multiple features, including nature sounds, accelerometer-derived "sleep cycle" tracking; snoring detection and "antisnoring" (e.g., Option: the phone vibrates or emits tongue-click sounds to rouse the patient to stop snoring); smart alarm with fail-safes to prevent the user from accidentally falling back asleep in the morning (e.g., the app will require the user to complete a mentally or physically engaging task such as answering arithmetic questions before the wake-up alarm will terminate); sleep graphs illustrating sleep duration, "sleep debt," and "light" and "deep" sleep percentages. They do claim that they can assess deep sleep with the use of actigraphy. They do have a disclaimer that that claim is not as precise as a polysomnogram. The accelerometer does record the movements at night. However, until they can conduct a clinical study using PSG as a gold standard, the value in clinical management is uncertain (Urbandroid Team n.d.; Newman 2013).

Entrain: This Android and Apple app encourages timed light exposure to reduce jet lag. "*Entrain* is a free app that connects users to lighting schedules developed by researchers at the University of Michigan. These schedules are mathematically demonstrated to facilitate rapid adjustment to new time zones, it has been found that light serves as the main stimulus for regulating the circadian rhythm. By documenting the individual's exposure to light over time, the therapist could simulate the body's clock and make recommendations for behavior." The user specifies the time zone change, arrival time, and arrival date for an upcoming trip, and the app creates a personalized pre-trip schedule of timed light (including intensity) and dark exposure to shift the user's circadian rhythm preemptively (Forger et al. 1999; Arcascope n.d.). The circadian clock is modeled using the equations in Forger et al. (1999). Sleep drive is modeled as a flip-flop system using a modified form of equations developed by Dr. Victoria Booth and Dr. Cecilia Diniz Behn.

Go! to Sleep: Created by Academicians at Cleveland Clinic Sleep Disorders Center, this app uses a lifestyle and sleep habit questionnaire to create a sleep score. It also provides daily sleep advice to improve one's score. 'It assigns you a "sleep score," rating the quality of sleep. Survey questions relate to caffeine, alcohol, exercise, and relaxation. The patient's answers are weighted by importance to create an overall score based on their lifestyle and morning and nighttime behaviors. The app includes a timer to track sleep and record interrupted sleep. If the app is upgraded, past results and chart progress can be reviewed, and daily sleep tips can be received" (Cleveland Clinic n.d.). Long-term outcome studies may be necessary to find their place in the clinical management of sleep patients using gold-standard PSG/actigraphy.

Wearable platforms: These devices attach to the person's body/clothing. It is easier to track the subject's movements. Disadvantages include subjective discomfort, and device displacement during sleep.

Fitbit: Moreno-Pino et al. studied two models of Fitbit wristbands: the Fitbit Charge 2, released in 2016, and the Fitbit Alta HR, released in 2017. Both devices are equipped with a three-axis accelerometer and a heart-rate monitor. They have similar operative systems and are expected to provide comparable results when used under the same conditions. The decision to use either one of the two models was made by the engineers prior to meeting the participant. Under ideal conditions, the Fitbit devices can record four stages of sleep: awake, light sleep, deep sleep, and

REM sleep. The provider refers to light sleep (usually stages N1 and N2) as the first stage of sleep, which is succeeded by periods of deep sleep and rapid eye movement (REM) sleep. To compare PSG and Fitbit devices, they decided to equate light sleep with N1 and N2 stages measured with the PSG, while deep sleep was considered N3. They studied 65 patients in sleep disorders clinics. Diagnosis of OSA was confirmed on 55 (84.6%). There were statistically significant differences between PSG and Fitbit measures for all sleep outcomes but rapid eye movement sleep. Fitbit devices overestimated total sleep time and underestimated the duration of wakefulness after initially falling asleep and the time taken to fall asleep (sleep onset). After correction of bias, Fitbit-delivered measures of sleep onset latency, which did not significantly differ from those provided by PSG. Fitbit wearable devices showed an acceptable sensitivity but poor specificity. Consumer sleep trackers still have insufficient accuracy for clinical settings, especially in clinical populations. It is time for the software developers and academicians to see if further refinement can be made to the performance of the device. If we can validate the reliability of sleep parameter outputs across various clinical centers, these methods may become useful for screening sleep disorders (Moreno-Pino et al. 2019).

Svensson et al. compared a wearable device, the Fitbit Versa (FV), to a validated portable single-channel EEG system across multiple nights in a naturalistic environment. Twenty participants (ten men and ten women) aged 25–67 years were recruited for the study. A total of 138 person-nights, equivalent to 76,539 epochs, were used in the validation process. Sleep measures were compared between the FV and portable EEG using Bland-Altman plots, paired *t*-tests, and epoch-by-epoch (EBE) analyses. The FV showed no significant bias with the EEG for the global sleep measures time in bed (TIB) and total sleep time (TST), and for calculated sleep efficiency (cSE = [TST/TIB] × 100). The FV had 92.1% sensitivity, 54.1% specificity, and 88.5% accuracy with a Cohen's kappa of 0.41, but a prevalence- and bias-adjusted kappa of 0.77. The predictive values for sleep and wakefulness were 95.0% and 42.0%, respectively. Although it has certain limitations in terms of specificity, this consumer sleep tracker has the potential to serve as a valuable tool for measuring sleep duration in longitudinal epidemiologic naturalistic studies. Even with the limitation of specificity, if the performance can be improved, this may be used as an adjunct test in the clinical evaluation of patients with sleep disorders (Svensson et al. 2019).

Liang and Martel conducted a study involving 23 participants, examining sleep data. They found that the sleep stage transition probabilities measured by the Fitbit Charge 2 significantly differed from those measured by the medical device, except for specific transitions such as deep sleep to wake, light sleep to REM sleep, and the probability of staying in REM sleep. Bland–Altman plots revealed a systematic bias ranging from 0% to 60%. The Fitbit device tends to overestimate the duration spent in any given sleep stage, while it underestimates the frequency of transitions between different stages. Additionally, a sleep efficiency (SE) greater than 90% was associated with a notable increase in measurement error. The Pittsburgh sleep quality index (PSQI) score below five and wake after sleep onset (WASO) less than 30 min were linked to significant reductions or increases in errors, depending on the specific sleep metrics being assessed. The study by Liang and Chapa-Martell (2019) suggests that the Fitbit Charge 2 yields less accurate estimates of sleep stage transitions compared to standard medical devices. This potential discrepancy should be carefully considered when utilizing data from wearable technology in the clinical assessment of sleep disorders (Liang and Chapa-Martell 2019).

Garmin: Stevens et al. studied 55 subjects wearing a Garmin wearable device (vivosmart 3) on the nondominant wrist. EEG data were simultaneously collected using a takehome device (sleep profiler) that recorded three-channel EEG for sleep staging. EEG data were scored by a registered polysomnographic technologist using AASM guidelines. "An automated algorithm to distinguish sleep stages (wake, light (N1 and N2), deep, and REM was developed using a neural network approach. Per epoch, the accuracy of 69.7% was achieved, with a Cohen's kappa of 0.54 ± 0.12. A sleep sensitivity of 95.8% and a wake specificity of 73.4% were obtained." Their study demonstrates that a wearable device employing optical PPG and accelerometry signals may be used as a valid method to estimate sleep stages in a cohort of young adults. The wake specificity is not high, and this limitation may need to be considered in the evaluation of patients with sleep-disordered patients (Stevens and Siengsukon 2019).

The same investigators conducted another study using the same device. Ninety (58 M, 32 F) patients scheduled for PSG at either KUMC or CGMH were recruited to participate in the study. Each participant underwent an overnight PSG monitored and scored by a registered PSG technologist (RPSGT). Garmin Fenix 5X+ watch was also instrumented on the participant's wrist to log raw photoplethysmography (PPG) and accelerometer signals. The wearable sensor data were then post-processed on PC to obtain SpO_2 and used to detect sleep apnea events throughout the entire sleep by a proprietary algorithm as an automatic apnea detector. The severity of sleep apnea was assessed for each participant by apnea-hypopnea index (AHI) based on RPSGT's scoring. The risk assessment provided by the detector aims to detect patients with AHI $\geq$15 as considerable risk. The AHI of the cohort was 8.0 (2.6 to 25.7). When correlated with clinical AHI, the R2 values of RPSGT-scored ODI and the automatic detector's ODI are 0.95 and 0.84, respectively. The detector's risk assessment reaches an accuracy of 89% with a sensitivity and specificity of 96% and 98%, respectively, against the PSG AHI $\geq$15. With consumer wearable sensor data, automatic apnea detection may differentiate patients with significant OSA (AHI $\geq$15) from the unafflicted population with high specificity and accuracy. If this can be repeated in a larger sample in multicenter settings and varying severity. In that case, this may be of value in screening patients with OSAS and be used in professional settings (Stevens et al. 2019).

Mouritzen et al. studied the validation of the Garmin Vivosmart 4 (GV4) sleep tracker by comparing it with polysomnography (PSG), as well as assessing the consistency of measurements within the GV4 device itself. A total of 18 healthy adults (13 females, with a mean age of 56.1 ± 12.0 years) without reported sleep disorders were enrolled in the study. Sleep data was collected simultaneously using both GV4 and PSG devices for one night, while intra-device reliability was assessed in one participant over 23 consecutive nights. An intra-device agreement was considered sufficient (observed agreement = 0.85 ± 0.13). The Garmin Vivosmart 4 (GV4) sleep tracker demonstrated high accuracy (0.90) and sensitivity (0.98) in detecting sleep, but its specificity was low (0.28). Cohen's kappa coefficient was calculated for sleep/wake detection (0.33) and sleep stage detection (0.20). GV4 significantly underestimated wakeful time, including wake after sleep onset (WASO), and overestimated light sleep and total sleep time (TST) according to paired t-tests ($p < 0.001$ for time awake and WASO, $p = 0.045$ for light sleep, and $p = 0.001$ for TST). The discrepancies in sleep onset and sleep end between GV4 and PSG values were not statistically significant. These findings suggest that while GV4 may not provide reliable information about sleep architecture, it could still be useful for detecting changes in sleep onset, sleep end, and TST (ICC ≥ 0.825) when monitoring groups over a long period of time.However, generalizations are difficult due to our sample limitations and may need to be replicated with a larger sample done in different clinical settings (Mouritzen et al. 2020).

Ōura Ring: The ŌURA ring is a commercially available "sleep tracker" measuring and processing information from several users' bio-signals.

Zamboti et al. evaluated 41 healthy adolescents and young adult sleep data using the ŌURA ring and standard PSG in a single laboratory overnight. There were no significant differences in the summary variables of sleep onset latency (SOL), total sleep time (TST), and wake after sleep onset (WASO) between the ŌURA ring and polysomnography (PSG).

The ŌURA ring exhibited a discrepancy in sleep stage measurement when compared to polysomnography (PSG), with an underestimation of approximately 20 min in PSG N3 sleep stage and an overestimation of approximately 17 min in PSG REM sleep stage. In terms of sleep detection, the ŌURA ring demonstrated a high sensitivity of 96%. The agreement between the ŌURA ring and PSG was 65% for "light sleep" (N1 + N2), 51% for "deep sleep" (N3), and 61% for REM sleep. The specificity in detecting wake was **48%**. Like PSG-N3, "deep sleep" detected with the Oura ring was negatively correlated with advancing age. ŌURA ring correctly categorized 90.9%, 81.3%, and 92.9% into PSG-defined TST ranges of <6 h, 6–7 h, >7 h, respectively. The low specificity in detecting the wake stage may be a concern. If the performance can be improved with improvement in technology, this may be of value in the initial management of sleep disorders like insomnia (de Zambotti et al. 2019).

Chee et al. evaluated 53 adolescent patients. They underwent overnight PSG monitoring while wearing both an Ōura ring and Actiwatch 2 (Philips Respironics, USA). Measurements were made over multiple nights and across three levels of sleep opportunity (five nights with either 6.5 or 8 h, and three nights with 9 h). Actiwatch data at two sensitivity settings were analyzed. Compared with PSG, Oura consistently underestimated TST by an average of 32.8 to 47.3 min across the different TIB conditions; Actiwatch 2, at its default setting, underestimated TST by 25.8 to 33.9 min. The Ōura significantly overestimated WASO by an average of 30.7 to 46.3 min. It was comparable to Actiwatch 2 at default sensitivity in the 6.5 and 8 h TIB conditions. The Ōura significantly underestimated REM sleep (12.8 to 19.5 min) and light sleep (51.1 to 81.2 min) relative to PSG but overestimated N3 by

31.5 to 46.8 min). EBE analyses demonstrated excellent sleep-wake accuracies, specificities, and sensitivities—between 0.88 and 0.89 across all TIBs. The Oura ring yielded comparable sleep measurement to research-grade actigraphy at the latter's default settings. Sleep staging needs improvement. There is a significant limitation in the performance of the disease. When the patients come with these devices' sleep data, these limitations must be considered (Chee et al. 2021).

10.4 Conclusions

As many devices are available in the market today, it is necessary to compare the device performances and use the specific device to assess and manage sleep disorders. Stone et al. have already compared many of these devices. A total of eight popular commercial sleep devices were assessed, which included the *Apple Watch Series 3* (Apple, Cupertino, California, United States), *Beddit Sleep Monitor* 3.0 (previously Beddit, Espoo, Finland, manufacturing now owned by Apple), *Fatigue Science* Readiband (Fatigue Science, Vancouver, British Columbia), *Fitbit Ionic* (Fitbit, San Francisco, California, United States), *Garmin* Vivosmart 4 (Garmin, Olathe, Kansas, United States), second generation *Ōura smart ring* (ŌURA, Oulu, Finland), *Polar A370* (Polar, Kempele, Finland), and the *WHOOP Strap 2.0* (WHOOP, Boston, Massachusetts, United States.

- Total sleep time (TST), total wake time (TWT), and sleep efficiency (SE) were measured with greater accuracy (lower percent errors) and limited bias by Fitbit Ionic [mean absolute percent error, bias (95% confidence interval); TST: 9.90%, 0.25 (−0.11, 0.61); TWT: 25.64%, −0.17 (−0.28, −0.06); SE: 3.49%, 0.65 (−0.82, 2.12)] and Oura smart ring [TST: 7.39%, 0.19 (0.04, 0.35); TWT: 36.29%, −0.18 (−0.31, −0.04); SE: 5.42%, 1.66 (0.17, 3.15)], whereas all other devices demonstrated a propensity to over or underestimate at least one if not all of the aforementioned sleep metrics. No commercial sleep technology accurately quantified sleep stages.
- Commercial sleep technologies displayed lower error and bias values when quantifying sleep/wake states than sleep staging durations.
- End-users may then be able to determine more accurately which sleep device can be used as an adjunct to other sleep devices. However, as recommended by the AASM, this needs to be in conjunction with a comprehensive sleep evaluation.
- Significant TST overestimations and TWT underestimations by the Garmin Vivosmart were observed. Additionally, negligible biases for TST and SE were observed for the Fitbit Ionic wrist device in the recent study, whereas previous models of Fitbit devices overestimated both metrics.
- As movement patterns vary with different states of wakefulness, sleep monitoring technologies often utilize accelerometers. Of the devices assessed in this study, the Apple Watch Series 3 (measured via Sleep++ and SleepWatch apps), Fatigue Science Readiband, Fitbit Ionic, Garmin Vivosmart 4, Oura Smart Ring, Polar

A370, and WHOOP Strap 2.0 were all equipped with accelerometers. However, their clinical value may be limited unless augmented by other reliable sleep devices with known performance for clinical management purposes.
- WHOOP reported the more accurate results among commercial devices analyzed for light and deep sleep stages durations.
- In general, the SenseWear, Fitbit Charge HR, Jawbone UP3, and Garmin Vivosmart devices can accurately measure total sleep time (TST) and time in bed (TIB) when compared to a sleep diary in healthy adults in a real-life environment. However, these trackers cannot be considered reliable in measuring wake times during a night of sleep.

Unfortunately, which device is "acceptable enough for clinical management purposes" remains a question. In the future, this would need well-designed, large-scale clinical studies with each of these devices using gold-standard polysomnogram as a comparator.

References

Al-Mardini M, Aloul F, Sagahyroon A, et al. (2013) On the use of smartphones for detecting obstructive sleep apnea. In 13th IEEE International Conference on BioInformatics and BioEngineering, pp 1–4

Arcascope (n.d.) Biology in the fourth dimension. http://entrain.math.lsa.umich.edu/

Arlotto P, Grimaldi M, Naeck R, Ginoux J-M (2014) An ultrasonic contactless sensor for breathing monitoring. Sensors 14(8):15371–15386. https://doi.org/10.3390/s140815371

Berryhill S, Morton CJ, Dean A et al (2020) Effect of wearables on sleep in health individuals: a randomized crossover trial and validation study. J Clin Sleep Med 16(5):775–783

Chee Y, Han J, Youn J, Park K (2005) Air mattress sensor system with balancing tube for unconstrained measurement of respiration and heartbeat movements. Physiol Meas 26:413

Chee NIYN, Ghorbani S, Golkashani HA, Leong RLF, Ong JL, Chee MWL (2021) Multi-night validation of a sleep tracking ring in adolescents compared with a research actigraph and polysomnography. Nat Sci Sleep 13:177–190. https://doi.org/10.2147/NSS.S286070

Chen Z et al. (2013) Unobtrusive sleep monitoring using smartphones. In 2013 7th International Conference on Pervasive Computing Technologies for Healthcare and Workshops, Venice, pp 145–152

Cleveland Clinic (n.d.) GO! TO SLEEP®. https://my.clevelandclinic.org/mobile-apps/go-to-sleep-app

de Zambotti M, Rosas L, Colrain IM, Baker FC (2019) The sleep of the ring: comparison of the ŌURA sleep tracker against polysomnography. Behav Sleep Med 17(2):124–136. https://doi.org/10.1080/15402002.2017.1300587

Devine JK, Schwartz LP, Hursh S (2021) What do researchers want in a consumer sleep technology? Sleep 44(5):zsab078. https://doi.org/10.1093/sleep/zsab078

Edouard P, Campo D, Bartet P, Marais L, Petitjean M, Roisman G, Bruyneel M, Escourrou P (2020) 0579 non-intrusive and unattended sleep analyzer effectively screens patients suspected of sleep apnea: a comparison with polysomnography. Sleep 43(Suppl_1):A222. https://doi.org/10.1093/sleep/zsaa056.576

Forger DB, Jewett ME, Kronauer RE (1999) A simpler model of the human circadian pacemaker. J Biol Rhythm 14(6):533–538. https://doi.org/10.1177/074873099129000867

Garde A, Dehkordi P, Karlen W, Wensley D, Ansermino JM, Dumont GA (2014) Development of a screening tool for sleep disordered breathing in children using the phone Oximeter™. PLoS One 9(11):e112959. https://doi.org/10.1371/journal.pone.0112959

Khosla S, Deak MC, Gault D, Goldstein CA, Hwang D, Kwon Y, O'Hearn D, Schutte-Rodin S, Yurcheshen M, Rosen IM, Kirsch DB, Chervin RD, Carden KA, Ramar K, Aurora RN, Kristo DA, Malhotra RK, Martin JL, Olson EJ, Rosen CL, Rowley JA, American Academy of Sleep Medicine Board of Directors (2018) Consumer sleep technology: an American Academy of Sleep Medicine position statement. J Clin Sleep Med 14(5):877–880

Liang Z, Chapa-Martell MA (2019) Accuracy of fitbit wristbands in measuring sleep stage transitions and the effect of user-specific factors. JMIR Mhealth Uhealth 7(6):e13384. https://doi.org/10.2196/13384

Miller DJ, Lastella M, Scanlan AT et al (2020) A validation study of the WHOOP strap against polysomnography to assess sleep. J Sports Sci 38(22):2631–2636. https://doi.org/10.1080/02640414.2020.1797448

Moreno-Pino F, Porras-Segovia A, López-Esteban P, Artés A, Baca-García E (2019) Validation of Fitbit Charge 2, and Fitbit Alta HR against polysomnography for assessing sleep in adults with obstructive sleep apnea. J Clin Sleep Med 15(11):1645–1653

Mouritzen NJ, Larsen LH, Lauritzen MH, Kjær TW (2020) Assessing the performance of a commercial multisensory sleep tracker. PLoS One 15(12):e0243214. https://doi.org/10.1371/journal.pone.0243214

Newman J (2013) 50 best android apps for 2013. Time Magazine. http://techland.time.com/2013/07/01/50-best-android-apps-for-2013/slide/sleep-as-android

Perez-Pozuelo I, Zhai B, Palotti J et al (2020) The future of sleep health: a data-driven revolution in sleep science and medicine. NPJ Digit Med 3:42. https://doi.org/10.1038/s41746-020-0244-4

Sadek I, Bellmunt J, Kodyš M, Abdulrazak B, Mokhtari M (2020) Novel unobtrusive approach for sleep monitoring using fiber optics in an ambient assisted living platform. In International Conference on Smart Homes and Health Telematics, Springer, pp 48–60

Sleep Cycle (2014). http://sleepcycle.com

Stevens S, Siengsukon C (2019) Commercially available wearable provides valid estimate of sleep stages (P3.6-042). Neurology 92(15 Suppl):P3.6-042

Stevens S, Siengsukon CF, Yu S, Fehler M, Chen N-H, Chuang L-P, Lin S-W, Burgett S (2019) 1007 validation of consumer wearable technology to assess sleep apnea risk. Sleep 42(Suppl_1):A405. https://doi.org/10.1093/sleep/zsz067.1004

Stone JD, Rentz LE, Forsey J, Ramadan J, Markwald RR, Finomore VS Jr, Galster SM, Rezai A, Hagen JA (2020) Evaluations of commercial sleep technologies for objective monitoring during routine sleeping conditions. Nat Sci Sleep 12:821–842. https://doi.org/10.2147/NSS.S270705

Svensson T, Chung U, Tokun S et al (2019) A validation study of a consumer wearable sleep tracker compared to a portable EEG system in naturalistic conditions. J Psychosom Res 126:109822

Tuominen J, Peltola K, Saaresranta T, Valli K (2019) Sleep parameter assessment accuracy of a consumer home sleep monitoring ballistocardiograph beddit sleep tracker: a validation study. J Clin Sleep Med 15(3):483–487

Urbandroid Team (n.d.) Sleep as android. http://urbandroid.org/

Yang R-Y et al (2019) Pneumatic sensor for cardiorespiratory monitoring during sleep. Biomed Phys Eng Express 5:055014

Chapter 11
Traffic Safety in Sleep Deprivation, Sleepiness, and Sleep Disorders

Ahmed S. BaHammam, Mahadevappa Hunasikatti, and Seithikurippu R. Pandi-Perumal

Abstract Every year, a substantial number of people are killed or injured due to motor vehicle accidents (MVAs). A large number of these collisions are caused by driving errors. Vehicles travel at fast speeds, necessitating the driver's constant attention and alertness. However, scientific evidence suggests that tired drivers or drivers who drive at unusual hours are more prone to make mistakes. When drivers have sleep disorders such as obstructive sleep apnea (OSA) or central hypersomnolence disorders, the situation worsens. Patients with OSA and narcolepsy have been found to have an increased risk of falling asleep behind the wheel and crashing. Narcolepsy and OSA treatment has increased driver vigilance and minimized the likelihood of MVAs. Motorists susceptible to dozing off while driving can benefit from tests that determine the level of daytime sleepiness or drowsiness while driving. As a result, unambiguous standards must be developed to oversee the periodic assessment of drivers' risk of falling asleep behind the wheel, particularly commercial drivers.

Keywords Drivers · Drowsiness · Sleepiness · Legislation · Narcolepsy · Obstructive sleep apnea

---

A. S. BaHammam (✉)
The University Sleep Disorders Center, Department of Medicine, College of Medicine, King Saud University, Riyadh, Saudi Arabia

King Saud University Medical City, King Saud University, Riyadh, Saudi Arabia
e-mail: ashammam@ksu.edu.sa

M. Hunasikatti
Division of Respiratory, Sleep and Anesthesiology (DSRA), OHT1, Food and Drug Administration (FDA), Silver Spring, MD, USA

S. R. Pandi-Perumal
Somnogen Canada Inc., College Street, Toronto, ON, Canada

Saveetha Medical College and Hospitals, Saveetha Institute of Medical and Technical Sciences, Saveetha University, Chennai, Tamil Nadu, India

A. S. BaHammam, M. Hunasikatti (eds.), *Sleep Apnea Frontiers*, Progress in Sleep Research, https://doi.org/10.1007/978-981-99-7901-1_11

Key Points

- One of the leading causes of car accidents is drowsiness.
- Sleep deprivation, obstructive sleep apnea, and central hypersomnias are all common sleep issues linked to hypersomnolence and MVAs.
- Effective treatment of sleep problems and sleep deprivation can boost vigilance, lowering the risk of an accident.
- Sleepiness can be measured with diagnostic tests, which can be used to determine driving fitness.
- Physicians, especially family physicians and general practitioners, should be taught about the dangers of sleepy driving and the significance of assessing patients' fitness to drive regularly.
- Before giving driver's licenses, legislation should be drafted to measure hypersomnolence.

11.1 Introduction

The number of automobiles sold has increased in recent years, implying that the number of people driving or riding in vehicles has also increased. As the number of vehicles and drivers has grown, so has the number of accidents. Motor vehicle accidents (MVAs) are a prominent source of death and morbidity worldwide, and they are anticipated to be the fourth greatest cause of death by 2030 (World Health Organization 2011, 2018). MVAs claim the lives of approximately 1.3 million people globally each year, with another 20 to 50 million suffering nonfatal injuries that result in severe disability (World Health Organization 2011). According to the current research, weariness, and sleepiness are substantial risk factors for MVAs (Gupta et al. 2017). Although it is difficult to assess the frequency of MVAs involving drowsy drivers, several modeling studies have suggested that the fatal accident rate ranges from 15% to 33% (Centers for Disease Control and Prevention 2013). Drowsiness results in various neurological problems, including slower reaction times, decreased concentration, and poor decision-making abilities (Centers for Disease Control and Prevention 2013). According to French research of 4774 drivers, 11.8% had Epworth sleepiness scores (ESS) of 11, 29% reported having sleepiness at the wheel severe enough to need halting, and 47% and 39% reported feeling sleepy while driving at night and day, respectively (Sagaspe et al. 2010).

Furthermore, 10.7% of respondents said they had a near-miss accident in the preceding year (46% of which were sleep-related), and 6% said they had an accident (5% of which were sleep-related) (Sagaspe et al. 2010). Globally, nearly 1.3 million people die each year due to MVAs, and road accidents are the leading cause of death in children and young adults aged 5–29 years (World Health Organization 2021). In 2013, drowsy drivers were accountable for more than 72,000 collisions, approximately 44,000 injuries, and there were 800 reported deaths in addition to the 72,000 accidents mentioned above and 44,000 injuries caused by drowsy drivers, as per the Centers for Disease Control and Prevention (CDC) data (Centers of Disease Control and Prevention 2020). However, according to the CDC analysis, this important risk factor is vastly underreported (Centers of Disease Control and Prevention 2020). Furthermore, the survey claims that some kinds of drivers are more likely to become drowsy while driving. Those who do not get enough sleep (for whatever reason), drivers of commercial vehicles, such as trucks and buses, drivers who use substances that cause drowsiness (e.g., sleeping pills, alcohol, and opiates), and those with sleep problems are all at danger (Centers of Disease Control and Prevention 2020). People who sleep less than 6 h per day and snore are more likely to fall asleep behind the wheel, according to one study (Centers of Disease Control and Prevention 2020). Short periods of sleep during the night and extended periods of driving have been observed to enhance the likelihood of rear-end collisions and single-vehicle accidents (Abe et al. 2010). Furthermore, driving at night increases the risk of an accident (Rice et al. 2003).

The available evidence linking MVAs to tiredness, sleep loss, or sleep problems is discussed in this chapter. Furthermore, we discuss the legal status of sleep-related elements in the issue of driver's licenses in various nations. Finally, we dig deeper into the evidence of the impact of sleep disorders therapy on MVAs.

11.2 Sleep Deprivation

An Individual, either a long-haul truck driver or a long-haul or ultra-long-haul pilot, might be subjected to acute or chronic sleep deprivation. Acute sleep deprivation is a condition in which a person is unable to sleep for the entire night or a significant portion of it. Chronic sleep deprivation is characterized by a persistent lack of adequate sleep, whereby an individual fails to obtain the necessary amount of daily sleep for a few days to several weeks. This sort of sleep deprivation also has a cumulative impact, so the individual becomes chronically sleep-deprived after a few days. Both sorts of sleep deprivation are harmful to one's health. Sleep deprivation is estimated to cause 56,000 accidents in the United States each year (National Highway Traffic Safety Administration 2011). Other countries have observed a similar association between sleep deprivation and an increased risk of MVAs (BaHammam et al. 2014; Leechawengwongs et al. 2006). It has been discovered that after 17 h of continuous awake, cognitive skills (e.g., reaction time and hand-eye coordination) suffer substantially, and an increasing number of errors occur (Williamson and Feyer 2000). Attention, as well as the ability to comprehend information, is severely hampered (Williamson and Feyer 2000). After 17 h of continuous awake, cognitive and psychomotor performance deteriorates to a level equal to the impairment seen in persons with a blood alcohol concentration of 0.05%, which is unsuitable for industrial or driving employment (Williamson and Feyer 2000). A recent study of a representative sample of crashes of about 7000 drivers involved in the United States reported that driving after sleeping less than 7 h. in a 24-h period is associated with an elevated risk of culpable crash involvement (Tefft 2018). An inverse relationship was observed between sleep duration and the risk of MVAs, with the most significant risk for drivers who have slept less than 4 h, 3.4 (95% CI = 2.1 to 5.6) times the increase in odds of culpable involvement in single-vehicle crashes compared to those who slept 7–9 h (Tefft 2018).

Another study looked at the effect of awake time on performance impairment during the simulated afternoon and midnight driving in a group of healthy persons who had a normal sleep duration (7 h) before taking the test (Akerstedt et al. 2013). The findings indicated that midnight driving was associated with increased tiredness and an increased risk of accidents due to the prolonged period of wakefulness, despite regular sleep duration on previous nights (Abe et al. 2010; Akerstedt et al. 2013; Anund et al. 2009). Most adults need at least 7 h of sleep per 24 h, while adolescents need at least 8 h (Alfawaz et al. 2021). When the driving test was administered during the day, adolescents who received 6.5 h of nightly sleep for 5 days performed worse on simulated driving than those who received 10 h of nocturnal sleep (Garner et al. 2015).

Long-haul truck drivers commonly limit their sleep to 4 h per day while on duty, resulting in sleep deprivation and an increased risk of an MVA (Perez-Chada et al. 2005). According to another survey, over half of the truck drivers worked unrealistic schedules, with more than 50 h of driving each week, and 27% of drivers reported poor sleep quality (Braeckman et al. 2011). Professional drivers were found to have a higher prevalence of sleep debt than control participants (Carter et al. 2003). These examples demonstrate the consequences of prolonged partial sleep loss on MVA risk. When paired with alcohol use, an increased length of wakefulness has been shown to impair driving ability, even throughout the elimination phase of ethanol use (Arnedt et al. 2000).

The effect of sleep loss on the risk of MVAs appears to be age-dependent. A counterbalanced design study that assessed the effect of sleep restriction for the previous 5 h on prolonged (2 h) afternoon simulated driving in 20 younger (mean age 23 years) and 19 older (mean age 67 years) healthy drivers found that after sleep restriction, young drivers had significantly more sleepiness-related lane deviations and higher low-frequency EEG power (4–11 H) (Filtness et al. 2012).

11.2.1 Obstructive Sleep Apnea and Hazards of Driving Risks

Snoring serves as an indication or sign of obstructive sleep apnea (OSA), a prevalent sleep disease that affects 6% to 17% of the population (Senaratna et al. 2017). Poor performance on the psychomotor vigilance test is associated with OSA (Pack et al. 2006). Several studies have found that, when compared to controls, OSA patients are three times more likely to fall asleep while driving and are three times as likely to cause MVAs (Amra et al. 2012; Rodenstein 2009; Tregear et al. 2009; Bonsignore et al. 2021). However, most studies were observational and used screening questionnaires with poor sensitivity or specificity, resulting in an unacceptably high number of missed cases (false-negative) and false-positive cases (Bonsignore et al.

2021); therefore, there is an urgent need for good quality studies to describe the best ways to diagnose OSA and detect sleepiness among drivers.

A high-risk driver has moderate to severe daytime sleepiness and a recent unintentional motor vehicle crash or near miss due to sleepiness, weariness, or inattention, according to the American Thoracic Society (ATS) (Strohl et al. 2013). Table 11.1 summarizes research conducted in various countries to examine OSA, drowsiness, and the risk of MVAs. Using studies published between 1980 and 2003 and data from the National Safety Council, it has been estimated that OSA was responsible for roughly 810,000 crashes in the United States, resulting in the loss of 1400 lives and $16 billion in damages per year (Sassani et al. 2004). However, the link between OSA and poor performance is not ubiquitous, and some research found no link between OSA and MVAs (Pack et al. 2006; de Mello et al. 2013; Ozoh et al. 2013). A recent study used both before-after and case-control study design to assess the risk of MVAs among CPAP-treated, and untreated patients matched for gender, age, and apnea-hypopnea index (AHI) and reported that the MVA incidence did not change after CPAP treatment (Myllyla et al. 2020). Nonetheless, it has been stated that reporting MVAs as a result of OSA may have a detrimental impact on an individual's employment and ability to keep their driver's license (Kales and Straubel 2014). As a result, drivers suffering from OSA may underreport MVAs to their treating physicians.

The effect of OSA on driving performance is most likely related to the severity of the OSA (Pack et al. 2006; Ozoh et al. 2013). Accident rates are proportional to the severity of OSA, and the risk rises as the severity rises due to increased daytime sleepiness (Cui et al. 2006; Horstmann et al. 2000). AHI values (AHI of 40) and Epworth Sleepiness Scale (ESS) scores (ESS of 11 and ESS of 16) of 616 Japanese drivers were shown to be associated with an increased chance of dozing off at the wheel as well as a higher probability of being involved in MVAs in a study (Komada et al. 2009). According to a systematic evaluation of published data, noncommercial drivers with OSA had a considerably elevated probability of being involved in MVAs (two to three times higher accident rate) (Ellen et al. 2006). In OSA patients, increased drowsiness is a risk factor for MVAs. The European sleep apnea database of 8476 drivers with OSA found that an ESS score of >16, driving more than 15,000 km/year, sleeping fewer than 5 h/day, and usage of hypnotics were the main risk factors for MVAs, with an increasing number of risk variables across the OSA severity spectrum (Karimi et al. 2014). Other evidence suggests that daytime sleepiness raises the risk of MVAs in both OSA patients and those with simple snoring (Perez-Chada et al. 2005; Akkoyunlu et al. 2013; Lloberes et al. 2000; Masa et al. 2000; Shiomi et al. 2002; Alahmari et al. 2019). Early diagnosis and treatment of OSA and education of the patient and family on the dangers of sleepy driving are likely to reduce the prevalence of sleepiness-related crashes in OSA patients (Strohl et al. 2013).

Table 11.1 A summary of major studies that addressed obstructive sleep apnea and motor vehicle accidents among drivers

Authors	Country	Type of study	Study population	Results
Hanning and Welsh (1996)	United Kingdom	Questionnaire-based survey	2247 drivers with "no claim" history. Questioned for a history of accidents, snoring, and daytime sleepiness	Snorers were more likely to report daytime sleepiness and having had to pull off from the road while driving
Pack et al. (2006)	USA	Questionnaire followed by actigraphy and in-lab PSG; an experimental study	1329 drivers; 247 at high risk for OSA and 159 at low risk for OSA	Short sleep duration and severe OSA were associated with sleepiness and poor performance on the road. Of the study population, 29% had OSA, with severe OSA in 5%
Ozoh et al. (2013)	Lagos	Descriptive cross-sectional study using questionnaires	500 male commercial drivers.	Of the study population, 49% had a high risk for OSA and 14.4% had daytime sleepiness. No relationship was observed between OSA, daytime sleepiness, and past history of MVAs
Dogan et al. (2006)	Turkey	Descriptive cross-sectional study using a questionnaire	340 drivers	Of the study population, 41.2% had habitual snoring. Habitual snorers had OR = 1.6 (1.03–2.53) for accidents
Lioberes et al. (2000)	Spain	A case-control study with PSG	172 patients with OSA and 40 controls	Self-reported sleepiness among OSA patients increased the chance of accidents (OR = 5; 95% CI 2.3–10.9)
Perez-Chada et al. (2005)	Argentina	Questionnaire-based study	738 long-haul truck drivers	Snoring more than 3 times a week (OR 1.73, 95% CI = 1.23–2.44), sleepiness while driving (OR 1.92, 95% CI = 1.08–1.96), and ESS score > 10 (odds ratio 2.53, 95%

(continued)

Table 11.1 (continued)

Authors	Country	Type of study	Study population	Results
				CI = 1.61–3.97) predicted accidents or near-miss accidents
Akkoyunlu et al. (Cui et al. 2006)	Turkey	Questionnaire-based study	520 city drivers	The accident rate was predicted by daytime sleepiness among those at high risk for OSA
Masa et al. (2000)	Spain	A case-control study using PSG	145 habitually sleepy drivers and controls	Those with a history of MVA had a higher respiratory disturbance index (OR = 8.5; 95% CI = 1.2–59)
Shiomi et al. (2009)	Japan	A case-control study using PSG	448 patients with OSA and 106 simple snorers	The accident rates were higher in patients with severe OSA. Sleepiness among OSA and snorers predicted risk of MVA
Braekman et al. (Alfawaz et al. 2021; Braeckman et al. 2011)	Belgium	A cross-sectional cohort study using questionnaires	476 truck drivers	Of the study population (21.5%) were at high risk for OSA and 18% had scores >10 on ESS
BaHammam et al. (2014)	Saudi Arabia	A cross-sectional study using questionnaires	1219 city drivers	The prevalence of high risk for OSA was higher among drivers who had MVAs in the past 6 months

11.2.2 MVAs and Central Hypersomnolence Disorders

MVAs are more likely in patients with central causes of hypersomnolence (Bartels and Kusakcioglu 1965; George et al. 1996; Grubb 1969; Kotterba et al. 2004). However, just a few research have been conducted on this subject. A French study found that central diseases producing hypersomnolence (narcolepsy and primary hypersomnia) enhanced the likelihood of having MVAs (Pizza et al. 2015) and that the likelihood of having MVAs was related to daytime drowsiness (Pizza et al. 2015). At study inclusion, the risk of MVAs was high in both treated and untreated participants (untreated, OR = 2.21; 95% CI = [1.30–3.76], treated OR = 2.04; 95% CI = [1.26–3.30]). However, the risk of MVAs for patients treated for at least 5 years was not different from the risk for healthy participants (OR = 1.23; 95%

CI = [0.56–2.69]), implying that long-term treatment may potentially reverse the risk of MVAs (Pizza et al. 2015).

11.2.3 Medication and the Dangers of MVAs

Several studies have found that people with insomnia and those on various sleep drugs are more prone to accidents as a result of weariness, poor attentiveness, and sleepiness behind the wheel (Daley et al. 2009; Leger et al. 2006; Williamson et al. 2011). Several studies investigated the long-term impact of hypnotic medications on driving performance. Participants were given hypnotic medication before going to bed, and their driving abilities were assessed the next day. According to the findings of these investigations, benzodiazepine hypnotics and zopiclone considerably decreased driving performance, although zolpidem and zaleplon had less of an effect (Verster et al. 2004, 2006, 2011). Ramelteon, a melatonin receptor antagonist, has been demonstrated to dramatically decrease driving performance the following morning (Mets et al. 2011).

Current evidence corroborates epidemiological data indicating an increased risk of accidents among people who take benzodiazepine hypnotics and zopiclone (Barbone et al. 1998).

11.2.4 Countermeasures That May Help to Reduce Drowsiness When Driving

Given that tiredness is a major cause of MVAs, it is critical to evaluate the countermeasures that may prevent sleepiness and lower the risk of MVAs. A variety of countermeasures can be used to decrease tiredness among sleep-deprived drivers or while driving at strange hours. The installation of rumble strips in the middle of the road, particularly on small roads, has been shown to rouse sleeping drivers (Anund et al. 2011). When evaluated 2 h after administration, modafinil (300 mg) was found to reduce lane deviation in sleep-deprived subjects, with no effect on vehicle speed beyond permitted limits (Gurtman et al. 2008). However, modafinil can cause exhilaration, impairing one's ability to properly appraise the risk of driving in such settings (Gurtman et al. 2008). Caffeine (200 mg) and napping have been shown to reduce errors while driving at night without affecting future sleep (Philip et al. 2006; Sagaspe et al. 2007a). After 64 h of continuous wakefulness, the effects of three stimulants, modafinil (400 mg), caffeine (600 mg), and dextroamphetamine (20 mg), on alertness and vigilance were examined. Caffeine had an instant impact, while dextroamphetamine had a longer effect, and modafinil was found to be equal to placebo (Killgore et al. 2008). Caffeine had the highest rate of adverse effects, although dextroamphetamine negatively influenced post-test sleep (Killgore et al. 2008). A recent study suggested that a small dose of caffeine has positive effects on alertness in patients with narcolepsy (Aldosari et al. 2020). Some studies have suggested that driving at night with a blue light in the automobile can be beneficial (Taillard et al. 2012). Scheduled naps while driving at night have also increased attentiveness and reduced errors (Signal et al. 2009).

11.2.5 The Effect of Treatment for Sleep Disorders on the Risk of MVAs

One important concern is whether treating sleep disturbances can help to prevent MVAs. It has been stated numerous times in the literature that optimal and effective treatment of sleep disturbances can enhance awareness and minimize the likelihood of accidents (George 2001; Gurubhagavatula et al. 2008). It has been discovered that treating narcolepsy and primary hypersomnia lowers the risk of MVAs (Pizza et al. 2015; Philip et al. 2014). A comprehensive review and meta-analysis of OSA, CPAP, and accidents discovered that using CPAP reduced the probability of an accident in drivers with moderate to severe OSA (RR = 0.278, 95% CI: 0:22 to 0:35, p 0.001); self-reported tiredness, one of the symptoms of OSA, was also reduced (Tregear et al. 2009) (Figs. 11.1 and 11.2).

A poll was conducted to ascertain sleep physicians' perspectives on fitness-to-drive in individuals with sleep disorders (Mets et al. 2012). Most sleep doctors agreed that patients with untreated OSA, narcolepsy, or insomnia should not drive

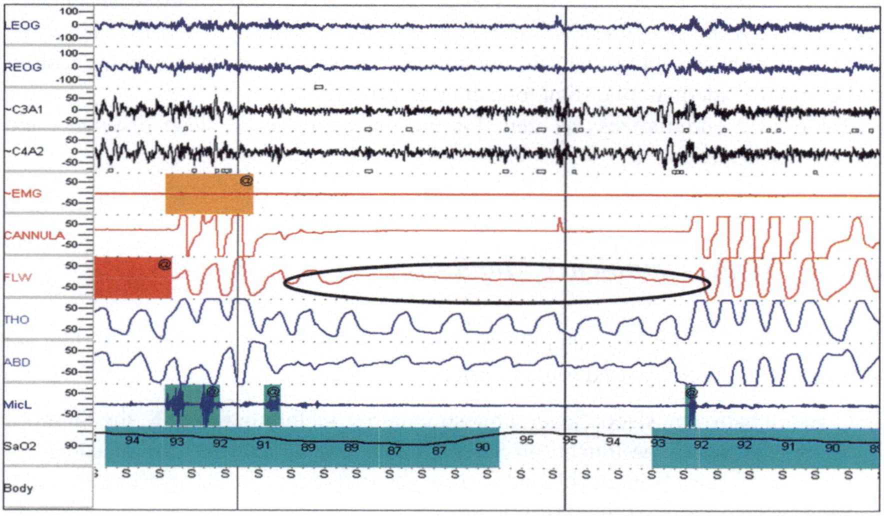

Fig. 11.1 A 2-min window showing an example of obstructive apnea. Airflow is absent despite persistent respiratory paradoxical effort. The event is followed by arousal and desaturation

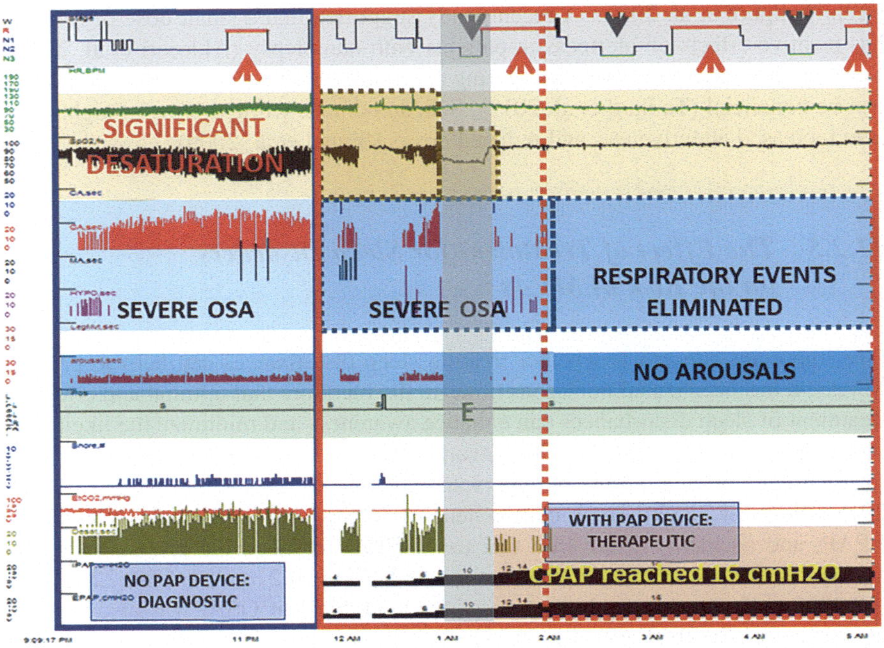

Fig. 11.2 A histogram of a split-night study of a patient with obstructive sleep apnea. The first part of the study is diagnostic, and in the second part, continuous positive airway pressure (CPAP) therapy was titrated. Obstructive events were eliminated when the optimal pressure was reached

unless they receive treatment (Mets et al. 2012). According to these specialists, excessive daytime sleepiness was the biggest limiting factor for driving (Mets et al. 2012). Nevertheless, physicians do not seem to assess their patients' driving history and eligibility adequately (Alkharboush et al. 2017).

Table 11.2 summarizes the data for the management of sleep disorders and driving performance.

11.2.6 Hypersomnolence Monitoring

Several techniques for assessing hypersomnolence have been developed. Below are some of the commonly used tools in brief.

1. Questionnaires on sleep: Several questionnaires, including the ESS, the Stanford Sleepiness Scale, the Pittsburgh Sleep Quality Index, and the Karolinska Sleepiness Scale, have been developed to assess hypersomnolence (Buysse et al. 1989; Johns 1991; Kaida et al. 2006). Unfortunately, subjective drowsiness ratings do not correspond with objective sleepiness tests. Furthermore, none of those tools have been validated for MVA risk.

Table 11.2 A summary of major studies that assessed the effects of treatment on the risk of motor vehicle accidents in patients with hypersomnolence

Authors	Type of study	Intervention	Study population	Results
Philip et al. (2006)	Randomized, crossover, double-blind placebo-controlled trial	Modafinil 400 mg/day or placebo for 5 days before a driving test	13 patients with narcolepsy and 14 with idiopathic hypersomnia	Modafinil improved wakefulness and reduced inappropriate line crossing and the standard deviation of the lateral position of the vehicle
Hirshkowitz et al. (2007)	Randomized double-blind study	Armodafinil 150 mg/day	Patients with OSA on nCPAP having residual EDS	Armodafinil improved wakefulness and reduced fatigue
Hack et al. (2001)	Case-control study	nCPAP for OSA and sleep for sleep-deprived patients	26 patients with OSA before and after nCPAP; 12 healthy subjects after sleep deprivation and after normal sleep or before and after alcohol intoxication	nCPAP for OSA, normal sleep in sleep-deprived subjects, and removal of alcohol reversed impaired driving. The effects of OSA are related to sleep fragmentation/ sleep deprivation, while alcohol induces impairment through impaired cognitive performance
Cassel et al. (Hack et al. 2001; Cassel et al. 1996)	Cohort study	nCPAP for OSA patients	59 patients using nCPAP for 1 year	Spells of sleepiness, sleep onset latency, fatigue, and attention are reduced with nCPAP, thus reducing the number of MVAs
Horstmann et al. (2000)	Cohort study	nCPAP for OSA	85 patients with OSAHS using nCPAP	nCPAP reduced the number of accidents from 10.6 to 2.7 per million kilometers

2. OPTALERT (System of Infrared Reflectance Oculography): OPTALERT is a new infrared reflectance oculography method for monitoring eye and eyelid movements (Krahn et al. 2021). During driving operations, transducers are connected to a glass frame and continually monitor drowsiness using a new scale (Johns Drowsiness Scale-JDS) (Johns et al. 2007, 2008).
3. The Multiple Sleep Latency Test (MLST) (Krahn et al. 2021): This test examines a person's proclivity to sleep in a comfortable position, a criterion widely employed in diagnosing narcolepsy. Patients with mean sleep latencies of 5 min or less are deemed pathologically sleepy and should be advised about

their elevated risk of industrial or motor vehicle accidents (Richardson et al. 1978).
4. The Maintenance of Wakefulness Test (MWT) (Krahn et al. 2021) evaluates a person's capacity to remain awake. According to one study, an MWT mean sleep latency of 0–19 min is connected with simulated driving impairment (Sagaspe et al. 2007b). More research is needed to confirm that abnormal MWT scores are linked to actual driving impairment.
5. The Oxford Sleep Resistance Test (OSLER) (74 points): This test incorporates aspects of the MWT as well as psychomotor testing.
6. Pupillography: This test assesses the diameter and variability of the pupil, as well as their association with subjective sleepiness complaints (Balkin et al. 2004).
7. Actigraphy: This instrument analyses shift workers' sleep duration and circadian rhythm abnormalities, and it can be used to detect hypersomnia (Smith et al. 2018).
8. The most generally used test to measure sleep and associated abnormalities is polysomnography (PSG), also referred to as the benchmark or the most reliable method for diagnosing OSA (Caples et al. 2021).

These survey instruments and physiological tests are critical for effectively diagnosing sleep problems and their symptoms, thereby avoiding driver errors and accidents that may negatively impact public health. Nonetheless, none of the methods mentioned above can reliably predict the degree of impairment and the probability of MVAs (George et al. 1996; Kotterba et al. 2004). Furthermore, present methods of simulated driving cannot reliably estimate the likelihood of accidents, but they may predict who is unlikely to be involved in an MVA (Turkington et al. 2001; Vakulin et al. 2014).

11.3 The Treatment: Physician's Role

Several studies have demonstrated that daytime sleepiness is connected with an increased risk of MVAs (Gupta et al. 2017). Drivers are frequently unable to forecast their degree of awareness prior to driving or their performance while driving (Verster and Roth 2012). MVAs impose a high economic and social cost. As a result, routine health examinations should be performed prior to obtaining or renewing a driver's license, especially in patients with sleep disorders. Physicians should consider this risk when evaluating patients with sleep disorders, medical illnesses, or drugs that may cause excessive daytime sleepiness (Alkharboush et al. 2017). It is debatable whether physicians have a legal obligation to evaluate their patients' medical fitness to drive (Adams 2010; Laycock 2010). While some nations hold doctors legally responsible for failing to notify specific medical issues local traffic authorities (Berger et al. 2000; Reuben and St George 1996), others do not (Alkharboush et al. 2017). Furthermore, physicians in some countries have a high level of awareness about the relevance of this issue, but physicians in other nations have a lack of felt responsibility. According to a Canadian survey, 72% of family physicians say they should be held legally accountable for alerting traffic authorities about patients who are unable to drive (Jang et al. 2007). In contrast, a study in Saudi Arabia found that just 15% of those polled thought it was their responsibility to report patients who were unable to drive to local traffic authorities (Jang et al. 2007).

11.3.1 Legislative Concerns

Numerous European countries, as well as Canada, Australia, and several US states, have detailed traffic rules that restrict persons suffering from sleep disorders such as OSA and narcolepsy from driving or acquiring a driver's license until the disease is medically treated (de Mello et al. 2013; Krieger 2007; Masullo et al. 2014). Several European nations (Belgium, the Netherlands, Finland, Spain, Sweden, France, and the United Kingdom) have placed legislative restrictions on the issuance of a driving license to patients suffering from OSA and narcolepsy (Masullo et al. 2014). However, in other nations (particularly developing countries), no traffic laws govern sleepy driving. To limit the risk of sleepy driving, regulations must be put in place. Sleepy drivers should be held liable for the dangers they take and the damage they create if they do not seek medical attention (Powell and Chau 2011; National Conference of State Legislatures 2018). Furthermore, clear regulations are required to identify the legal duties of treating physicians in the case of hypersomnolence patients.

11.4 Conclusions

Given that evidence worldwide shows hypersomnolence, sleep deprivation, and sleep disorders as major causes of industrial and road accidents, sleep-related measures should be included in screening at least commercial drivers (if not all drivers). This screening should be part of the first medical work-up before a commercial driver is issued a license. Furthermore, these drivers' health should be monitored on an annual basis. Drivers should be encouraged to seek the advice of a sleep physician and follow that advice. Additionally, drivers should be informed of the necessity of getting enough sleep before driving and employing effective sleep-related countermeasures. Furthermore, physicians, particularly family physicians and general practitioners, should be educated about the dangers of sleepy driving and the significance of routine fitness-to-drive exams.

Funding This work was partially supported by a grant from the National Plan for Science and Technology Program by the King Saud University Project in Saudi Arabia. The sponsor had no role in the design or conduct of this research.

References

Abe T, Komada Y, Nishida Y, Hayashida K, Inoue Y (2010) Short sleep duration and long spells of driving are associated with the occurrence of Japanese drivers' rear-end collisions and single-car accidents. J Sleep Res 19(2):310–316. https://doi.org/10.1111/j.1365-2869.2009.00806.x

Adams AJ (2010) Rebuttal: should family physicians assess fitness to drive? YES. Can Fam Physician 56(12):e411

Akerstedt T, Hallvig D, Anund A, Fors C, Schwarz J, Kecklund G (2013) Having to stop driving at night because of dangerous sleepiness—awareness, physiology and behaviour. J Sleep Res 22(4):380–388. https://doi.org/10.1111/jsr.12042

Akkoyunlu ME, Kart L, Uludag M, Bayram M, Alisha G, Ozcelik H, Karakose F, Sezer M (2013) [Relationship between symptoms of obstructive sleep apnea syndrome and traffic accidents in the city drivers]. Tuberkuloz ve toraks 61(1):33–37

Alahmari MD, Alanazi TM, Batawi AA, Al-Osaimi EA, Alrabeeah S, Jebakumar Z, Olaish AH, Almeneessier AS, BaHammam AS (2019) Sleepy driving and risk of obstructive sleep apnea among truck drivers in Saudi Arabia. Traffic Inj Prev 20(5):498–503. https://doi.org/10.1080/15389588.2019.1608975

Aldosari MS, Olaish AH, Nashwan SZ, Abulmeaty MMA, BaHammam AS (2020) The effects of caffeine on drowsiness in patients with narcolepsy: a double-blind randomized controlled pilot study. Sleep Breath 24(4):1675–1684. https://doi.org/10.1007/s11325-020-02065-6

Alfawaz R, Aljuraiban G, AlMarzooqi M, Alghannam A, BaHammam A, Dobia A, Alothman S, Aljuhani O, Aljaloud K (2021) The recommended amount of physical activity, sedentary behavior, and sleep duration for healthy Saudis: a joint consensus statement of the Saudi Public Health Authority. Ann Thorac Med 16:239–244. https://doi.org/10.4103/atm.atm_4133_4121

Alkharboush GA, Al Rashed FA, Saleem AH, Alnajashi IS, Almeneessier AS, Olaish AH, Bhatti JA, BaHammam AS (2017) Assessment of patients' medical fitness to drive by primary care physicians: a cross-sectional study. Traffic Inj Prev 18(5):488–492. https://doi.org/10.1080/15389588.2016.1274029

Amra B, Dorali R, Mortazavi S, Golshan M, Farajzadegan Z, Fietze I, Penzel T (2012) Sleep apnea symptoms and accident risk factors in Persian commercial vehicle drivers. Sleep Breath 16(1):187–191. https://doi.org/10.1007/s11325-010-0473-x

Anund A, Kecklund G, Kircher A, Tapani A, Akerstedt T (2009) The effects of driving situation on sleepiness indicators after sleep loss: a driving simulator study. Ind Health 47(4):393–401

Anund A, Ahlstrom C, Kecklund G, Akerstedt T (2011) Rumble strips in centre of the lane and the effect on sleepy drivers. Ind Health 49(5):549–558

Arnedt JT, Wilde GJ, Munt PW, MacLean AW (2000) Simulated driving performance following prolonged wakefulness and alcohol consumption: separate and combined contributions to impairment. J Sleep Res 9(3):233–241

BaHammam AS, Alkhunizan MA, Lesloum RH, Alshanqiti AM, Aldakhil AM, Pandi-Perumal SR, Sharif MM (2014) Prevalence of sleep-related accidents among drivers in Saudi Arabia. Ann Thorac Med 9(4):236–241. https://doi.org/10.4103/1817-1737.140138

Balkin TJ, Bliese PD, Belenky G, Sing H, Thorne DR, Thomas M, Redmond DP, Russo M, Wesensten NJ (2004) Comparative utility of instruments for monitoring sleepiness-related performance decrements in the operational environment. J Sleep Res 13(3):219–227. https://doi.org/10.1111/j.1365-2869.2004.00407.x

Barbone F, McMahon AD, Davey PG, Morris AD, Reid IC, McDevitt DG, MacDonald TM (1998) Association of road-traffic accidents with benzodiazepine use. Lancet 352(9137):1331–1336

Bartels EC, Kusakcioglu O (1965) Narcolepsy: a possible cause of automobile accidents. Lahey Clin Found Bull 14(1):21–26

Berger JT, Rosner F, Kark P, Bennett AJ (2000) Reporting by physicians of impaired drivers and potentially impaired drivers. J Gen Intern Med 15(9):667–672. https://doi.org/10.1046/j.1525-1497.2000.04309.x

Bonsignore MR, Randerath W, Schiza S, Verbraecken J, Elliott MW, Riha R, Barbe F, Bouloukaki I, Castrogiovanni A, Deleanu O, Goncalves M, Leger D, Marrone O, Penzel T, Ryan S, Smyth D, Teran-Santos J, Turino C, McNicholas WT (2021) European Respiratory Society statement on sleep apnoea, sleepiness and driving risk. Eur Respir J 57(2):2001272. https://doi.org/10.1183/13993003.01272-2020

Braeckman L, Verpraet R, Van Risseghem M, Pevernagie D, De Bacquer D (2011) Prevalence and correlates of poor sleep quality and daytime sleepiness in Belgian truck drivers. Chronobiol Int 28(2):126–134. https://doi.org/10.3109/07420528.2010.540363

Buysse DJ, Reynolds CF 3rd, Monk TH, Berman SR, Kupfer DJ (1989) The Pittsburgh Sleep Quality Index: a new instrument for psychiatric practice and research. Psychiatry Res 28(2): 193–213

Caples SM, Anderson WM, Calero K, Howell M, Hashmi SD (2021) Use of polysomnography and home sleep apnea tests for the longitudinal management of obstructive sleep apnea in adults: an American Academy of Sleep Medicine clinical guidance statement. J Clin Sleep Med 17(6): 1287–1293. https://doi.org/10.5664/jcsm.9240

Carter N, Ulfberg J, Nystrom B, Edling C (2003) Sleep debt, sleepiness and accidents among males in the general population and male professional drivers. Accid Anal Prev 35(4):613–617

Cassel W, Ploch T, Becker C, Dugnus D, Peter JH, von Wichert P (1996) Risk of traffic accidents in patients with sleep-disordered breathing: reduction with nasal CPAP. Eur Respir J 9(12): 2606–2611

Centers for Disease Control and Prevention (2013) Drowsy driving - 19 states and the District of Columbia, 2009–2010. MMWR Morb Mortal Wkly Rep 61(51–52):1033–1037

Centers of Disease Control and Prevention (2020) Drowsy driving: asleep at the wheel CDC. https://www.cdc.gov/sleep/features/drowsy-driving.html

Cui R, Tanigawa T, Sakurai S, Yamagishi K, Iso H (2006) Relationships between sleep-disordered breathing and blood pressure and excessive daytime sleepiness among truck drivers. Hypertens Res 29(8):605–610. https://doi.org/10.1291/hypres.29.605

Daley M, Morin CM, LeBlanc M, Gregoire JP, Savard J, Baillargeon L (2009) Insomnia and its relationship to health-care utilization, work absenteeism, productivity and accidents. Sleep Med 10(4):427–438. https://doi.org/10.1016/j.sleep.2008.04.005

de Mello MT, Narciso FV, Tufik S, Paiva T, Spence DW, Bahammam AS, Verster JC, Pandi-Perumal SR (2013) Sleep disorders as a cause of motor vehicle collisions. Int J Prev Med 4(3): 246–257

Dogan OT, Dal U, Ozsahin SL, Akkurt I, Seyfikli Z (2006) [The prevalence of sleep related disorders among the drivers and it's relation with traffic accidents]. Tuberkuloz ve toraks 54(4):315–321

Ellen RL, Marshall SC, Palayew M, Molnar FJ, Wilson KG, Man-Son-Hing M (2006) Systematic review of motor vehicle crash risk in persons with sleep apnea. J Clin Sleep Med 2(2):193–200

Filtness AJ, Reyner LA, Horne JA (2012) Driver sleepiness-comparisons between young and older men during a monotonous afternoon simulated drive. Biol Psychol 89(3):580–583. https://doi.org/10.1016/j.biopsycho.2012.01.002

Garner AA, Miller MM, Field J, Noe O, Smith Z, Beebe DW (2015) Impact of experimentally manipulated sleep on adolescent simulated driving. Sleep Med 16(6):796–799. https://doi.org/10.1016/j.sleep.2015.03.003

George CF (2001) Reduction in motor vehicle collisions following treatment of sleep apnoea with nasal CPAP. Thorax 56(7):508–512

George CF, Boudreau AC, Smiley A (1996) Comparison of simulated driving performance in narcolepsy and sleep apnea patients. Sleep 19(9):711–717

Grubb TC (1969) Narcolepsy and highway accidents. JAMA 209(11):1720

Gupta R, Pandi-Perumal SR, Almeneessier AS, BaHammam AS (2017) Hypersomnolence and traffic safety. Sleep Med Clin 12(3):489–499. https://doi.org/10.1016/j.jsmc.2017.03.018

Gurtman CG, Broadbear JH, Redman JR (2008) Effects of modafinil on simulator driving and self-assessment of driving following sleep deprivation. Hum Psychopharmacol 23(8):681–692. https://doi.org/10.1002/hup.983

Gurubhagavatula I, Nkwuo JE, Maislin G, Pack AI (2008) Estimated cost of crashes in commercial drivers supports screening and treatment of obstructive sleep apnea. Accid Anal Prev 40(1): 104–115. https://doi.org/10.1016/j.aap.2007.04.011

Hack MA, Choi SJ, Vijayapalan P, Davies RJ, Stradling JR (2001) Comparison of the effects of sleep deprivation, alcohol and obstructive sleep apnoea (OSA) on simulated steering performance. Respir Med 95(7):594–601. https://doi.org/10.1053/rmed.2001.1109

Hanning CD, Welsh M (1996) Sleepiness, snoring and driving habits. J Sleep Res 5(1):51–54

Hirshkowitz M, Black JE, Wesnes K, Niebler G, Arora S, Roth T (2007) Adjunct armodafinil improves wakefulness and memory in obstructive sleep apnea/hypopnea syndrome. Respir Med 101(3):616–627. https://doi.org/10.1016/j.rmed.2006.06.007

Horstmann S, Hess CW, Bassetti C, Gugger M, Mathis J (2000) Sleepiness-related accidents in sleep apnea patients. Sleep 23(3):383–389

Jang RW, Man-Son-Hing M, Molnar FJ, Hogan DB, Marshall SC, Auger J, Graham ID, Korner-Bitensky N, Tomlinson G, Kowgier ME, Naglie G (2007) Family physicians' attitudes and practices regarding assessments of medical fitness to drive in older persons. J Gen Intern Med 22(4):531–543. https://doi.org/10.1007/s11606-006-0043-x

Johns MW (1991) A new method for measuring daytime sleepiness: the Epworth sleepiness scale. Sleep 14(6):540–545

Johns M, Tucker A, Chapman R, Crowley K, Michael N (2007) Monitoring eye and eyelid movements by infrared reflectance oculography to measure drowsiness in drivers. Somnologie 11:234–242

Johns M, Chapman R, Crowley K, Tucker A (2008) A new method for assessing the risks of drowsiness while driving. Somnologie 12:66–74

Kaida K, Takahashi M, Akerstedt T, Nakata A, Otsuka Y, Haratani T, Fukasawa K (2006) Validation of the Karolinska sleepiness scale against performance and EEG variables. Clin Neurophysiol 117(7):1574–1581. https://doi.org/10.1016/j.clinph.2006.03.011

Kales SN, Straubel MG (2014) Obstructive sleep apnea in North American commercial drivers. Ind Health 52(1):13–24

Karimi M, Hedner J, Lombardi C, McNicholas WT, Penzel T, Riha RL, Rodenstein D, Grote L (2014) Driving habits and risk factors for traffic accidents among sleep apnea patients--a European multi-centre cohort study. J Sleep Res 23(6):689–699. https://doi.org/10.1111/jsr.12171

Killgore WD, Rupp TL, Grugle NL, Reichardt RM, Lipizzi EL, Balkin TJ (2008) Effects of dextroamphetamine, caffeine and modafinil on psychomotor vigilance test performance after 44 h of continuous wakefulness. J Sleep Res 17(3):309–321. https://doi.org/10.1111/j.1365-2869.2008.00654.x

Komada Y, Nishida Y, Namba K, Abe T, Tsuiki S, Inoue Y (2009) Elevated risk of motor vehicle accident for male drivers with obstructive sleep apnea syndrome in the Tokyo metropolitan area. Tohoku J Exp Med 219(1):11–16

Kotterba S, Mueller N, Leidag M, Widdig W, Rasche K, Malin JP, Schultze-Werninghaus G, Orth M (2004) Comparison of driving simulator performance and neuropsychological testing in narcolepsy. Clin Neurol Neurosurg 106(4):275–279. https://doi.org/10.1016/j.clineuro.2003.12.003

Krahn LE, Arand DL, Avidan AY, Davila DG, DeBassio WA, Ruoff CM, Harrod CG (2021) Recommended protocols for the Multiple Sleep Latency Test and Maintenance of Wakefulness Test in adults: guidance from the American Academy of Sleep Medicine. J Clin Sleep Med 17(12):2489–2498. https://doi.org/10.5664/jcsm.9620

Krieger J (2007) Sleep apnoea and driving: how can this be dealt with? Eur Respir Rev 16(106): 189–195

Laycock KM (2010) Rebuttal: should family physicians assess fitness to drive? NO. Can Fam Physician 56(12):e412

Leechawengwongs M, Leechawengwongs E, Sukying C, Udomsubpayakul U (2006) Role of drowsy driving in traffic accidents: a questionnaire survey of Thai commercial bus/truck drivers. J Med Assoc Thail 89(11):1845–1850

Leger D, Massuel MA, Metlaine A (2006) Professional correlates of insomnia. Sleep 29(2): 171–178

Lloberes P, Levy G, Descals C, Sampol G, Roca A, Sagales T, de la Calzada MD (2000) Self-reported sleepiness while driving as a risk factor for traffic accidents in patients with obstructive sleep apnoea syndrome and in non-apnoeic snorers. Respir Med 94(10):971–976

Masa JF, Rubio M, Findley LJ (2000) Habitually sleepy drivers have a high frequency of automobile crashes associated with respiratory disorders during sleep. Am J Respir Crit Care Med 162(4 Pt 1):1407–1412. https://doi.org/10.1164/ajrccm.162.4.9907019

Masullo A, Feola A, Marino V, Iadevaia C, Trabucco Aurilio M, Marsella LT (2014) Sleep disorders and driving licence: the current Italian legislation and medico-legal issues. Clin Ter 165(5):e368–e372. https://doi.org/10.7417/CT.2014.1766

Mets MA, de Vries JM, de Senerpont Domis LM, Volkerts ER, Olivier B, Verster JC (2011) Next-day effects of ramelteon (8 mg), zopiclone (7.5 mg), and placebo on highway driving performance, memory functioning, psychomotor performance, and mood in healthy adult subjects. Sleep 34(10):1327–1334. https://doi.org/10.5665/SLEEP.1272

Mets MA, Alford C, Verster JC (2012) Sleep specialists' opinion on sleep disorders and fitness to drive a car: the necessity of continued education. Ind Health 50(6):499–508

Myllyla M, Anttalainen U, Saaresranta T, Laitinen T (2020) Motor vehicle accidents in CPAP-compliant obstructive sleep apnea patients-a long-term observational study. Sleep Breath 24(3): 1089–1095. https://doi.org/10.1007/s11325-020-02023-2

National Conference of State Legislatures (2018) Summaries of current drowsy driving laws. NCSL. https://www.ncsl.org/research/transportation/summaries-of-current-drowsy-driving-laws.aspx

National Highway Traffic Safety Administration (2011) An analysis of the significant decline in motor vehicle traffic fatalities in 2008. Ann Emerg Med 58(4):377–378. https://doi.org/10.1016/j.annemergmed.2011.08.008

Ozoh OB, Okubadejo NU, Akanbi MO, Dania MG (2013) High-risk of obstructive sleep apnea and excessive daytime sleepiness among commercial intra-city drivers in Lagos metropolis. Niger Med J 54(4):224–229. https://doi.org/10.4103/0300-1652.119607

Pack AI, Maislin G, Staley B, Pack FM, Rogers WC, George CF, Dinges DF (2006) Impaired performance in commercial drivers: role of sleep apnea and short sleep duration. Am J Respir Crit Care Med 174(4):446–454. https://doi.org/10.1164/rccm.200408-1146OC

Perez-Chada D, Videla AJ, O'Flaherty ME, Palermo P, Meoni J, Sarchi MI, Khoury M, Duran-Cantolla J (2005) Sleep habits and accident risk among truck drivers: a cross-sectional study in Argentina. Sleep 28(9):1103–1108

Philip P, Taillard J, Moore N, Delord S, Valtat C, Sagaspe P, Bioulac B (2006) The effects of coffee and napping on nighttime highway driving: a randomized trial. Ann Intern Med 144(11): 785–791

Philip P, Chaufton C, Taillard J, Capelli A, Coste O, Leger D, Moore N, Sagaspe P (2014) Modafinil improves real driving performance in patients with hypersomnia: a randomized double-blind placebo-controlled crossover clinical trial. Sleep 37(3):483–487. https://doi.org/10.5665/sleep.3480

Pizza F, Jaussent I, Lopez R, Pesenti C, Plazzi G, Drouot X, Leu-Semenescu S, Beziat S, Arnulf I, Dauvilliers Y (2015) Car crashes and central disorders of hypersomnolence: a French study. PLoS One 10(6):e0129386. https://doi.org/10.1371/journal.pone.0129386

Powell NB, Chau JK (2011) Sleepy driving. Sleep Med Clin 6:117–124. https://doi.org/10.1016/j.smc.2010.12.006

Reuben DB, St George P (1996) Driving and dementia California's approach to a medical and policy dilemma. West J Med 164(2):111–121

Rice TM, Peek-Asa C, Kraus JF (2003) Nighttime driving, passenger transport, and injury crash rates of young drivers. Inj Prev 9(3):245–250

Richardson GS, Carskadon MA, Flagg W, Van den Hoed J, Dement WC, Mitler MM (1978) Excessive daytime sleepiness in man: multiple sleep latency measurement in narcoleptic and control subjects. Electroencephalogr Clin Neurophysiol 45(5):621–627

Rodenstein D (2009) Sleep apnea: traffic and occupational accidents--individual risks, socioeconomic and legal implications. Respiration 78(3):241–248. https://doi.org/10.1159/000222811

Sagaspe P, Taillard J, Chaumet G, Moore N, Bioulac B, Philip P (2007a) Aging and nocturnal driving: better with coffee or a nap? A randomized study. Sleep 30(12):1808–1813

Sagaspe P, Taillard J, Chaumet G, Guilleminault C, Coste O, Moore N, Bioulac B, Philip P (2007b) Maintenance of wakefulness test as a predictor of driving performance in patients with untreated obstructive sleep apnea. Sleep 30(3):327–330

Sagaspe P, Taillard J, Bayon V, Lagarde E, Moore N, Boussuge J, Chaumet G, Bioulac B, Philip P (2010) Sleepiness, near-misses and driving accidents among a representative population of French drivers. J Sleep Res 19(4):578–584. https://doi.org/10.1111/j.1365-2869.2009.00818.x

Sassani A, Findley LJ, Kryger M, Goldlust E, George C, Davidson TM (2004) Reducing motor-vehicle collisions, costs, and fatalities by treating obstructive sleep apnea syndrome. Sleep 27(3):453–458

Senaratna CV, Perret JL, Lodge CJ, Lowe AJ, Campbell BE, Matheson MC, Hamilton GS, Dharmage SC (2017) Prevalence of obstructive sleep apnea in the general population: a systematic review. Sleep Med Rev 34:70–81. https://doi.org/10.1016/j.smrv.2016.07.002

Shiomi T, Arita AT, Sasanabe R, Banno K, Yamakawa H, Hasegawa R, Ozeki K, Okada M, Ito A (2002) Falling asleep while driving and automobile accidents among patients with obstructive sleep apnea-hypopnea syndrome. Psychiatry Clin Neurosci 56(3):333–334. https://doi.org/10.1046/j.1440-1819.2002.01004.x

Signal TL, Gander PH, Anderson H, Brash S (2009) Scheduled napping as a countermeasure to sleepiness in air traffic controllers. J Sleep Res 18(1):11–19. https://doi.org/10.1111/j.1365-2869.2008.00702.x

Smith MT, McCrae CS, Cheung J, Martin JL, Harrod CG, Heald JL, Carden KA (2018) Use of actigraphy for the evaluation of sleep disorders and circadian rhythm sleep-wake disorders: an American Academy of Sleep Medicine clinical practice guideline. J Clin Sleep Med 14(7):1231–1237. https://doi.org/10.5664/jcsm.7230

Strohl KP, Brown DB, Collop N, George C, Grunstein R, Han F, Kline L, Malhotra A, Pack A, Phillips B, Rodenstein D, Schwab R, Weaver T, Wilson K (2013) An official American Thoracic Society Clinical practice guideline: sleep apnea, sleepiness, and driving risk in noncommercial drivers. An update of a 1994 statement. Am J Respir Crit Care Med 187(11):1259–1266. https://doi.org/10.1164/rccm.201304-0726ST

Taillard J, Capelli A, Sagaspe P, Anund A, Akerstedt T, Philip P (2012) In-car nocturnal blue light exposure improves motorway driving: a randomized controlled trial. PLoS One 7(10):e46750. https://doi.org/10.1371/journal.pone.0046750

Tefft BC (2018) Acute sleep deprivation and culpable motor vehicle crash involvement. Sleep 41(10):zsy144. https://doi.org/10.1093/sleep/zsy144

Tregear S, Reston J, Schoelles K, Phillips B (2009) Obstructive sleep apnea and risk of motor vehicle crash: systematic review and meta-analysis. J Clin Sleep Med 5(6):573–581

Turkington PM, Sircar M, Allgar V, Elliott MW (2001) Relationship between obstructive sleep apnoea, driving simulator performance, and risk of road traffic accidents. Thorax 56(10):800–805

Vakulin A, Catcheside PG, Baulk SD, Antic NA, Banks S, Dorrian J, McEvoy RD (2014) Individual variability and predictors of driving simulator impairment in patients with obstructive sleep apnea. J Clin Sleep Med 10(6):647–655. https://doi.org/10.5664/jcsm.3792

Verster JC, Roth T (2012) Drivers can poorly predict their own driving impairment: a comparison between measurements of subjective and objective driving quality. Psychopharmacology 219(3):775–781. https://doi.org/10.1007/s00213-011-2400-7

Verster JC, Veldhuijzen DS, Volkerts ER (2004) Residual effects of sleep medication on driving ability. Sleep Med Rev 8(4):309–325. https://doi.org/10.1016/j.smrv.2004.02.001

Verster JC, Veldhuijzen DS, Patat A, Olivier B, Volkerts ER (2006) Hypnotics and driving safety: meta-analyses of randomized controlled trials applying the on-the-road driving test. Curr Drug Saf 1(1):63–71

Verster JC, Spence DW, Shahid A, Pandi-Perumal SR, Roth T (2011) Zopiclone as positive control in studies examining the residual effects of hypnotic drugs on driving ability. Curr Drug Saf 6(4):209–218

Williamson AM, Feyer AM (2000) Moderate sleep deprivation produces impairments in cognitive and motor performance equivalent to legally prescribed levels of alcohol intoxication. Occup Environ Med 57(10):649–655

Williamson A, Lombardi DA, Folkard S, Stutts J, Courtney TK, Connor JL (2011) The link between fatigue and safety. Accid Anal Prev 43(2):498–515. https://doi.org/10.1016/j.aap.2009.11.011

World Health Organization (2011) Global plan for the decade of action for road safety 2011–2020. https://www.who.int/roadsafety/decade_of_action/plan/plan_english.pdf?ua=2021. Accessed 29 Aug 2021

World Health Organization (2018) Global status report on road safety. https://www.who.int/publications/i/item/9789241565684. Accessed 29 Aug 2021

World Health Organization (2021) Road traffic injuries. World Health Organization. https://www.who.int/news-room/fact-sheets/detail/road-traffic-injuries. Accessed 28 Aug 2021

Chapter 12
Advances in the Surgical Treatments for Snoring and Obstructive Sleep Apnea

Anjum Khan and Mahadevappa Hunasikatti

Abstract The first-line therapy for OSA is usually behavioral modification, including weight loss if appropriate, followed by gold standard therapy of continuous positive airway pressure (CPAP). However, some patients fail or do not tolerate CPAP therapy. For those patients who are not able to use or tolerate CPAP therapy, oral appliances, mandibular advancement devices (MAD), and various other surgical options are available.

The general aim of the various surgical options is to enlarge and/or stabilize the upper airway by directing modifications to the nasal septum, inferior nasal turbinate (unilaterally/bilaterally), adenoids, upper and lower pharyngeal structures, and base of the tongue.

The chapter gives a synopsis of the current perspectives on various modalities of surgical intervention. This is divided into various sections, as sleep apnea is a multistaging disease. Advocated here is the value of least invasive intervention to most invasive intervention and the impact on the general and cardiovascular consequences on the patients. Recently described DISE procedures have been described. From commonly performed procedures like UVPP (uvulopalatopharyngoplasty) to the most invasive procedures like tracheostomy and their place in the management of OSA is discussed. We have attempted conflicting data on the same procedures, and the current perspectives as defined by the otolaryngologist community of physicians are described here.

Keywords Behavioral modification · Weight loss · Continuous positive airway pressure (CPAP) · Oral appliances · Surgical options

A. Khan
ENT Division, US Food and Drug Administration, Silver Spring, MD, USA

M. Hunasikatti (✉)
Division of Respiratory, Sleep and Anesthesiology (DSRA), OHT1, Food and Drug Administration (FDA), Silver Spring, MD, USA

A. S. BaHammam, M. Hunasikatti (eds.), *Sleep Apnea Frontiers*, Progress in Sleep Research, https://doi.org/10.1007/978-981-99-7901-1_12

12.1 Introduction

Obstructive sleep apnea (OSA) is identified by recurrent occurrences of partial or total blockage in the upper airway while asleep. The characteristic features of OSA are excessive daytime sleepiness, and snoring, which may also be associated with gasping. These obstructions result in the desaturation of the blood's normal oxygen levels and are usually terminated by brief arousals from sleep. Multiple sleep arousals result in sleep fragmentation, which impairs daytime activity. It is well known that adults with OSA-associated daytime somnolence are at higher risk for motor vehicle accidents. Drivers with known obstructive sleep apnea (OSA) have a two- to threefold higher likelihood of being involved in motor vehicle accidents (Mozafari et al. 2014). OSA and snoring are associated with systemic hypertension, resistant hypertension, cardiac arrhythmias, myocardial infarction, and cerebrovascular accidents.

The first-line therapy for OSA is usually behavioral modification, including weight loss if appropriate, followed by gold standard therapy of continuous positive airway pressure (CPAP). However, some patients fail or do not tolerate CPAP therapy. For those patients who are not able to use or tolerate CPAP therapy, oral appliances, mandibular advancement devices (MAD), and various other surgical options are available.

The general aim of the various surgical options is to enlarge and/or stabilize the upper airway by directing modifications to the nasal septum, inferior nasal turbinate (unilaterally/bilaterally), adenoids, upper and lower pharyngeal structures, and base of the tongue.

12.2 Procedures Involving Soft Palate

Soft palatal procedures affect the reduction or reconstruction of the collapsible portions of the soft palate.

12.2.1 Uvulopalatopharyngoplasty (UPPP)

Uvulopalatopharyngoplasty, first introduced in the United States by Fujita in 1981 as an alternative to tracheostomy, remains the most common surgical procedure of the upper airway. During a UPPP procedure submucosal excision of the tonsils (if present), removal of posterior soft palate/uvula, and pharyngeal tissue is followed by closure of the tonsillar pillars. Even though the UPPP procedure usually results in the reduction of the apnea hypopnea index (AHI), it does not normalize it. The amount of resection is individualized for each patient as determined by the potential space and the size of the mucosa of the tonsillar pillar between the two palatal arches.

The reduction of AHI measures the success of the surgery. The AHI, however, is an imperfect, surrogate metric for OSA. Therefore, most clinicians use additional surrogate measures like the oxygen desaturation index (ODI) and Epworth sleepiness scale (ESS) to document improvement in clinical status.

To demonstrate the utility of the conventional UPPP procedure, many authors have performed studies comparing the UPPP procedure to CPAP, modified radiofrequency tissue ablation (MRFTA), nasal surgery, etc.

Weaver et al. conducted a retrospective cohort database study that included all sleep apnea patients treated with CPAP or UPPP. They showed that 1339 (7.1%) of 18,754 CPAP patients and 71 (3.4%) of 2072 UPPP patients were dead. After adjustment, CPAP patients had a 31% higher probability of being dead at any time than UPPP patients. Authors concluded that UPPP confers a survival advantage over CPAP, after adjustments for age, gender, race, year of treatment, and comorbidity (Weaver et al. 2004).

Amali et al. compared the efficacy of modified radiofrequency tissue ablation (MRFTA) with uvulopalatopharyngoplasty (UPPP) based on both subjective and objective outcome measures in individuals diagnosed with mild-to-moderate obstructive sleep apnea (OSA). Compared to conventional UPPP, where excision of the soft palate and tonsils is performed, MFRTA uses radiofrequency technology to puncture and coagulate at 5 points, 1 cm apart around the midline of the soft palate and four to six lesions in the tonsils on each side. The resection of excess mucosa is limited to a triangular shape on both sides of the uvula, leaving the uvular musculature in place. While the overall target area and goal of enlarging the airway remain, the FMRTA procedure limits tissue resection. In addition, the procedure can be performed under local/topical anesthesia compared to UPPP, which is performed under general anesthesia. Forty patients with mild-to-moderate OSA were randomly divided into UPPP and MRFTA groups. The evaluation was based on the AHI, sleep apnea quality of life index (SAQLI), Epworth sleepiness scale (ESS), immediately before surgery and 6 months postoperatively. The authors of the study concluded that for mild OSA, the effectiveness of MRFTA in lowering AHI was comparable to that of UPPP. In addition, MFRTA was noted to be more effective than UPPP in enhancing the quality of life because of less morbidity and fewer treatment-related symptoms. The study is limited because of the short follow-up time of 6 months (Amali et al. 2017).

Wadhera et al. reported their experience of 98 patients who had a UPPP performed by the same surgeon in a regional Australian ear, nose, and throat practice between 2012 and 2017. Forty-seven patients participated in structured phone interviews into their daytime symptoms, sleep quality, partner's experience with snoring, postoperative pain, and quality of life as measured by the Functional Outcome of Sleep Questionnaire 10 (FOSQ-10). Data were qualitatively and quantitatively analyzed. Postoperatively, there was an improvement in the mean quality of life, from 16.5 (SD ±3) to 18.1 (SD ±1.7). The majority of patients (54.8%) reported that their daytime symptoms had completely resolved postoperatively, and 71.4% had no further complaints of snoring from their partner (Wadhera and Fogarty 2020).

The UPPP procedure has evolved to reduce complications and improve airway outcomes. Associated side effects with current techniques include postoperative pain, acute and chronic dysphagia, velopharyngeal incompetence, voice change, taste change, long-term problems of nasopharyngeal stenosis, and altered pharyngeal sensation (Kezirian et al. 2004). To minimize the risk of palatal dysfunction, newer techniques avoid transection of muscles, and create neo-uvula by not cutting palatal mucosal margins, and preservethe anterior uvula. Friberg et al. showed that the vast majority of OSA patients were satisfied 24 months after surgery, even though a third of them were experiencing side effects. Younger patients had fewer side effects than older patients. Patients with better OSA outcomes were also more satisfied. After 9 years, the side effects were no longer problematic (Friberg et al. 2020).

12.2.2 Laser-Assisted Uvulopalatoplasty (LAUP)

This outpatient surgical technique is an alternative to the standard UPPP and involves a series of laser incisions and vaporizations designed to shorten the uvula and modify and tighten the soft palatal tissue. The procedure sequentially reshapes superficial palatal tissue using a carbon dioxide laser. There is variation in how LAUP laser incisions may be made using a beam of CO_2 laser: (1) Excision of the uvular base through full palatal depth and extension bilaterally to anterior and posterior tonsillar pillars including tonsillectomy (if tonsils present), and (2) through and through full thickness vertical trenches created in the free edge of the soft palate on either side of the uvula removing its core.

Camacho et al. reported A Systematic Review and Meta-Analysis regarding laser-assisted uvulopalatoplasty for obstructive sleep apnea. Twenty-three adult studies (717 patients) reported outcomes. The mean (M) $\pm$ standard deviations (SDs) of the apnea-hypopnea index (AHI) before and after LAUP were calculated was 28 $\pm$ 13 and 19 $\pm$ 12 events/h (32% reduction). In their meta-analysis, LAUP reduced AHI by 32% among all patients, while the LSAT (lowest oxygen saturation) only changed minimally. Among individual patients, the data showed a success rate of 23%, a cure rate of 8%, and a worsening of the apnea-hypopnea index (AHI) in 44% of the patients. Significant heterogeneity and inconsistency were observed among LAUP studies except for studies that included LAUP with tonsillectomy. Based on the unfavorable results of the analysis, the authors reached the conclusion that caution should be exercised when performing LAUP or that it should be avoided altogether (Camacho et al. 2017). However, Schiffman et al. reported the efficacy of Minimally Invasive Combined Nd:YAG and Er:YAG Laser-Assisted Uvulopalatoplasty for the Treatment of Obstructive Sleep Apnea. This method involves the utilization of lasers, specifically Nd:YAG and Er:YAG, which are composed of neodymium-doped yttrium aluminum garnet and erbium-doped yttrium aluminum garnet materials, respectively. These lasers are employed to apply controlled heat and treat the soft palate and neighboring tissues that have experienced relaxation. As a result, collagen shrinkage and the formation of new collagen fibers occur, which are considered beneficial outcomes. The efficacy of the minimally invasive LAUP procedure, combining Nd:YAG laser ($\lambda = 1064$ nm) and the effectiveness of treating obstructive sleep apnea (OSA) using erbium-doped yttrium aluminum garnet (Er:YAG) laser (wavelength of 2940 nm) on the soft palate was evaluated in a study involving 27 patients with varying degrees of OSA. The patients received three 20-min laser sessions in an outpatient setting over a period of 45–60 days. After the treatment, measurements of the apnea-hypopnea index (AHI) were taken before and after the sessions. The study found that all patients experienced a decrease in AHI, with an average improvement of 66.3% (ranging from 32% to 100%). Furthermore, 78% (21) of the patients achieved an improvement of 50% or more. Although the NightLase LAUP treatment of OSA support short-term favorable results and a safe therapeutic method, the authors of the study acknowledge the need for further research and long-term prospective trial to improve the evidence (Shiffman et al. 2021).

12.2.3 UPPP Modifications and Variants

(i) To minimize some of the long-term complications of nasopharyngeal stenosis, velopharyngeal insufficiency, and globus sensation following UPPP, modifications like preservation of uvular muscle/and or preservation of uvular mucosa only, removal of soft palate mucosa with adjacent adipose tissue, etc. are regularly practiced.

(ii) Expansion pharyngoplasty and expansion sphincter pharyngoplasty (ESP) can be considered variants of UPPP. In expansion pharyngoplasty, palatopharyngeus muscles on both sides are dissected and horizontally transected. The muscle is then rotated, lifted, and fixated to the hamulus pterygoideus. These procedures are good for retropalatal obstruction and lateral pharyngeal wall collapse with easy applicability, rapid recovery, and without significant complication (Hong et al. 2019).

(iii) Functional expansion pharyngoplasty in the treatment of OSA is a conservative modification of expansion sphincter pharyngoplasty (ESP), as described by Pang and Woodson. In the Functional modification of ESP, the key part following bilateral tonsillectomy is the separation of the palatopharyngeus muscle (PPM) from the superior pharyngeus constrictor (SPC) and mucosa of tonsillar pillars. The PPM is transected bilaterally, and the upper portion is sutured to the hamulus and the lower portion is sutured to SPC. The authors believe the functional technique is more physiologic than expansion sphincter pharyngoplasty (Sorrenti and Piccin 2013).

12.3 Role of Nasal Surgery in Sleep Apnea (Adjunct)

Snoring is caused by the vibration of several structures in the oral cavity, like the tissues targeted in this procedure include the soft palate, uvula, tonsils, base of the tongue, and epiglottis and pharyngeal walls. Relief of nasal obstruction alone rarely cures OSA. However, various nasal surgical procedures (septoplasty, septorhinoplasty, nasal valve surgery, and turbinate reduction) can be used as adjuncts and are performed to maximize the effectiveness of CPAP use by patients, when correction of structural problems is needed to reduce nasal resistance, and as part of multilevel surgery used to treat sleep apnea (O'Halloran 2003; Pang 2013).

12.4 Tonsillectomy and Adenoidectomy

12.4.1 Pediatric

Most children with sleep apnea are noted to have tonsillar hypertrophy and/or adenoid hypertrophy. Other most common causes of childhood sleep apnea include childhood obesity, small jaw, Down syndrome, or cerebral palsy. A family history of OSA and allergic rhinitis also increases the risk of OSA in children. Tonsillectomy with adenoidectomy is a combination surgery that has been used to treat pediatric sleep apnea. Tonsillectomy as an isolated treatment modality is rarely performed in pediatric patients with OSA. Even though these two procedures can be performed on many patients as outpatient surgery (i.e., discharged to home the same day of surgery), it is important to note that high-risk children may need to be monitored closely in PICU to avoid/minimize postoperative complications. The American Academy of Pediatrics and "American Academy of Otolaryngology and Head and Neck Surgery (AAO-HNS)" recommend inpatient management for pediatric patients who are obese, have complicated medical histories, have severe OSA (defined as an apnea-hypopnea index of >10 by the AAOHNS and >24 by the American Academy of Pediatrics), and are younger than 3 years of age. This list relates to a higher rate of postoperative complications seen in younger children undergoing adenotonsillectomy who have little respiratory and hemodynamic reserve (Baugh et al. 2011).

The timing of the adenotonsillectomy procedure needs to be individualized due to the issue of reported events of spontaneous resolution of OSA in children. Chervin et al. conducted a CHAT study (The Childhood Adenotonsillectomy Trial), which randomized 50% of eligible children with OSAS (obstructive sleep apnea syndrome) to a control arm (watchful waiting), with 7-month follow-up symptom inventories, physical examinations, and polysomnography. Polysomnographic and symptomatic resolution were defined respectively by an apnea/hypopnea index (AHI) <2 and obstructive apnea index (OAI) <1 and by an OSAS symptom score (Pediatric Sleep Questionnaire [PSQ]) < 0.33 with ≥25% improvement from baseline. After

194 children aged 5 to 9 years underwent 7 months of watchful waiting, 82 (42%) no longer met polysomnographic criteria for OSAS. Among 167 children with baseline PSQ scores ≥0.33, only 25 (15%) experienced symptomatic resolution. Baseline predictors were low PSQ and PSQ snoring subscale scores; absence of habitual snoring, loud snoring, observed apneas, or a household smoker; higher quality of life; fewer attention-deficit/hyperactivity disorder symptoms; and female sex. Only lower PSQ and snoring scores were independent predictors. Many candidates for AT no longer have OSAS on polysomnography after 7 months of watchful waiting, whereas meaningful symptom improvement is not common. In practice, a low baseline AHI and normal waist circumference, or low PSQ and snoring score, may help identify an opportunity to avoid adenotonsillectomy (AT) (Chervin et al. 2015). Although most patients will have a resolution of OSA, some patients may have residual OSA. In patients with residual symptoms, additional treatment like intranasal steroids and oral leukotriene therapy or CPAP may be needed together with repeat PSG to ensure adequate resolution.

12.4.2 Adults

There are not many large clinical studies of tonsillectomy as the sole therapy for OSA. Camacho et al. performed a single systemic review and meta-analysis of tonsillectomy alone for OSA in the adult population in 2016. This meta-analysis included 17 studies totaling 216 patients. Adult patients in the study ($n = 203$), were grouped based on their tonsil sizes grouped as the condition of the tissues can be described as hypertrophied, large, enlarged, significantly enlarged, or falling within grades 2 to 4. Analysis recorded showed a decrease in apnea-hypopnea index by 65% (from 40.5 ± 29/h to 14 ± 17/h). For individuals with an apnea-hypopnea index (AHI) of less than 30/h before the surgery, the tonsillectomy success rate was 100% (25 of 25), and the cure rate was 84% (21 of 25) with a mean postoperative AHI of 2.4 ± 2.1/h; in contrast, patients with a preoperative apnea-hypopnea index (AHI) of 30 or higher per hour had a tonsillectomy success rate of 72.4% (21 out of 29), a cure rate of 34.4% (10 out of 29), and a mean post-operative AHI of 14.3 ± 13.9/h. These findings suggest that isolated tonsillectomy can effectively treat obstructive sleep apnea (OSA) in adults, particularly among those with enlarged tonsils and mild to moderate OSA (AHI < 30/h) (Camacho et al. 2016). Of the 17 articles identified, 12 did not describe the palate position and tonsil size, which is essential to patient selection for treating OSA based on tonsillectomy alone. Therefore, it remains unclear whether enough literature supports tonsillectomy alone as a surgical modality for adults with OSA.

Interestingly, Smith et al. performed a single-group pretest-posttest study from January 2011 to January 2016 on a total of 46 consecutive adult subjects. Seven patients were excluded, and 39 subjects were selected for intervention by tonsillectomy alone. This study aimed to determine if tonsillectomy alone is an effective treatment for improving obstructive sleep apnea in adult subjects with tonsillar hypertrophy. Inclusion criteria included AHI >5, hypertrophic tonsils (3+ or 4+), and a modified Mallampati grade 1 or 2. Exclusion criteria included BMI > 40, craniofacial abnormalities, and diagnosis of parasomnias. A surgical response to treatment was defined by >50% decrease in AHI and a decrease in the overall AHI to <20. Postoperative sleep studies were performed at 8 weeks after surgery. Patient-reported outcomes were measured using patients' subjective symptoms, insomnia severity index (ISI), Epworth sleepiness scale (ESS), and Functional Outcome of Sleep Questionnaire-10 (FOSQ-10). This prospective study reported quality of life indicators improved significantly after tonsillectomy as the single treatment for OSA in a unique population of tonsillar Hypertrophy. However, the study was uncontrolled, and had short-term follow-up with many patients not seen back at the study site (Smith et al. 2017).

12.5 Multilevel Procedures for Treatment of OSA

12.5.1 Lingual Tonsillectomy with Palatal Surgery (Combined Procedure)

Patients with OSA often have multiple site upper airway obstructions that may not be adequately addressed with UPPP alone. Several tongue-directed procedures have been developed to reduce the size of the tongue and/or prevent it from falling backward. Studies have shown that combining these tongue region procedures with palate procedures achieves better results than palate procedures alone, and the removal of tonsils on the back of the tongue is one such procedure. Lingual tonsillectomy involves the removal of lymphoid tissue, mainly localized on either side of the tongue base, present in some cases during adulthood. Different surgical treatments aim to widen the retrolingual space using coblator or lasers. Samutsakorn et al. conducted a systematic review and meta-analysis regarding the value of lingual tonsillectomy in conjunction with palatal surgery for the treatment of obstructive sleep apnea in adults. A total of four studies (107 patients) were analyzed. The combined analysis using random effects models showed a significant statistical improvement of polysomnographic respiratory events postoperatively, with an average AHI reduction of 18.51 events/h, and an increase in minimal arterial oxygen saturation of 5.26. The result also showed that the procedures significantly improved sleepiness, resulting in an average Epworth sleepiness scale (ESS) reduction of 5.44. A combination of lingual tonsillectomy with palatal surgery may offer benefits in a select group of OSA adults, in terms of apnea/hypopnea index, minimal arterial oxygen saturation, and Epworth sleepiness scale. Data were based on a meta-analysis of all case series with 2–6 months follow-up. These combined procedures may need to be individualized (Samutsakorn et al. 2018).

12.5.2 Midline Glossectomy for OSA

In a few selected pediatric and adult patients with significant macroglossia and moderate-to-severe obstructive sleep apnea with significant comorbid illness, midline partial glossectomy is an effective surgical modality and can also be used adjunctively as a part of other multilevel pharyngeal surgical procedures for OSA. Midline glossectomy involves the removal of the central portion of the tongue base, which is usually accomplished using laser, radiofrequency ablation probes, or coblation (American Academy of Otolaryngology Head and Neck Surgery 2021).

Fajita et al. described midline glossectomy (MLG) using the carbon dioxide laser as a procedure that directly enlarges the hypopharyngeal airspace. They presented their initial experience of 12 patients by performing midline glossectomy as the sole procedure on 11 patients who had failed uvulopalatopharyngoplasty (UPPP) and who were felt to have significant hypopharyngeal collapse on physical examination as well as Müller's maneuver. Five (42%) were considered responders with a respiratory disturbance index (RDI) decreasing from 60.6/h to 14.5/h. In seven nonresponders, there was no significant change in the RDI (62.6 to 48.4 events/h). Cephalometric analysis showed that responders tended to be more retrognathic and significantly less obese (body mass index (BMI = 30.6)) than nonresponders (BMI = 37.9) (Fujita et al. 1991).

Lin et al. evaluated the efficacy of base of tongue (BOT) resection in the management of obstructive sleep apnea/hypopnea syndrome (OSAHS) through the use of transoral robotic surgery (TORS). They described (between June 2010 and May 2012), BOT resection via TORS on 27 patients with OSAHS. Twelve patients who underwent BOT resection alone were included in this study. The median age for these 12 patients was 48.5 (range, 19–64) and included nine females and three males. The mean apnea-hypopnea index (AHI) was 43.9 ± 41.1 preoperatively and 17.6 ± 16.2 postoperatively. This difference in AHI was statistically significant and reflected an average AHI reduction of 56.2 ± 28.3%. Statistically significant reductions in daytime somnolence level, as measured by Epworth sleepiness scale (13.7 ± 5.2 preoperatively vs. 6.4 ± 4.5 postoperatively, $P < 0.001$), and snoring intensity, as reported by a bed partner using a visual analog scale (8.6 ± 1.2 preoperatively vs. 4.2 ± 1.9 postoperatively), were achieved (Lin et al. 2013).

12.5.3 TORS (Trans-Oral Robotic Surgery)

Trans-oral robotic surgery (TORS) is a minimally invasive surgical technique that has been increasingly utilized for procedures such as base of tongue (BOT) resection and lingual tonsillectomy. TORS offers several advantages over traditional tongue reduction methods, including superior magnification, precise manipulation of tissues, better control of bleeding, and careful avoidance of vital vessels and nerves (Gangwani et al 2019).TORS uses sophisticated technology to access hard-to-reach areas in the back of the throat, which were traditionally accessed through a mandibulotomy, a procedure that involves cutting through the jaw and requires extensive reconstruction. This traditional approach often resulted in patients having difficulty talking, swallowing, and eating after recovery (Scott et al 2018). In contrast, TORS allows for a less invasive approach, reducing post-operative discomfort and shortening hospital stays (Scott et al 2018).The use of TORS in BOT resection and lingual tonsillectomy has been shown to result in less acute morbidity compared to other therapeutic methods (Patel et al 2021). A study found that patients treated with TORS base of tongue reduction for obstructive sleep apnea had diminished swallow function at 27 months, indicating a potential long-term benefit of this approach (Patel et al 2021). Moreover, TORS has been found to provide a wider surgical view and a more consistent removal of lingual tissue compared to other techniques, such as coblation surgery (Cammaroto et al. 2017). However, it's important to note that TORS also has a slightly higher rate of minor complications and significant costs (Cammaroto et al. 2017).

12.6 Maxillomandibular Surgery/Jaw Realignment Surgery

(a) Maxillomandibular Advancement (MMA)

In selected patients, the use of jaw realignment surgery in persons with OSA who are unresponsive to other therapies has been demonstrated to be an effective treatment. Maxillomandibular advancement achieves enlargement of the nasopharyngeal, retropalatal, and hypopharyngeal airway by physically expanding the facial skeletal framework by repositioning upper and lower jaw via Le Fort I maxillary and sagittal split mandibular osteotomies. The procedure also suspends the attached airway muscles in an anterior position together with an increase in pharyngeal soft tissue tension, affecting the nose and hypopharynx, and achieving expansion in the medial-lateral and anteroposterior dimensions of the upper airway (Gokce et al. 2014). MMA is an effective treatment of OSA but not without risk of injuries to the palatine arteries, infra-orbital nerve, and tooth roots, around osteotomy for the LeFort procedure. Similarly, the bilateral sagittal split osteotomy can damage the Inferior alveolar nerve with subsequent loss of feeling to lower teeth, chin, and lip. Above all is the change in facial skeleton and esthetics.

Zaghi et al. conducted a metanalysis to identify criteria associated with surgical outcomes of MMA using aggregated individual patient data from multiple studies. Inclusion criteria consisted of studies in (1) adult patients who underwent MMA as a treatment for OSA, (2) report of preoperative and post-operative quantitative outcomes for the apnea-hypopnea index (AHI) and/or respiratory disturbance index (RDI), and (3) report of individual patient data. The primary outcomes were changes in the AHI and RDI after MMA for each patient. Secondary outcomes included surgical success, defined as the percentage of patients with more than 50% reduction of the AHI to fewer than 20 events/h, and OSA cure, defined as a post-MMA AHI of fewer than 5 events/h. Forty-five studies with individual data from 518 unique patients/interventions were included. Among patients for whom data were available, 197 of 268 (73.5%) had undergone prior surgery for OSA. Mean (SD) postoperative changes in the AHI and RDI after MMA were −47.8 (25.0) and −44.4 (33.0), respectively; mean reductions of AHI and RDI outcomes were 80.1% (1.8%) and 64.6% (4.0%), respectively; and 512 of 518 patients (98.8%) showed improvement. Significant improvements were also seen in the mean (SD) postoperative oxygen saturation nadir 70.1% to 87.0% and Epworth sleepiness scale score of 13.5 to 3.2. Rates of surgical success and cure were 389 (85.5%) and 175 (38.5%), respectively, among 455 patients with AHI data and 44 (64.7%) and 13 (19.1%), among 68 patients with RDI data. Preoperative AHI of fewer than 60 events/h was the factor most strongly associated with the highest incidence of surgical cure. Nevertheless, patients with a preoperative AHI of more than 60 events/h experienced substantial net improvements despite modest surgical cure rates. Maxillomandibular advancement is an effective treatment for OSA. Most patients with high residual AHI and RDI after other unsuccessful surgical procedures for OSA are likely to benefit from MMA (Zaghi et al. 2016).

Romano et al. published a long-term follow-up of patients who had maxillomandibular advancement for the treatment of obstructive sleep apnea syndrome (retrospective study). A total of 19 patients with severe OSAS were treated during the period 2007–2016. They were followed up 6 months postoperatively (T1) and after a mean follow-up of 6.7 (range 4–10) years (T2, between November 2017 and February 2018) for the polysomnography variables, and the Epworth sleeping scale (ESS) and body mass index (BMI) were recorded. All polysomnographic variables improved significantly from baseline to T1 and from baseline to T2. The ESS significantly improved from baseline to T2. Nearly all patients at T1 and over two-thirds at T2 had a 50% reduction of the apnea/hypopnea index when compared with baseline, and a value lower than 20 of the same indexes. The BMI did not change significantly from T0 to T2. Their long-term follow-up has documented the stability of the outcomes of the MMA for the treatment of OSAS (Romano et al. 2020).

Camacho et al. performed a meta-analysis to examine outcomes in the intermediate-term (1 to <4 years), long-term (4 to <8 years), and very long-term (≥8 years) for maxillomandibular advancement (MMA) as a treatment for OSA. The authors systematically reviewed the international literature through July 26, 2018. Out of 445 studies screened, 6 met the criteria (120 patients). Thirty-one patients showed a reduction in AHI from a mean 48.3 events/h (95% CI, 42.1–54.5) pre-MMA to 8.4 (95% CI 5.6, 11.2) in the intermediate term. Fifty-four patients showed a reduction in AHI from a mean 65.8 events/h (95% CI, 58.8–72.8) pre-MMA to 7.7 (95% CI 5.9, 9.5 in the long term. Thirty-five showed a reduction in AHI from a mean 53.2 events/h (95% CI 45, 61.4) pre-MMA to 23.1 (95% CI 16.3, 29.9) in the very long term. Improvement in sleepiness was maintained at all follow-up periods. The lowest oxygen saturation improvement was maintained in the long term. Authors observed that while patients treated with MMA maintained improvements in AHI, sleepiness, and lowest oxygen saturation in the long term; mean AHI increased to moderate OSA in the long term (Camacho et al. 2019). MMA is an effective surgical option for patients with OSA. However, the nasal changes associated with significant maxillary advancement are an important concern. In a study of 379 subjects 71 (18.7%) underwent corrective nasal surgery with a median time to surgery of 10.5 months. Ten patients (2.6%) required a second corrective nasal operation (Liu et al. 2017a).

(b) Distraction osteogenesis maxillary expansion (DOME)

(c) This surgical technique is recommended for adult individuals who have OSA who are identified to have narrow nasal floors and high-arched palate without soft tissue redundancy. Stanley Yung-Chuan Liu, MD, DDS et al., conducted a prospective cohort of 20 subjects, who underwent DOME (Liu et al., 2017b). This technique involves placement of a maxillary expander with mini-implants to custom fit the narrow palate under local anesthesia, followed by LeFort level 1 osteotomy via small incisions. After wedging the mid-palatal suture, an expander is turned to ensure separation of the maxilla (1 mm diastema). Patients then turn the expander daily by 0.25 mm. One month after the surgery, orthodontic treatment is required to close the gap between maxillary incisors. The key feature of distraction osteogenesis is the incremental traction to the reparative callus that then initiates adaptive changes in the soft tissues. Authors report that "separation of mid-palatal suture and subsequent maxillary and nasal floor expansion was successfully achieved in all subjects." Authors also noted significant reductions in outcome measures of ESS: Before 12.3 ± 4.10, After 7.8 ± 4.8; NOSE: Before 11.7 ± 5.3, After 3.85 ± 3.23; AHI: Before 30.9 ± 27.1, After 14.2 ± 9.3; oxygen desaturation index (ODI): Before 23.0 ± 28.4, After 8.7 ± 6.9, and nasal airflow resistance. These results need to be interpreted with the knowledge that expanders remain in place for 8 months and longer-term data is not available to observe the rate of relapse (Liu et al. 2017a)

A systemic review by Wai Kin Tsui et al. (published online in 2016) showed that mandibular distraction osteogenesis (MDO) was effective in resolving OSA in adults with the retrognathic mandible. From this review, authors concluded that MMA could enlarge the airways in three dimensions by expanding the skeletal framework, thus rendering the pharyngeal soft tissues and tongue more resistant to collapse during inspiration. Out of 534 studies, only 98 were found to be relevant to the treatment of OSA with MDO. Only 12 articles fulfilled the five criteria and entered the final round of appraisal. All 12 studies were retrospective and were on adults with unilateral or bilateral temporomandibular joint (TM J) ankylosis with associated retrognathic mandible and OSA. All 12 studies used pre and postintervention polysomnograms (PSGs). Significant improvement was reported for AHI/RDI in the adult group from 51.7/h (43.0–58.0/h) pre-operatively to 2.9/h (2.1–3.4/h) postoperatively. The lowest oxygen saturation improved in adults from 66% to 77% preoperatively to 90.3% to 98.2% postoperatively. Although this review demonstrates the effectiveness of the procedure, the evidence-based data on MDO is believed to be weak (Tsui et al. 2016).

12.7 Upper Airway Stimulation for Obstructive Sleep Apnea

A relatively new development in the armamentarium of treatments for OSA is upper airway stimulation which involves unilateral respiration synchronized stimulation of the hypoglossal nerve delivered via an implanted neuromodulation technology. The procedure requires placing a cuff electrode on the distal branches of the hypoglossal nerve to stimulate the tongue protrusers (genioglossus and transverse/vertical muscles), inserting an implanted pulse generator in the right upper chest, and placing a respiratory sensing lead between external and internal intercostal muscles of the ribs.

From July 2014 through October 2015, 60 patients were enrolled in a prospective single-arm study conducted by Heiser, MD et al. The study enrolled subjects with moderate-to-severe OSA, AHI: 20–65, failure of CPAP therapy, body mass index (BMI) $\leq$ 32 kg/m^2, and absence of complete circumferential collapse on drug-induced sleep endoscopy. All patients had failed CPAP as first-line therapy, 14 had attempted oral appliance therapy, and 15 had prior upper airway OSA operations (UPPP, uvulopalatal flap, tongue base reduction, etc.). Patients who completed the 6-month follow-up achieved an AHI reduction of 61 $\pm$ 24% compared with baseline. Per Sher criteria (AHI $<$ 20 with at least 50% reduction), 68% were deemed responders. The four patients, who had UPPP and missed the 6 month visit, a success rate of 63% was observed. The authors reported a statistically significant reduction in ODI, apnea index, hypopnea index, and minimal SpO_2 nadir from baseline to 6 months. Patient-reported outcome measures (ESS and FOSQ) compared with baseline and further improved at the 6-month visit. Stimulation-related morbidity was low. Therapy adherence was high after 6 months. The authors concluded that this multicenter study further strengthened the evidence that upper airway stimulation can be "successfully transferred from the previous controlled setting (STAR TRIAL) into routine clinical practice" (Heiser et al. 2017).

Tucker Woodson et al. conducted a 5-year outcome from a multicenter, prospective cohort study of patients with OSA treated with upper airway stimulation (UAS) via a unilateral hypoglossal nerve implant. From a cohort of 126 patients, 97 completed the protocol and 71 consented to a polysomnogram. Those having CPAP failure with moderate to severe OSA, body mass index <32 kg/m (Weaver et al. 2004), and no unfavorable collapse were enrolled in the phase 3 study. Significant improvements were noted in sleepiness levels, as assessed by the Epworth sleepiness scale, with scores returning to normal ranges for 33% to 78% of participants. Similarly, there were positive changes in the quality of life, with scores improving from 15% to 67% of their initial values. The rate of positive response in terms of AHI (apnea-hypopnea index) improvement, defined as AHI less than 20/h and a reduction of more than 50%, was 75% among the study participants ($n = 71$). When the last observation carried forward analysis was used, the responder rate at the 5-year mark was 63%. Six percent of the patients experienced serious adverse events related to the device. Results support that in a select group of participants, upper airway stimulation therapy can provide improvement in objective and quality-of-life measures across 5 years (Woodson et al. 2018).

12.7.1 Drug-Induced Sleep Endoscopy (DISE)

Drug-induced sleep endoscopy is an evaluation technique that requires nasoendoscopic examination of the upper airway under conditions of spontaneous ventilation and pharmacologic unconscious sedation. Since the recognition of OSA as a condition that can present with multiple levels of collapse, increased interest has also developed in using DISE to facilitate the appropriate selection of the best surgical procedures and maximize the success rate. DISE has been recommended as a patient selection tool for upper airway stimulation (UAS) due to its ability to select patients with an increased probability of therapeutic success. DISE is also used for prescribing mandibular advancement devices and positional therapy. However, administration techniques vary worldwide partly because it is hard to mimic natural sleep and there is no best way at present to perform DISE despite many classification systems.

DISE is typically indicated for children with persistent OSA after tonsillectomy, those with OSA without tonsillar hypertrophy, children with risk factors predisposing them to multiple sites of obstruction, or when sleep-state-dependent laryngomalacia is suspected.

Many different scoring systems have been used to report both adult and pediatric DISE findings, but none is universally accepted (Wilcox et al. 2017).

12.7.1.1 Scoring Systems for Adults

VOTE (Kezirian 2011): The areas of assessment include the velum, oropharynx, tongue base, and epiglottis, commonly referred to as VOTE. A grading scale ranging from 0 to 2 is utilized to evaluate the extent of obstruction: 0 indicating no obstruction, 1 indicating partial obstruction, and 2 indicating total obstruction. Furthermore, it is valuable to characterize the pattern of collapse, which can be classified as lateral, anterior-posterior, or concentric.

NOHL (Vicini 2012): NOHL is focused on four anatomical zones: nose, oropharynx, hypopharynx, and larynx. The severity of obstruction was categorized in conjunction with the Müller maneuver grading system. Grade 1 is an obstruction between 0% and 25% airway collapse (no collapse during Müller maneuver). Grade

2 is an obstruction between 25% and 75% airway collapse. Grade 3 is an obstruction between 50% and 75% airway collapse. Grade 4 is obstruction greater than 75% airway collapse (total pharyngeal wall collapse during Müller maneuver).

12.7.1.2 Scoring System for Children

Six different scoring systems (VOTE, SERS, Chan, Bachar, Fishman, Boudewyns) have been used to report pediatric DISE findings. Unlike the adult scoring systems, obstruction at the nasopharynx (including adenoids) and supraglottis are commonly incorporated into these scales.

VOTE (velum, oropharynx, tongue base, and epiglottis): Although this scoring system is frequently used, it excludes the evaluation of the nasopharynx and supraglottis.

SERS (sleep endoscopy rating scale): Bachar provides similar information but incorporates an overall severity score for upper airway obstruction. There are no published reproducible results for the rating system.

CHAN: Chan and VOTE scoring system evaluate each obstruction level individually.

Boudewyns: This classification system includes six sites (adenoids, tonsils, tongue base, palate, epiglottis, laryngomalacia) characterized as fixed or dynamic.

Fishman: This grading system evaluates five sites (nasal, nasopharynx, lateral walls, tongue base, and supraglottis.

Kezirian reports insights from drug-induced sleep endoscopy in nonresponders to pharyngeal surgery for OSA. This retrospective cohort study of 33 subjects included consecutive subjects who were nonresponders to previous palate surgery (and tonsillectomy, whether performed concurrently or not) to treat OSA, and who underwent DISE. DISE findings were summarized with three analyses: Analysis-1 was a global dichotomous (yes/no) assessment of obstruction at the level of palate/velum and hypopharynx, Analysis II for the degree of palatal and hypopharyngeal obstruction and Analysis III evaluated specific structures with a determination of which structure at the level of palate and hypopharynx was the primary factor in airway obstruction and also a dichotomous evaluation of whether each of the individual structures contributed to airway obstruction structures were grouped as those at the level of the palate (velum/soft palate, tonsils when present) and hypopharynx (oropharyngeal lateral walls, tongue, and epiglottis). Additional evaluation of mouth opening was recorded (degree of opening <1 cm between maxillary and mandibular incisors, mild (1–2 cm), moderate (2–3 cm), and severe (>3 cm). Over half of the subjects had PSG. During DISE, the majority of subjects demonstrated global obstruction (Analysis I) in the palatal region and almost all subjects demonstrated hypopharyngeal obstruction, largely severe (>75%) obstruction (Analyses I and II). Subjects with residual palatal obstruction were older and trended toward higher BMI than those without. There was a notable diversity of specific structures that contributed to airway obstruction during DISE (Analysis III). These were not systematically associated with previous hypopharyngeal procedures overall or with specific procedures. Tongue contribution of obstruction was common. Like those without previous OSA Surgery, a diversity of factors may contribute to residual upper airway obstructions in surgical non-responders. The author concludes that "residual upper airway obstruction in OSA surgery non-responders likely occurs due to multiple mechanisms and evaluation techniques like DISE may enhance the understanding of these mechanisms." The authors also observed that oropharyngeal lateral wall collapse was associated with poorer outcomes after secondary surgery and may explain a portion of nonresponders to both primary and secondary surgery (Kezirian 2011).

12.8 Trachesotomy

Tracheostomy is the only definitive therapy and most invasive of all treatments of OSA as it bypasses the upper airway. The definitive nature of the tracheostomy procedure, the appearance following the establishment of tracheotomy, and the change in the ability to vocalize results in significant quality of life/psychosocial impact for the individual. However, the procedure does result in a decrease in AHI and improves ESS when compared to untreated OSA cohorts. As many effective therapies are available at this time, the tracheostomy procedure should be reserved only as a last resort for morbidly obese patients with concurrent obesity hypoventilation syndrome, patients with significant craniofacial anomalies, and OSA patients who have failed or are intolerant of other nonsurgical and surgical treatments.

12.8.1 Tracheostomy for Adults

Camacho et al. published a systematic review and meta-analysis on tracheostomy as a treatment for adult obstructive sleep apnea. They reviewed outcomes for polysomnography, sleepiness, and mortality in patients undergoing tracheostomy to treat adult obstructive sleep apnea (OSA). Using retrospective review, they identified 18 relevant studies that were primarily case series. Post-tracheostomy AI (apnea index) improved from 73.0 ± 27.1 to 0.2 ± 1.2/h and an AI mean difference of -83.47. The mean AHI decreased from 92.0 ± 34.8 to 17.3 ± 20.5/h and an AHI mean difference of -79.82 compared with preoperative status was observed. Post-surgically, central apneas developed; however, the central AI demonstrated near normalization to a mean of 2.1 ± 3.5/h after 14 weeks. ODI decreased from 78.2 ± 25.8/h to 20.8 ± 25.5/h. Four studies demonstrated a statistically significant improvement in subjective sleepiness post-tracheostomy. Postoperative statistically significant overall and cardiovascular mortality reductions compared with untreated historical cohorts weres reported (Camacho et al. 2014).

12.8.2 Tracheostomy for Pediatric OSA

In children, tracheostomy is infrequently performed for the treatment of obstructive sleep apnea (OSA). Typically, patients in need of this procedure have an underlying syndromic condition that leads to the obstruction of the upper airway.

Rizzi et al. examined a group of 29 children below 18 years old, who received surgical treatment tracheostomy due to the presence of severe obstructive sleep apnea (OSA) during the period of January 1, 2010 to December 31, 2015. Of 29 patients, 13 had craniofacial abnormalities, 10 had a neurologic disorder with hypotonia or neuromuscular dysfunction, and 5 had a diagnosis of laryngomalacia on a fiberoptic exam. The mean follow-up was 30.6 months. All patients were admitted to PICU with a mean stay of 23.2 days. Most children who undergo tracheostomy for OSA will remain dependent at 24 months (Rizzi et al. 2017).

References

Amali A, Motiee-Langroudi M, Saedi B, Rahavi-Ezabadi S, Karimian A, Amirzargar B (2017) A comparison of uvulopalatopharyngoplasty and modified radiofrequency tissue ablation in mild to moderate obstructive sleep apnea: a randomized clinical trial. J Clin Sleep Med 13(9): 1089–1096

American Academy of Otolaryngology Head and Neck Surgery (2021) Position statement: midline glossectomy for OSA. American Academy of Otolaryngology Head and Neck Surger, Alexandria

Baugh RF, Archer SM, Mitchell RB et al (2011) Clinical practice guideline: tonsillectomy in children. Otolaryngol Head Neck Surg 144(1 Suppl):S1–S30

Camacho M, Certal V, Brietzke SE et al (2014) Tracheostomy as treatment for adult obstructive sleep apnea: a systematic review and meta-analysis. Laryngoscope 124:803–811

Camacho M, Li D, Kawai M, Zaghi S, Teixeira J, Senchak AJ, Brietzke SE, Frasier S, Certal V (2016) Tonsillectomy for adult obstructive sleep apnea: a systematic review and meta-analysis. Laryngoscope 126(9):2176–2186. https://doi.org/10.1002/lary.25931

Camacho M, Nesbitt NB, Lambert E, Song SA, Chang ET, Liu SY, Kushida CA, Zaghi S (2017) Laser-assisted uvulopalatoplasty for obstructive sleep apnea: a systematic review and meta-analysis. Sleep 40(3):zsx004. https://doi.org/10.1093/sleep/zsx004

Camacho M, Noller MW, Del Do M, Zaghi S, Wei JM, Gouveia CJ, Boyd SB, Guilleminault C (2019) Long-term results for maxillomandibular advancement to treat obstructive sleep apnea: a meta-analysis. Otolaryngol Head Neck Surg 160(4):580–593

Cammaroto G, Montevecchi F, D'Agostino G, Zeccardo E, Bellini C, Galletti B, Shams M, Negm H, Vicini C (2017) Tongue reduction for OSAHS: TORSs vs coblations, technologies vs techniques, apples vs oranges. Eur Arch Otorhinolaryngol 274(2):637–645. https://doi.org/10.1007/s00405-016-4112-4. Epub 2016 May 24. PMID: 27221388

Chervin RD, Ellennerg SS, Hou X, Garetz SL, Katz ES, Hodges EK, Mitchell RB, Jones DT, Arens R, Amin R, Redline S, Rosen CL (2015) Prognosis for spontaneous resolution of OSA in children. Chest 148(5):1204–1213

Friberg D, Sundman J, Browaldh N (2020) Long-term evaluation of satisfaction and side effects after modified uvulopalatopharyngoplasty. Laryngoscope 130:263–268

Fujita S, Woodson BT, Clark JL, Wittig R (1991) Laser midline glossectomy as a treatment for obstructive sleep apnea. Laryngoscope 101(8):805–809

Gangwani K, Shetty L, Seshagiri R, Kulkarni D (2019) Comparison of TORS with conventional surgery for oropharyngeal carcinomas in T1-T4 lesions. Ann Maxillofac Surg 9(2):387–392. https://doi.org/10.4103/ams.ams_33_18. PMID: 31909020; PMCID: PMC6933975

Gokce SM, Gorgulu S, Gokce HS, Bengi AO, Karacayli U, Ors F (2014) Evaluation of pharyngeal airway space changes after bimaxillary orthognathic surgery with a 3-dimensional simulation and modeling program. Am J Orthod Dentofac Orthop 146(4):477–492

Heiser C, Mauer JT, Hofauer B, Sommer JU, Seitz A, Steffen A (2017) Outcomes of upper airway stimulation for OSA in a multicenter German postmarket study. Otolaryngol Head Neck Surg 156(2):378–384

Hong S-N, Kim HG, Han S-Y, Ji J-Y, Kim M-K, Han DH, Won T-B, Kim D-Y, Kim HJ (2019) Indications for and outcomes of expansion sphincter pharyngoplasty to treat lateral pharyngeal collapse in patients with obstructive sleep apnea. JAMA Otolaryngol Head Neck Surg 145(5): 405–412

Kezirian EJ (2011) Non-responders to pharyngeal surgery for OSA: insights from drug-induced sleep endoscopy. Laryngoscope 121(6):1320–1326. https://doi.org/10.1002/lary.21749

Kezirian EJ, Weaver EM, Yueh B, Deyo RA, Khuri SF, Daley J et al (2004) Incidence of serious complications after uvulopalatophryngoplasty. Laryngoscope 114(3):450–453

Lin HS, Rowley JA, Badr MS et al (2013) Transoral robotic surgery for treatment of obstructive sleep apnea-hypopnea syndrome. Laryngoscope 123(7):1811–1816. https://doi.org/10.1002/lary.23913

Liu SY-C, Lee P-J, Awad M, Riley RW, Zaghi S (2017a) Corrective nasal surgery after maxillomandibular advancement for OSA: experience from 379 cases. Otolaryngol Head Neck Surg 157(1):156–159

Liu SY-C, Guilleminualt C, Huon L-K, Yoon A (2017b) Distraction osteogenesis Maxillary Expansion (DOME) for adult obstructive sleep apnea patients with high arched palate. Otolaryngol Head Neck Surg 157(2):345–348

Mozafari A, Zand N, Hoseini SAA, Mohebi S, Fard RG, Rasouli A, Beiglo MG, Hatami S (2014) Relationship between road accidents with sleep apnea and sleep quality among truck drivers in Iran. Eur Respir J 44(Suppl 58):2296

O'Halloran LR (2003) The lateral crural J-flap repair of nasal valve collapse. Otolaryngol Head Neck Surg 128(5):640–649

Pang KP (2013) Role of nasal surgery in the treatment of OSA. Curr Otorhinolaryngol Rep 1:20–24

Patel MR, Ottenstein L, Ryan M, Farrell A, Studer M, Baddour HM, Magliocca K, Griffith C, Stokes W, Switchenko J, Aiken A, El-Deiry M, Solares CA, Steuer C, Saba N, Beitler J (2021) TORS elective lingual tonsillectomy has less acute morbidity than therapeutic base of tongue TORS. Oral Oncol 117:105294. https://doi.org/10.1016/j.oraloncology.2021.105294. Epub 2021 Apr 17. PMID: 33878679; PMCID: PMC8281337

Rizzi CJ, Amin JD, Isiah A, Valdez TA, Jeyakumar A, Smart SE, Pereira KD (2017) Tracheostomy for severe pediatric OSA: indications and outcomes. Otolaryngol Head Neck Surg 157(2): 309–313

Romano M, Karanxha L, Baj A et al (2020) Maxillomandibular advancement for the treatment of obstructive sleep apnoea syndrome: a long-term follow-up. Br J Oral Maxillofac Surg 58(3): 319–323

Samutsakorn P, Hirunwiwatkul P, Chaitusaney B et al (2018) Lingual tonsillectomy with palatal surgery for the treatment of obstructive sleep apnea in adults: a systematic review and meta-analysis. Eur Arch Otorhinolaryngol 275:1005–1013. https://doi.org/10.1007/s00405-018-4887-6

Scott SI, Madsen AKØ, Rubek N, Channir HI, Kehlet H, von Buchwald C (2018) Time course of subacute pain after transoral robotic surgery (TORS) for oropharyngeal squamous cell carcinoma versus traditional bilateral tonsillectomy in adults - a case-control study. Acta Otolaryngol 138(9):837–842. https://doi.org/10.1080/00016489.2018.1461238. Epub 2018 May 9. PMID: 29741427

Shiffman HS, Khorsandi J, Cauwels NM (2021) Minimally invasive combined Nd:YAG and Er: YAG laser-assisted uvulopalatoplasty for treatment of obstructive sleep apnea. Photobiomodul Photomed Laser Surg 39(8):550–557

Smith M, Peterson E, Yaremchuk KL (2017) The role of tonsillectomy in adults with tonsillar hypertrophy and OSA. Otolaryngol Head Neck Surg 157(2):331–335

Sorrenti G, Piccin O (2013) Functional expansion pharyngoplasty in the treatment of obstructive sleep apnea. Laryngoscope 123:2905–2908

Tsui WK, Yang Y, Cheung LK, Leung YY (2016) Distration osteogenesis as a treatment of obstructive sleep apnea syndrome. Medicine (Baltimore) 95(36):e4674. https://doi.org/10.1097/MD0000000000004674

Wadhera V, Fogarty G (2020) Uvulopalatopharyngoplasty: a specific surgical technique aimed at minimizing patient morbidity, with outcomes over 5 years in a regional centre. Aust J Otolaryngol 3:11. https://doi.org/10.21037/ajo.2020.03.03

Weaver EM, Maynard C, Yeuh B (2004) Survival of veterans with sleep apnea: continous positive airway pressure vs. surgery. Otolaryngol Head and Neck Surg 130(6):659–652. https://doi.org/10.1016/jotohns.2003.12.012. Erratum in: Otolaryngology Head and Neck Surg.2004 Jul;131(1):144

Wilcox LJ, Bergeron M, Reghunathan S, Isman SL (2017) An updated review of pediatric drug induced sleep endoscopy. Laryngoscope Investig Otolaryngol 2(6):423–431

Woodson T, Strohl KP, Soose RJ, Gillespie MB, Maurer JT, de Vries N, Padhya TA, Badr MS, Lin H-s, Vanderveken OM, Mickelson S, Strollo PJ Jr (2018) Upper airway stimulation for obstructive sleep apnea: 5-year outcomes. Otolaryngol Head Neck Surg 159(1):194–202

Zaghi S, Holty JC, Certal V et al (2016) Maxillomandibular advancement for treatment of obstructive sleep apnea: a meta-analysis. JAMA Otolaryngol Head Neck Surg 142(1):58–66. https://doi.org/10.1001/jamaoto.2015.2678

Chapter 13
Impact of REM Sleep on Sleep Disorders: Current Perspectives

Ahmed S. BaHammam, Mahadevappa Hunasikatti, and Seithikurippu R. Pandi-Perumal

Abstract Sleep is a complex process characterized by different stages, including rapid eye movement (REM) and non-rapid eye movement (NREM) sleep. Throughout the night, the body cycles through these stages, with REM sleep episodes increasing in duration. REM sleep is associated with active dreaming, while the body experiences muscle atonia and impaired breathing. Research on REM sleep continues to unravel its mysteries. The regulation of sleep cycles is influenced by the circadian rhythm and various brain regions. Cholinergic and glutamatergic activities in specific areas of the brain contribute to the occurrence of REM sleep. Parameters such as REM latency and REM density are used to measure REM sleep and have been linked to sleep deprivation, depression, and respiratory abnormalities. REM-predominant obstructive sleep apnea (OSA) is a condition where apneas and hypopneas occur mainly during REM sleep periods. It is more common in younger individuals, women, and patients with mild or moderate OSA. Treatment with positive airway pressure therapy has shown improvements in functional outcomes for patients with REM OSA. REM sleep behavior disorder (RBD) is a parasomnia characterized by dream-enacting behaviors during REM sleep without muscle atonia. It can be idiopathic or comorbid with neurodegenerative disorders. RBD has a higher prevalence in males and can lead to sleep-related injuries. Psychiatric disorders such as major depressive disorder (MDD) and schizophrenia affect sleep architecture. MDD is associated with disrupted sleep parameters, including

A. S. BaHammam
The University Sleep Disorders Center, Department of Medicine, College of Medicine, King Saud University, Riyadh, Saudi Arabia

King Saud University Medical City, King Saud University, Riyadh, Saudi Arabia

M. Hunasikatti (✉)
Division of Respiratory, Sleep and Anesthesiology (DSRA), OHT1, Food and Drug Administration (FDA), Silver Spring, MD, USA

S. R. Pandi-Perumal
Somnogen Canada Inc., College Street, Toronto, ON, Canada

Saveetha Medical College and Hospitals, Saveetha Institute of Medical and Technical Sciences, Saveetha University, Chennai, Tamil Nadu, India

A. S. BaHammam, M. Hunasikatti (eds.), *Sleep Apnea Frontiers*, Progress in Sleep Research, https://doi.org/10.1007/978-981-99-7901-1_13

decreased sleep efficiency and altered REM sleep. Antidepressant medications can influence sleep quality and may exacerbate or induce sleep disorders. Understanding the effects of antidepressants on sleep is crucial for effective treatment of depression. Different classes of antidepressants can disrupt sleep architecture and REM sleep latency. In summary, REM sleep plays a significant role in sleep physiology and is associated with various neurophysiological and pathological conditions. Further research is needed to fully understand the mechanisms and clinical implications of REM sleep disorders.

Keywords Sleep stages · REM sleep · Sleep disorders · Obstructive sleep apnea (OSA) · REM sleep behavior disorder (RBD) · Antidepressant medications

13.1 Introduction

During sleep, our body cycles through three distinct stages: rapid eye movement (REM) and non-rapid eye movement (NREM) sleep. The body usually cycles through these stages on average four to six times, 90 min in each stage. As the night progresses, fewer NREM stages occur, and the duration of REM sleep episodes increases with time (Memar and Faradji 2018). Sleep may be categorized into wake, N1, N2, N3, and R (REM). Sleep is scored in sequential 30-s epochs, and each of these epochs is assigned a specific sleep stage based on the American Academy of Sleep Medicine (AASM) scoring rules. Most of the sleep is spent in the N2 stage in normal subjects (Malik et al. 2018). REM stage is associated with active dreaming. The EEG is like an awake individual, but the skeletal muscles are atonic and without movement. The only exception is the eye and diaphragmatic breathing muscles, which remain functional. The breathing rate is impaired, and more erratic and irregular. This stage usually starts about 90 min after falling asleep, and each of the REM cycles gets longer throughout the night. The first period typically lasts a few minutes, and the final one can last almost an hour (Ferri et al. 2017).

13.2 AASM Scoring Criteria for Stage R (REM) Sleep

- EEG demonstrates a low-voltage, mixed EEG pattern. Sawtooth waves are common during REM sleep; these 2–6 Hz wave patterns are sharply contoured and occur in brief bursts.
- The characteristic that defines this stage is the rapid movement of the eyes. These are defined on EOG by conjugate, irregular, sharply peaked eye movements with an initial phase of fewer than 500 ms.
- EMG demonstrates atonia, the inhibition of alpha motor neurons directly leads to the absence of voluntary muscle activity, except for the muscles controlling eye movement (extraocular muscles) and breathing (diaphragm) (American Academy of Sleep Medicine n.d.).

13.3 Significance of REM Staging

Despite many well-done clinical and preclinical studies, the mystery regarding the REM stage continues.

Basic science: The sleep cycle is regulated by the circadian rhythm, a process controlled by the suprachiasmatic nucleus (SCN) of the hypothalamus (España and Scammell 2011). The extraocular movements after which REM sleep is named are mainly due to the cholinergic activity of the pedunculopontine tegmentum (PPT) center and glutamatergic activity of the medial pontine reticular formation (mPRF) (https://www.statpearls.com/articlelibrary/viewarticle/28329/). Datta et al. showed that "selective RSD (REM sleep deprivation) induced increased expression of BDNF (brain-derived neurotrophic factor (BDNF) expression in the PPT (pedunculopontine tegmentum) and SubCD (subcoeruleus nucleus) are determinant factors in the development of the homeostatic drive for REM sleep" (Datta et al. 2015).

Many parameters are used to measure REM sleep, including the time to onset, length of the REM stage, and amount/intensity of eye movement that occurs during REM. REM latency refers to the time elapsed between the onset of sleep to the first REM stage. REM density refers to the number of eye movements during REM sleep, and changes have been associated with sleep deprivation and depression. REM sleep is associated with more frequent and longer duration of apneas, hypopneas, and severe hypoxemia (American Academy of Sleep Medicine n.d.; Shrivastava et al. 2014; Pirzada and BaHammam 2021).

13.4 Neurophysiological and Pathological Conditions Concerning REM Sleep

13.4.1 REM-Predominant Obstructive Sleep Apnea

The reported prevalence of REM-related OSA is quite variable, ranging from as little as 10% to as much as 36% of the patient population with OSA undergoing PSG (Conwell et al. 2012). The variability in the reported prevalence is related to the lack of an accepted definition of REM-predominant OSA. The currently accepted definition is refined by considering REM sleep duration in terms of percentage or minutes and excluding significant NREM-OSA.

OSA can present with or provoke various psychological symptoms. BaHammam et al. reviewed studies that have examined dreams, dream recall, and dream content in patients with OSA. Obstructive events induce recurrent sleep fragmentation and intermittent desaturations in patients with OSA, which may trigger different parasomnias, including nightmares. Contradictory results have been reported concerning dreams in patients with OSA. A clear association between post-traumatic stress disorders, comorbid OSA, and nightmares has been reported in several studies.

Furthermore, an improvement in nightmare frequency with continuous positive airway pressure (CPAP) treatment has been shown. An inverse relationship between the severity of OSA reflected by the apnea-hypopnea index and dream recall has been demonstrated in several studies. (BaHammam and Almeneessier 2019; BaHammam et al. 2018)

REM-predominant OSA is a sleep-breathing abnormality during which apneas and hypopneas occur predominantly or exclusively during REM sleep periods. Body position has an important effect on breathing abnormalities during REM sleep. During this sleep stage, the supine posture has a worsening effect, facilitating the occurrence of breathing abnormalities, but the duration of these events is like the duration of events occurring in the lateral posture (Oksenberg et al. 2010). As REM sleep is accompanied by smaller lung volumes, a decrease in airway muscle tone, reduced hypoxic and hypercapnic ventilatory effort, and an increase in blood pressure at the termination of the obstructive events, obstructive events tend to worsen in REM sleep (Pirzada and BaHammam 2021).

Studies have consistently shown that REM-predominant OSA occurs more commonly in younger individuals, women, children, and patients with mild or moderate OSA despite the wide variation in prevalence (Goh et al. 2000). Functional outcomes in patients with REM OSA improve after treatment with PAP therapy comparable to that observed in patients with OSA not restricted to REM (Su et al. 2012). The sequelae of REM-predominant OSA are getting more clear now. Severe OSA that occurs primarily during REM sleep is associated with a higher incidence of composite cardiovascular and metabolic endpoints but in only those with prevalent cardiovascular disease (Iannella G et al. 2020 and Aurora et al. 2018).

Despite the overwhelming research related to REM sleep behavior disorder (RBD) and OSA as different entities, few papers paid attention to the coexistence of RBD and OSA. The salient differences are as follows. The nadir SpO_2 of OSA patients with RBD achieves a lower drop compared to those observed in REM-OSA patients (Giardino et al. 2021). The improvement observed in RBD patients on OSA severity, denoted by a lower drop in nadir SpO_2, was named by Giardino et al. “the respiratory RSWA (REM sleep without atonia-RSWA) benefit” (Giardino et al. 2021). These findings may suggest that a shorter duration for sleep events and a consequent minor reduction of nadir oxygen saturation were related to an enhancement of active Pcrit (active critical closing pressure of the upper airway) in OSA-RBD patients.

13.5 REM Sleep Behavior Disorder

REM sleep behavior disorder (RBD) is a particularly intriguing parasomnia. RBD is a parasomnia characterized by dream-enacting behaviors associated with REM sleep without muscle atonia (also known as REM sleep without atonia or RSWA). These patients have repeated episodes of sleep-related vocalization and/or complex motor behaviors, which may be documented by PSG. Alternatively, they may also be based on a clinical history of dream enactment during REM sleep (American Academy of Sleep Medicine 2014). RBD can be isolated, also known as idiopathic (iRBD), or comorbid (secondary) with other conditions related to neurodegenerative, autoimmune, or structural abnormalities or medications (Dauvilliers et al. 2018).

Muscle atonia during REM is a protective mechanism that involves both muscle atonia and suppression of motor-skeletal activity. This process entails active inhibition by neurons in the medulla, in addition to inhibition of the spinal motor neurons through feedback from the forebrain and thalamus (Fraigne et al. 2015). Several pontine regions, including the peri-locus coeruleus region, pedunculo-pontine nucleus (PPN), and laterodorsal tegmental nucleus (LDTN) excite neurons in the medulla, resulting in descending inhibitory projections to spinal alpha motor-neurons causing hyperpolarization and muscle atonia. The disinhibition of these neurons in RBD patients causes the appearance of muscle activity during REM sleep.

Longitudinal studies have shown that patients diagnosed with iRBD may eventually be diagnosed with a neurological disorder such as Parkinson's disease (PD), dementia with Lewy bodies (DLB), and multisystem atrophy. Most patients are eventually diagnosed with synucleinopathies PD, DLB, and, less frequently, MSA. iRBD represented the prodromal period of these conditions.

In RBD, there is a predilection for the male gender, and prevalence estimates are 0.38% to 0.5% in the general population. Prevalence in patients with psychiatric disorders is greater, related to medications prescribed for psychiatric disorders. In a report by Olson et al. involving 93 patients with RBD, only 12 patients (13%) were female (American Psychiatric Association 2013; Olson et al. 2000). Between 33% and 65% of RBD patients have been reported to have had sleep-related injury to self or bed partner (Comella et al. 1998).

Sleep clinicians should maintain an index of suspicion for RBD regardless of the presenting sleep complaint in patients greater than 50 years of age, or those at any age with narcolepsy and other neurological disorders (Schenck 2016; Iranzo et al. 2014). RBD in narcolepsy usually occurs in a younger age group, does not have sex preponderance, and has fewer occurrences of complex and violent movements (Cipolli et al. 2011; BaHammam et al. 2020).

All patients with RBD and their bed partners should be educated on ways to alter the sleeping environment to prevent injury. The AASM guidelines do recommend clonazepam for RBD treatment, but the drug should be used with caution in patients with dementia, gait disorders, or concomitant OSA. Melatonin is also recommended for the treatment of RBD with the advantage that there are few side effects, and this may be a preferred medication for the elderly who may be at risk of falling with the use of benzodiazepines. Other medication with limited evidence includes zopiclone, benzodiazepines other than clonazepam, Yi-Gan San, desipramine, clozapine, carbamazepine, and sodium oxybate (Aurora et al. 2010).

13.6 Psychiatric Disorders and REM Sleep

13.6.1 Major Depressive Disorder (MDD)

It is well known that major psychiatric disorders like MDD significantly impact sleep architecture. MDD is correlated with a decrease in REM latency, an increase in REM density, and a prolonged time in the first REM cycle (Anderson and Bradley 2013). This has been confirmed in several studies that have consistently shown a reduction in sleep efficiency, decreased slow-wave sleep (SWS), and disinhibited REM sleep with shortened REM latency, a longer first REM cycle, and increased REM density (the number of eye movements during REM). Many epidemiologic studies showed poor sleep quality in 50–90% of those with depression (Riemann et al. 2001; Kempenaers et al. 1988).

REM sleep normalizes after depression treatment, but many antidepressants are known to lengthen REM latency and improve sleep continuity in both patients and normal controls. Increased sleep efficiency and REM sleep improvement have also been demonstrated after nonpharmacological therapies such as cognitive behavioral therapy. Despite these paradoxical observations, there is a significant impact on the sleep stages, and needs further research (Nofzinger et al. 1994; Thase et al. 1998).

13.6.2 Schizophrenia

Sleep is disturbed in 30–80% of patients with schizophrenia, depending on the degree of psychotic symptomatology (Keshavan and Tandon 1993). Dream recall and eye movements during sleep were studied by Dement in 1955 in patients with schizophrenia and compared to controls. They showed a reduction in dream recall in patients with schizophrenia and abnormal REM parameters (Dement 1955). Memory consolidation is also impaired compared to controls and positively correlated with the amount of SWS and sleep efficiency. Manoach et al. showed that SWS and Stage 2 sleep in the last quartile of the night (S2q4) sleep, each predicted the initial level of overnight improvement in schizophrenia, and their product explained 77% of the variance, suggesting that both sleep stages are necessary for consolidation (Manoach et al. 2010).

However, the direct effect of antipsychotic medication is controversial. First-generation or typical antipsychotics are associated with increased total sleep time, increased sleep continuity, and REM latency, but the reported effects on other sleep stages are variable (Cohrs 2008). However, these effects are not consistent in subsequent studies. In contrast, studies of the effects of atypical antipsychotics (clozapine, olanzapine, quetiapine, risperidone, ziprasidone, and paliperidone) in healthy controls and schizophrenia patients show a consistent effect on measures of sleep continuity, with an increased total sleep time and/or sleep efficiency and individually varying effects on other sleep parameters, such as an increase in REM

Table 13.1 Effects of antidepressants on sleep

Drug	Sleep architecture disruption	Sleep-wave state	REM latency	REM sleep	Mechanism of action
Tricyclic antidepressants	Increased	Increase	Increase	Decrease	Serotonin
Sedative-like					Norepinephrine
Tricyclic	Decrease	Decrease	Increase	Decrease	Serotonin
Antidepressants					Norepinephrine
Stimulating					
Mono-amine Oxidase Inhibitors (MAOI)	Variable	Variable	Increase	Decrease	Monoamine oxidase
SSRI	Variable	Variable	Increase	Decrease	Serotonin
e.g. Fluoxetine					
SNRI	Decrease	Variable	Increase	Decrease	Serotonin
Venlafaxine					Norepinephrine
Bupropion	Variable	Variable	Variable	Variable	Norepinephrine
					Dopamine
Other	Increase	Increase	None	None	Norepinephrine
Antidepressants					Dopamine

latency. Newer antipsychotics like clozapine most frequently induce EEG slowing and epileptiform discharges (Jackson and Seneviratne 2019).

Schizophrenia also affects the circadian rhythm. A recent study included prolonged actigraphy and melatonin profiles in 20 schizophrenia patients compared to 21 unemployed controls and found marked circadian rhythm abnormalities among patients with schizophrenia, including both delayed sleep-phase disorder and free-running, the non-24-h circadian rhythm in 50% of those studied, with no relationship to the neuroleptic used (Wulff et al. 2012).

Effect of drugs on REM Sleep (Table 13.1): In order to effectively treat depression, it is crucial to comprehend the impact of antidepressants on sleep patterns. Certain types of antidepressant medications have the potential to exacerbate or trigger primary sleep disorders such as restless legs syndrome, sleep bruxism, REM sleep behavior disorder, nightmares, and sleep apnea. Patients with MDD show abnormalities in sleep parameters. MDD has been reported to be associated with disrupted sleep architecture, presents as a lengthened period before falling asleep, heightened frequency and duration of waking up during sleep, and is characterized by increased time awake after initially falling asleep. (WASO) time, decreased sleep efficiency, and early morning awakenings. Early morning awakening, together with the altered distribution of REM sleep, is considered a biological marker of circadian rhythm disturbances in depression and is a characteristic biological marker of depression with melancholic features (Wichniak et al. 2013).

Antidepressant drugs may deteriorate sleep quality due to the interaction of the receptors with serotonin and norepinephrine reuptake inhibitors (SNRI), norepinephrine reuptake inhibitors (NRI), monoamine oxidase inhibitors (MAOI), selective serotonin reuptake inhibitors (SSRI), and activating tricyclic antidepressants.

Tricyclic antidepressants. Most of them increase total sleep time and decrease wake time after sleep onset, while many selective serotonin reuptake inhibitors have the opposite effect. Withdrawal from antidepressants may cause nightmares and intensify dreaming (Wilson and Argyropoulos 2005).

Phenelzine, an MAOI drug, can safely eliminate REM sleep in depressed patients without altering the duration of visually scored slow-wave sleep (Landolt et al. 2001). Tolerance to the anticholinergic/sedating effects develops rapidly (usually within a few days), while the REM-suppressing effects may persist. A REM sleep rebound and sleep disturbance can be experienced with the abrupt discontinuation of REM sleep-suppressing antidepressants.

In normal subjects receiving lithium for 2 weeks, total sleep time did not vary, REM sleep decreased, and REM sleep latency increased. In depressed subjects, either on short-term therapy or long-term therapy with lithium, sleep stages 3 and 4 increased, REM sleep decreased, REM latency increased, and REM activity/time spent asleep (an index of REM intensity per minute of sleep) decreased (Billiard 1987).

Agomelatine is a nonsedative antidepressant drug exerting agonistic action at melatonergic M1 and M2 receptors, and antagonistic action at serotonergic 5-HT2c receptors. It should be considered an alternative drug of choice in treating depressed patients with marked insomnia symptoms. In comparison to escitalopram, agomelatine improves sleep latency after both short (after 2 weeks) and long (after 24 weeks) treatment. Both drugs differ significantly in their effect on sleep continuity. In the second week, agomelatine slightly improves sleep continuity (increased total sleep time and sleep efficiency), and escitalopram worsens it, and this effect is not related to the suppression of REM sleep. While agomelatine has not received approval for marketing from the U.S. Food and Drug Administration (FDA) in the United States, it is available in other countries (Quera-Salva et al. 2011).

Central nervous system stimulants include amphetamines and related compounds such as methylphenidate, modafinil, and pemoline. Xanthene derivatives such as caffeine and theophylline are also considered CNS stimulants. As a rule, stimulants increase wakefulness, decrease total sleep time and SWS, and increase sleep latency. All of them uniformly decrease sleep continuity, SWS, and REM sleep, and increase sleep latency (Sharma and Prasad 2015).

Opioids: In general, morphine decreases the number and duration of REM periods, delays the onset of the first REM period, and increases the time between REM period onsets. Many studies have shown that acute use of various opioids results in increased REM latency, decreased REM sleep time, increased stage 1 and stage 2 sleep, and decreased SWS. The short-term use of opioids also results in a longer time to fall asleep, heightened wakefulness after initially falling asleep, and a corresponding decrease in both the total duration of sleep and the efficiency of sleep (Kay et al. 1969).

Orexin receptor antagonists: The neuropeptides orexin A and orexin B (aka. hypocretin1 and hypocretin 2) are thought to have a key role in the regulation of sleep and wake states like narcolepsy. At least three studies administered selective orexin receptor-2 antagonists (2-SORA) in either a healthy control (one study) or a clinical sample (two studies). Their results show that DORAs (dual orexin receptor antagonists) increase total sleep time primarily by promoting REM sleep without affecting or even decreasing non-REM sleep (Clark et al. 2020). Studies in other species, like rats, demonstrated that intracerebroventricular injection of OX2R antagonist (TCS-OX2–29) increased total sleep time and REM sleep. The increase in REM sleep was due to an increase in the number of REM episodes (Kummangal et al. 2013).

Antiseizure drugs: Older antiseizure drugs, like *phenobarbital*, *carbamazepine*, and *phenytoin*, inhibit CNS activity, but the exact inhibition mechanisms are not known. The older antiseizure drugs have varying effects on sleep architecture. Placidi et al. study showed that the acute administration of CBZ-CR (carbamazepine) increased the number of stage shifts, reduced REM sleep, and increased REM sleep fragmentation. In the TLE (Temporal Lobe Epilepsy) group, these effects were reversed entirely after chronic treatment. LTG increased REM sleep, reduced the number of entries into REM sleep, decreased the number of phase shifts, and decreased the percentage of SWS (Placidi et al. 2000c). Gabapentin (GBP) increased REM sleep percentage, increased the mean duration of REM periods, reduced the number of awakenings, and reduced stage 1 sleep percentage. Moreover, it has been demonstrated that CBZ-CR disrupts REM sleep, but only during acute administration (Placidi et al. 2000a).

By attaching to the GABAA receptor and facilitating the entry of chloride ions, PBT (phenobarbitone) promotes a decreased frequency of awakenings and a shorter time to fall asleep. Additionally, it prolongs the duration of non-rapid eye movement (NREM) sleep while diminishing the length of rapid eye movement (REM) sleep. While phenytoin increases stage N1 sleep and decreases stage N3 sleep (Shvarts and Chung 2013).

Levetiracetam (LEV): is a new generation AED used to treat both partial and generalized epilepsy. In their study, Cicolin et al. conducted polysomnography and Multiple Sleep Latency Tests on 14 healthy adults who received a maximum dosage of 2000 mg/day of LEV. The results showed a notable rise in the duration of all non-rapid eye movement (NREM) sleep stages, accompanied by a relative reduction in the duration of rapid eye movement (REM) sleep. Overall, LEV exhibited a tendency to enhance the consolidation of sleep without adversely affecting daytime alertness (Cicolin et al. 2006).

Gabapentin (GBP) is an amino acid initially developed as a structural analog of GABA, an inhibitory neurotransmitter in the adult brain. Polysomnography findings revealed a notable increase in the proportion of REM sleep, along with an extended average duration of REM sleep, a decrease in the frequency of arousals, as well as a shorter duration of the initial stage of sleep (N1). Placidi et al. conducted a study to assess 18 patients with refractory partial seizures who were receiving stable treatment with a daily dosage of 1800 to 2400 mg of GBP for 4 months. The results showed a notable rise in the proportion of REM sleep and slow-wave sleep (SWS), accompanied by a decrease in the number of awakenings and a reduction in the duration of stage 1 sleep. Additionally, 16 patients reported experiencing enhanced sleep quality during the study period (Placidi et al. 2000b).

Melatonin and melatonin receptor agonists: Ramelteon acts as an activator of melatonin receptors, specifically targeting the MT1 and MT2 receptors. Inhibiting the wakefulness-promoting activity of the suprachiasmatic nucleus. REM sleep expression is strongly circadian modulated, and the impact of REM sleep on primary brain functions, metabolic processes, and immune system function is significant. Melatonin increases REM sleep percentage and causes improvements in the subjective measures of daytime dysfunction as well as clinical global impression score. Exogenous administration of Melatonin normalizes circadian variation in human physiology and may be helpful in elderly and shift workers (Kunz et al. 2004).

Beta-blockers: Beta-blockers are classified as lipophilic or hydrophilic. The lipophilic agents include propranolol, metoprolol, and pindolol, while the hydrophilic agents include atenolol and sotalol. Hydrophilic agents do not cross the blood-brain barrier readily and, thus, have no sleep effects. Both lipophilic and hydrophilic beta-blockers may negatively affect sleep architecture by suppressing REM sleep (Rosen and Kostis 1985).

Benzodiazepines and nonbenzodiazepine receptor agonists: Benzodiazepines such as diazepam, lorazepam, midazolam, and nonbenzodiazepine receptor agonists such as zolpidem, zaleplon, eszopiclone are commonly prescribed sedative-hypnotic medications used to treat sleep disorders. Finally, benzodiazepines and nonbenzodiazepine receptor agonists may modestly reduce REM sleep when used at higher than indicated doses.

13.7 Steroid and Sleep Disorders

Steroids are commonly used, and in the era of COVID-19, steroids have been extensively used in the ICU in patients with respiratory failure. Other common indications for steroids in the ICU include anaphylaxis, airway edema, septic shock, asthma and COPD exacerbations, *pneumocystis* pneumonia, adrenal crisis, antiemetic treatment, elevated intracranial pressure from tumors, autoimmune disorders, and stress doses needed for chronic steroid users before invasive procedures.

Steroid-induced sleep disruption/psychosis: The administration of external steroids is believed to induce disturbances in sleep, which can lead to additional psychostimulant effects like changes in mood, nervousness, the onset of psychoses, and delirium. There are three pathways responsible for the neurocognitive adverse events (AEs) of steroids: behavior changes through modification of the HPA axis, changes in natural sleep-wake cycles, and hyperarousal caused by modification in neuroinhibitory pathways. In a randomized prospective crossover study involving healthy participants, the administration of dexamethasone (3 mg every 8 h for 48 h) led to notable alterations in sleep patterns as assessed through polysomnography.

Compared with placebo, steroid treatment showed a significantly longer percentage (SD) of stage 0/awake times (11.7% [11.4] vs. 2.9% [1.8]), a longer percentage (SD) of REM sleep latency (363.8 [74.5] vs. 202.8 [79.6] min; $P < 0.01$), and a reduced number (SD) of REM periods (3.8 [2.6] vs. 9.7 [3.6]; $P < 0.01$) (Devlin et al. 2018).

Mitigating steroid-induced effects: Although steroid therapy often cannot be altered in the critically ill population, research showed that steroid overuse is common in the ICU setting. Minimizing dosage and duration are important ways via which clinicians can mitigate unwanted effects. CNS AEs associated with steroids often can be reversed once therapy is discontinued. Avoiding split-dose administration has been proposed, given the natural diurnal production of cortisol. A review by Flaherty discusses the importance of avoiding pharmacologic agents in hospitalized older patients if possible due to known risks (falls, dependency, hip fractures, rebound insomnia, and risk of delirium) (Bourne and Mills 2004; Cole 2020).

13.8 Conclusion

The increased severity of OSA during REM sleep contributes to an increased risk for and severity of endothelial dysfunction, cardiovascular morbidity, and metabolic disorders (Budhiraja et al. 2007; Budhiraja and Quan 2005). Dean et al. conducted a cross-sectional analysis utilizing data from 2040 participants who participated in polysomnography as part of the multiethnic study of atherosclerosis (MESA) during the MESA Exam. They showed that the arousal index in REM and central apnea indices were only higher associated with elevated DBP (Dean et al. 2015). What is clinically significant REM OSA? At what time does REM time need to be considered in clinical evaluation? Mokhlesi et al. argue that a minimum REM sleep time of 30 min would be required to reduce the possibility of overestimating the impact of REM OSA in individuals with a limited duration of REM sleep. The presence of REM OSA is linked to hypertension both in a cross-sectional and longitudinal manner. This has significant clinical implications as the treatment of OSA typically focuses on the first half of the sleep period, leaving a substantial portion of REM sleep unaddressed (Mokhlesi et al. 2014).

The potential neurological implications of OSA were not recognized until polysomnography was used to evaluate patients. It is important to consider OSA as a possibility in older individuals displaying common symptoms of sleep apnea during both nighttime and daytime. Polysomnography (PSG) is an essential diagnostic tool for identifying OSA in elderly patients. In a recent study by Chen et al., it was found that vertigo patients have a higher incidence of REM-related OSA, and compliance with continuous positive airway pressure (CPAP) therapy is low in this population. This may have practical implications in the management of elderly patients with REM-related OSA associated with hypertension (Chen et al. 2021). Further research is needed to fully understand the mechanisms and clinical implications of REM sleep disorders, particularly in the elderly population, as highlighted by Iannella G et al. (2020) (Iannella G et al. 2020).

Key Points/Impact of REM Sleep

- The terms "REM-predominant," "REM-related," "REM-isolated," or "REM-only" OSA have been used for OSA, which occurs predominantly or exclusively during REM sleep (Mokhlesi 2012).
- Classically, research studies employ a ratio of REM apnoea-hypopnea index (REM-AHI)/NREMAHI of more than 2 as a working definition of REM-OSA (Almeneessier et al. 2017).
- Some studies added additional criteria of a total AHI minimum (generally 5/h) and an NREM-AHI maximum (typically between 8 and 15/h).
- Some studies have used 30 min of REM sleep as the minimum time required to make a practical remark on REM-OSA.
- REM-OSA is more prevalent among women; the prevalence decreases with increasing age in both sexes, such that the odds of having REM-OSA decreases by 26.7% per decade (Koo et al. 2008).
- OSA is associated with a higher cardiovascular risk, particularly hypertension. REM sleep is characterized by increased sympathetic activity, decreased vagal tone, and heightened cardiovascular instability compared to NREM sleep (Almeneessier et al. 2020).
- Elevated REM-AHI has been independently associated with poor glycemic control and insulin resistance.
- As REM sleep increases in the early morning hours, using CPAP for only 4 h or less is insufficient. Patients generally start using CPAP at bedtime; hence, if the mean duration of CPAP use is approximately 4 h, it is likely that a significant proportion of early morning REM-OSA events remain untreated. This may partially explain the negative trials regarding the effects of CPAP on cardiovascular outcomes (Grimaldi et al. 2014).
- As adherence to CPAP therapy in patients with REM-OSA is low and the risk of not treating REM-OSA is high; it is critical to improve patients' and practitioners' awareness of the importance of adherence to CPAP therapy in REM-OSA and the potential benefits of using CPAP during the entire sleep period.

References

Almeneessier AS, Almousa Y, Hammad O et al (2017) Long-term adherence to continuous positive airway pressure in patients with rapid eye movement-only obstructive sleep apnea: a prospective cohort study. J Thorac Dis 9:3755–3765

Almeneessier AS, Alshahrani M, Aleissi S et al (2020) Comparison between blood pressure during obstructive respiratory events in REM and NREM sleep using pulse transit time. Sci Rep 10: 3342

American Academy of Sleep Medicine (2014) International classification of asleep disorders, 3rd edn. American Academy of Sleep Medicine, Darien

American Academy of Sleep Medicine (n.d.) The AASM manual for the scoring of sleep and associated events. Version 2.6. American Academy of Sleep Medicine, Darien

American Psychiatric Association (2013) Diagnostic and statistical manual of mental disorders, 5th edn. American Psychiatric Association, Arlington

Anderson KN, Bradley AJ (2013) Sleep disturbance in mental health problems and neurodegenerative disease. Nat Sci Sleep 5:61–75. https://doi.org/10.2147/NSS.S34842

Aurora RN, Zak RS, Maganti RK, Auerbach SH, Casey KR, Chowdhuri S, Karippot A, Ramar K, Kristo DA, Morgenthaler TI (2010) Best practice guide for the treatment of rem sleep behavior disorder (RBD). J Clin Sleep Med 6(1):85–95

Aurora RN, Crainiceanu C, Gottlieb DJ, Kim JS, Punjabi NM (2018) Obstructive sleep apnea during REM sleep and cardiovascular disease. Am J Respir Crit Care Med 197(5):653–660. https://doi.org/10.1164/rccm.201706-1112OC

BaHammam AS, Almeneessier AS (2019) Dreams and nightmares in patients with obstructive sleep apnea: a review. Front Neurol 10:1127. https://doi.org/10.3389/fneur.2019.01127

BaHammam AS, Almeneessier AS, Pandi-Perumal SR (2018) Medieval Islamic scholarship and writings on sleep and dreams. Ann Thorac Med 13:72–75

BaHammam AS, Alnakshabandi K, Pandi-Perumal SR (2020) Neuropsychiatric correlates of narcolepsy. Curr Psychiatry Rep 22(8):36

Billiard M (1987) Lithium carbonate: effects on sleep patterns of normal and depressed subjects and its use in sleep-wake pathology. Pharmacopsychiatry 20(5):195–196. https://doi.org/10.1055/s-2007-1017102

Bourne RS, Mills GH (2004) Sleep disruption in critically ill patients--pharmacological considerations. Anaesthesia 59(4):374–384. https://doi.org/10.1111/j.1365-2044.2004.03664.x

Budhiraja R, Quan SF (2005) Sleep-disordered breathing and cardiovascular health. Curr Opin Pulm Med 11(6):501–506. https://doi.org/10.1097/01.mcp.0000183058.52924.70

Budhiraja R, Parthasarathy S, Quan SF (2007) Endothelial dysfunction in obstructive sleep apnea. J Clin Sleep Med 3(4):409–415. https://doi.org/10.5664/jcsm.26864

Chen P-Y, Chen T-Y, Chao P-Z, Liu W-T, Bai C-H, Tsao S-T et al (2021) REM-related obstructive sleep apnea and vertigo: a retrospective case-control study. PLoS One 16(6):e0252844. https://doi.org/10.1371/journal.pone.0252844

Cicolin A, Magliola U, Giordano A, Terreni A, Bucca C, Mutani R (2006) Effects of levetiracetam on nocturnal sleep and daytime vigilance in healthy volunteers. Epilepsia 47(1):82–85

Cipolli C, Franceschini C, Mattarozzi K, Mazzetti M, Plazzi G (2011) Overnight distribution and motor characteristics of REM sleep behaviour disorder episodes in patients with narcolepsy-cataplexy. Sleep Med 12(7):635–640

Clark JW, Brian ML, Drummond SPA, Hoyer D, Jacobson LH (2020) Effects of orexin receptor antagonism on human sleep architecture: a systematic review. Sleep Med Rev 53:101332. https://doi.org/10.1016/j.smrv.2020.101332

Cohrs S (2008) Sleep disturbances in patients with schizophrenia: impact and effect of antipsychotics. CNS Drugs 22(11):939–962

Cole JL (2020) Steroid-induced sleep disturbance and delirium: a focused review for critically ill patients. Fed Pract 37(6):260–267

Comella CL, Nardine TM, Diederich NJ, Stebbins GT (1998) Sleep-related violence, injury, and REM sleep behavior disorder in Parkinson's disease. Neurology 51:526–529

Conwell W, Patel B, Doeing D, Pamidi S, Knutson KL, Ghods F, Mokhlesi B (2012) Prevalence, clinical features, and CPAP adherence in REM-related sleep-disordered breathing: a cross-sectional analysis of a large clinical population. Sleep Breath 16:519–526

Datta S, Knapp CM, Koul-Tiwari R, Barnes A (2015) The homeostatic regulation of REM sleep: a role for localized expression of brain-derived neurotrophic factor in the brainstem. Behav Brain Res 292:381–392. https://doi.org/10.1016/j.bbr.2015.06.038

Dauvilliers Y, Schenck CH, Postuma RB, Iranzo A, Luppi PH, Plazzi G, Montplaisir J, Boeve B (2018) REM sleep behaviour disorder. Nat Rev Dis Primers 4(1):19. https://doi.org/10.1038/s41572-018-0016-5

Dean DA, Wang R, Jacobs DR et al (2015) A systematic assessment of the association of polysomnographic indices with blood pressure: the Multi-Ethnic Study of Atherosclerosis (MESA). Sleep 38(4):587–596. https://doi.org/10.5665/sleep.4576

Dement W (1955) Dream recall and eye movements during sleep in schizophrenics and normals. J Nerv Ment Dis 122:263–269

Devlin JW, Skrobik Y, Gélinas C, Needham DM, Slooter AJC, Pandharipande PP, Watson PL, Weinhouse GL, Nunnally ME, Rochwerg B, Balas MC, van den Boogaard M, Bosma KJ, Brummel NE, Chanques G, Denehy L, Drouot X, Fraser GL, Harris JE, Joffe AM, Kho ME, Kress JP, Lanphere JA, McKinley S, Neufeld KJ, Pisani MA, Payen JF, Pun BT, Puntillo KA, Riker RR, Robinson BRH, Shehabi Y, Szumita PM, Winkelman C, Centofanti JE, Price C, Nikayin S, Misak CJ, Flood PD, Kiedrowski K, Alhazzani W (2018) Clinical practice guidelines for the prevention and management of pain, agitation/sedation, delirium, immobility, and sleep disruption in adult patients in the ICU. Crit Care Med 46(9):e825–e873. https://doi.org/10.1097/CCM.0000000000003299

España RA, Scammell TE (2011) Sleep neurobiology from a clinical perspective. Sleep 34(7): 845–858. https://doi.org/10.5665/SLEEP.1112

Ferri R, Rundo F, Silvani A, Zucconi M, Bruni O, Ferini-Strambi L, Plazzi G, Manconi M (2017) REM sleep EEG instability in REM sleep behavior disorder and clonazepam effects. Sleep 40(8):zsx080

Fraigne JJ, Torontali ZA, Snow MB, Peever JH (2015) REM sleep at its core - circuits, neurotransmitters, and pathophysiology. Front Neurol 6:123

Giardino DL, Fasano P, Garay A (2021) The "respiratory REM sleep without atonia benefit" on coexisting REM sleep behavior disorder - obstructive sleep apnea. Sleep Sci 14(2):181–185. https://doi.org/10.5935/1984-0063.20200054

Goh DY, Galster P, Marcus CL (2000) Sleep architecture and respiratory disturbances in children with obstructive sleep apnea. Am J Respir Crit Care Med 162:682–686

Grimaldi D, Beccuti G, Touma C et al (2014) Association of obstructive sleep apnea in rapid eye movement sleep with reduced glycemic control in type 2 diabetes: therapeutic implications. Diabetes Care 37:355–363

Iannella G, Maniaci A, Magliulo G, Cocuzza S, La Mantia I et al (2020) Current challenges in the diagnosis and treatment of obstructive sleep apnea syndrome in the elderly. Pol Arch Intern Med 130(7–8):649–654

Iranzo A, Fernández-Arcos A, Tolosa E, Serradell M, Molinuevo JL, Valldeoriola F et al (2014) Neurodegenerative disorder risk in idiopathic REM sleep behavior disorder: study in 174 patients. PLoS One 9(2):e89741. https://doi.org/10.1371/journal.pone.0089741

Jackson A, Seneviratne U (2019) EEG changes in patients on antipsychotic therapy: a systematic review. Epilepsy Behav 95:1–9. https://doi.org/10.1016/j.yebeh.2019.02.005

Kay DC, Eisenstein RB, Jasinski DR (1969) Morphine effects on human REM state, waking state and NREM sleep. Psychopharmacologia 14:404–416. https://doi.org/10.1007/BF00403581

Kempenaers C, Kerkhofs M, Linkowski P, Mendlewicz J (1988) Sleep EEG variables in young schizophrenic and depressive patients. Biol Psychiatry 24(7):833–838

Keshavan MS, Tandon R (1993) Sleep abnormalities in schizophrenia: pathophysiological significance. Psychol Med 23(4):831–835

Koo BB, Patel SR, Strohl K, Hoffstein V (2008) Rapid eye movement-related sleepdisordered breathing: influence of age and gender. Chest 134:1156–1161

Kummangal BA, Kumar D, Mallick HN (2013) Intracerebroventricular injection of orexin-2 receptor antagonist promotes REM sleep. Behav Brain Res 237:59–62. https://doi.org/10.1016/j.bbr.2012.09.015

Kunz D, Mahlberg R, Müller C, Tilmann A, Bes F (2004) Melatonin in patients with reduced REM sleep duration: two randomized controlled trials. J Clin Endocrinol Metabol 89(1):128–134. https://doi.org/10.1210/jc.2002-021057

Landolt H, Raimo EB, Schnierow BJ, Kelsoe JR, Rapaport MH, Gillin JC (2001) Sleep and sleep electroencephalogram in depressed patients treated with phenelzine. Arch Gen Psychiatry 58(3): 268–276. https://doi.org/10.1001/archpsyc.58.3.268

Malik J, Lo YL, Wu HT (2018) Sleep-wake classification via quantifying heart rate variability by convolutional neural network. Physiol Meas 39(8):085004

Manoach DS, Thakkar KN, Stroynowski E et al (2010) Reduced overnight consolidation of procedural learning in chronic medicated schizophrenia is related to specific sleep stages. J Psychiatr Res 44(2):112–120. https://doi.org/10.1016/j.jpsychires.2009.06.011

Memar P, Faradji F (2018) A novel multi-class EEG-based sleep stage classification system. IEEE Trans Neural Syst Rehabil Eng 26(1):84–95

Mokhlesi B (2012) REM-related obstructive sleep apnea: to treat or not to treat? J Clin Sleep Med 8(3):249–250. https://doi.org/10.5664/jcsm

Mokhlesi B, Finn LA, Hagen EW, Young T, Hla KM, Van Cauter E, Peppard PE (2014) Obstructive sleep apnea during REM sleep and hypertension: results of the Wisconsin sleep cohort. Am J Respir Crit Care Med 190(10):1158–1167. https://doi.org/10.1164/rccm.201406-1136OC

Nofzinger EA, Schwartz RM, Reynolds CF 3rd et al (1994) Affect intensity and phasic REM sleep in depressed men before and after treatment with cognitive-behavioral therapy. J Consult Clin Psychol 62(1):83–91

Oksenberg A, Arons E, Nasser K, Vander T, Radwan H (2010) REM-related obstructive sleep apnea: the effect of body position. J Clin Sleep Med 6(4):343–348

Olson EJ, Boeve BF, Silber MH (2000) Rapid eye movement sleep behaviour disorder: demographic, clinical and laboratory findings in 93 cases. Brain 123:331–339

Pirzada AR, BaHammam AS (2021) Rapid eye movement predominant obstructive sleep apnoea: prognostic relevance and clinical approach. Curr Opin Pulm Med 27(6):514–522

Placidi F, Diomedi M, Scalise A, Marciani MG, Romigi A, Gigli GL (2000a) Effect of anticonvulsants on nocturnal sleep-in epilepsy. Neurology 54(5 Suppl. 1):S25–S32

Placidi F, Mattia D, Romigi A, Bassetti MA, Spanedda F, Marciani MG (2000b) Gabapentin-induced modulation of interictal epileptiform activity related to different vigilance levels. Clin Neurophysiol 111(9):1637–1642

Placidi F, Marciani MG, Diomedi M, Scalise A, Pauri F, Giacomini P, Gigli GL (2000c) Effects of lamotrigine on nocturnal sleep, daytime somnolence and cognitive functions in focal epilepsy. Acta Neurol Scand 102(2):81–86. https://doi.org/10.1034/j.1600-0404.2000.102002081.x

Quera-Salva MA, Hajak G, Philip P, Montplaisir J, Keufer-Le GS, Laredo J, Guilleminault C (2011) Comparison of agomelatine and escitalopram on nighttime sleep and daytime condition and efficacy in major depressive disorder patients. Int Clin Psychopharmacol 26:252–262. https://doi.org/10.1097/YIC.0b013e328349b117

Riemann D, Berger M, Voderholzer U (2001) Sleep and depression – results from psychobiological studies: an overview. Biol Psychol 57(1–3):67–10

Rosen RC, Kostis JB (1985) Biobehavioral sequelae associated with adrenergic-inhibiting antihypertensive agents: a critical review. Health Psychol 4:579

Schenck CH (2016) Expanded insights into idiopathic REM sleep behavior disorder. Sleep 39(1): 7–9

Sharma S, Prasad A (2015) Drug induced REM disorder. J Sleep Med Disord 2(3):1021

Shrivastava D, Jung S, Saadat M, Sirohi R, Crewson K (2014) How to interpret the results of a sleep study. J Community Hosp Intern Med Perspect 4(5):24983. https://doi.org/10.3402/jchimp.v4.24983

Shvarts V, Chung S (2013) Epilepsy, antiseizure therapy, and sleep cycle parameters. Epilepsy Res Treat 2013:670682. https://doi.org/10.1155/2013/670682

Su CS, Liu KT, Panjapornpon K, Andrews N, Foldvary-Schaefer N (2012) Functional outcomes in patients with REM-related obstructive sleep apnea treated with positive airway pressure therapy. J Clin Sleep Med 8(3):243–247. https://doi.org/10.5664/jcsm.1902

Thase ME, Fasiczka AL, Berman SR, Simons AD, Reynolds CF 3rd. (1998) Electroencephalographic sleep profiles before and after cognitive behavior therapy of depression. Arch Gen Psychiatry 55(2):138–144
Wichniak A, Wierzbicka A, Jernajczyk W (2013) Sleep as a biomarker for depression. Int Rev Psychiatry 25(5):632–645. https://doi.org/10.3109/09540261.2013.812067
Wilson S, Argyropoulos S (2005) Antidepressants and sleep: a qualitative review of the literature. Drugs 65(7):927–947
Wulff K, Dijk DJ, Middleton B, Foster RG, Joyce EM (2012) Sleep and circadian rhythm disruption in schizophrenia. Br J Psychiatry 200(4):308–316

Chapter 14
COVID Pandemic Impact on OSA and Other Sleep Disorders

Arup Haldar

Abstract The COVID-19 pandemic is the deadliest one in human history. The unprecedented changes it brought in human lifestyle, health, economy, business, science, and technology are themselves are subject of future research. These changes also affected sleep habits. Those patients having sleep disorders may have worsening sleep problems and many patients also have developed new sleep problems. Various sleep disorders have a deep interaction with COVID-19. Though sleep disorders were placed far lower in the ranks as important comorbidities in COVID, such observation may be related to lower diagnosis of sleep disorders. The interaction between "acute happy hypoxia" of COVID-19 and "chronic happy hypoxia" of OSA is a subject of major debate. Studies have found overwhelming similarities in chronic pathophysiological derangement in OSA and some acute exacerbation of those derangements in COVID. Numerous studies revealed the true prevalence of sleep disorders in COVID-19. COVID-19 resulted in increased incidences of insomnia and other non-OSA sleep disorders. Healthcare workers and children suffered from specific sleep disorders during the COVID pandemic and is a subject of future research. Sleep practice and CPAP adherence all changed significantly than pre COVID era. Another factor that can lead to an increased incidence of OSA during lockdown is an increased chance of getting obese due to various factors. They are a sedentary lifestyle, overeating, decreased outdoor physical activity, and increased alcohol use. This type of obesity is given name as 'covibesity' or COVID-obesity.

Healthcare workers faced tremendous mental and physical workloads during this pandemic. That took a toll on their sleep health. They faced extreme work pressure, fast adaptation to intense critical care situations, an unseen amount of severe critical patients, numerous deaths of patients, and risks of infection. A recent meta-analysis from France on 70 cross-sectional studies across the globe involving 101,017 healthcare participants concluded the pooled prevalence of anxiety was 30%, depression was 31%, and posttraumatic stress disorder (PTSD) was 31.4%. The pooled prevalence of sleep disorders was 44%, and this was mainly measured for

A. Haldar (✉)
Consultant Pulmonologist, Calcutta Medical Research Institute (CMRI) Hospital, Kolkata, West Bengal, India

A. S. BaHammam, M. Hunasikatti (eds.), *Sleep Apnea Frontiers*, Progress in Sleep Research, https://doi.org/10.1007/978-981-99-7901-1_14

insomnia. The present chapter deals with the impact of the pandemic on sleep disorders, with special emphasis on obstructive sleep apnea (OSA).

Keywords COVID-19 pandemic · Sleep disorders · Obstructive sleep apnea (OSA) · COVID-19 and sleep · Covibesity

14.1 Introduction

The emergence of the SARS-CoV-2 virus in December 2019 in Wuhan, China marked the onset of the COVID-19 pandemic. WHO declared it a public health emergency on January 30, 2020. This pandemic is the deadliest one in human history. The unprecedented changes it brought in human lifestyle, health, economy, business, science, and technology are themselves are subject of future research. The lockdown measure is an effective way of handling the pandemic but may be an added fee for daily wage earners. The anxiety about disease, hospitalization and death, the suffering from the disease itself with long-term complications, the depression from losing a family member, the anxiety about earning, and overall, the mood changes associated with prolonged confinement at home have affected the rest-work schedule and thus have a profound influence on sleep habits. Those patients having sleep disorders may have worsening sleep problems and many patients also have developed new sleep problems.

14.2 Sleep Disorders as Comorbidity in COVID

Sleep disorders are diseases like other diseases, such as hypertension and diabetes, and are important comorbidities in COVID patients. Chances of hospitalization or death in COVID increase with the presence of comorbidities. An observational study by the Centers for Disease Control (CDC) from April 2020 to March 2021 has thrown some light on the important comorbidities of COVID-19 (Kompaniyets et al. 2021). The study was conducted in 800 hospitals in the USA on 5,40,447 subjects above 18 years. The two important observations in the study were

1. The three most common underlying diseases associated with hospitalization are essential hypertension, dyslipidemia, and obesity.
2. The three most common underlying diseases associated with deaths are obesity, anxiety-related disorders, and uncontrolled diabetes.

Though sleep disorders were placed far lower in the ranks as important comorbidities in COVID, still such observation may be related to lower diagnoses of sleep disorders in general. Obesity is an important component of obstructive sleep apnea (OSA), and anxiety-related disorders contribute to poor sleep. So, sleep disorders are associated with the COVID pandemic in a multifaceted way. The present chapter deals with the impact of the pandemic on sleep disorders, with special emphasis on obstructive sleep apnea (OSA).

14.3 OSA and COVID-19

OSA is a disease that manifests with snoring, excessive daytime sleepiness, and fatigue. In this disease, there are repeated upper airway collapses or closures during sleep, resulting in a drop in oxygen saturation, frequent arousals and repeated sympathetic stimulation, loss of deep stages of sleep, swing in arterial blood gas, and changes in intrathoracic pressures. All these lead to a pro-inflammatory state and invite many diseases like hypertension, cardiovascular disorders, neurological diseases, diabetes, metabolic syndrome, and obesity. The prevalence of clinically significant OSA, requiring an urgent intervention, was 6–17% in the adult population, whereas the prevalence was as high as 49% in the advanced age group (Reddy et al. 2009; Tufik et al. 2010; Heinzer et al. 2015). If the mild OSA groups are also included in the study, the prevalence rate increases further. Two studies have shown a prevalence rate of OSA as 88% in men aged 65–69 years and 90% in men aged 60–85 years (Heinzer et al. 2015; Lee et al. 2014). The prevalence of OSA is always more in adult men than in adult women.

Expressions of COVID-19 among different people are varied. They may be completely asymptomatic in a few people but may be severe in others. The underlying host response determines the outcome of the disease. The host response depends on multiple factors. But four main determinants are the following-

1. Renin–angiotensin–aldosterone system (RAAS)
2. Oxidative stress of the individual
3. Endothelial dysfunction
4. Immune responses

14.4 The Pathophysiological Linkage Between COVID-19 and OSA

All of these four systems are in a deranged state in OSA patients; they are in hyperactive states due to intermittent hypoxia, sympathetic activation, and poor sleep quality. So, any acute insult like COVID-19 may throw these systems out of control (Halder 2021).

14.4.1 Renin–Angiotensin–Aldosterone System(RAAS)

Renin–angiotensin–aldosterone system(RAAS) is one of the prime regulators of cardiovascular and various other functions (Ekiz et al. 2020). Dysregulation of the renin–angiotensin–aldosterone system (RAAS) is implicated in the development of various disorders including hypertension, myocardial infarction, heart failure, diabetes, and inflammatory lung disease. Renin acts on angiotensinogen and converts it to Angiotensinogen I. Angiotensinogen converting enzyme (ACE) converts Angiotensin I to Angiotensin II. Angiotensin II is a potent vasoconstrictor and pro-inflammatory (Fraga-Silva et al. 2010). But this effect is counterbalanced by another enzyme called Angiotensin Converting Enzyme 2 or ACE2. ACE2 also acts on Angiotensin I and converts into Angiotensin (1–7), which is a vasodilator and anti-inflammatory. So ACE2 counters the activity of ACE by reducing the amount of Angiotensin II and increasing Angiotensin (1–7), acting as a negative regulator of RAAS. So ACE2 exhibits a protective action on the cardiovascular and respiratory systems. So, any downregulation of ACE2 translates into more Angiotensin II.

The classical pathway is activated by OSA. It produces more Angiotensin II, which is pro-inflammatory. It causes vasoconstriction, vascular remodeling, and a prothrombotic state. This is often manifested as hypertension and cardiovascular disorders. The following evidence supports this idea:

1. In a meta-analysis, OSA was associated with higher levels of Angiotensin II and aldosterone, particularly in patients with hypertension (Jin and Wei 2016)
2. In another study, increased angiotensin converting enzyme (ACE) activity was highlighted in untreated OSA patients regardless of hypertension (Barceló et al. 2001)
3. Further, the severity of nocturnal hypoxemia in OSA augments RAAS activity (Jin and Wei 2016; Barceló et al. 2001)
4. Obesity is a significant comorbidity in OSA and affects RAAS activity (McSharry and Malhotra 2020)

Similarly, in COVID-19, the RAAS pathway is activated. After the initial engagement of COVID-19 spike protein with the ACE2 receptor, there is subsequent downregulation of ACE2 abundance on cell surfaces (Vaduganathan et al. 2020). Continued viral infection and replication contribute to reduced membrane ACE2 expression. This leads to increased activity of the vasoconstrictor and proliferative axis (Angiotensin II/ACE) and downregulation of the protective axis (ACE2/Angiotensin 1–17). This, in turn, is associated with higher risks of acute thrombosis, destabilization of atherosclerotic plaques, enhanced platelet activity, and coagulation.

14.4.2 Oxidative Stress

Oxidative stress is the body's defensive mechanism that is meant to kill the non-self-protein. It is mediated by superoxide radicals, also known as reactive oxygen species (ROS). But overwhelmed ROS can also damage any cells containing DNA or RNA. So, it is kept in check with the simultaneous action of two systems—the damaging pro-oxidant (PO) system and the balancing antioxidant system (AO). Various studies have shown that oxidative stress increase in OSA. This is due to multiple reasons.

1. **Enhanced release of superoxide from leukocytes:** Neutrophils and monocytes present in the bloodstream of individuals with obstructive sleep apnea (OSA) demonstrated significantly increased release of superoxide radical anions in laboratory conditions. This is mainly due to the activation of NADPH oxidase (NOX). This is a major superoxide-generating enzyme, and its expression is unregulated in OSA.
2. **Reduced bioavailability of nitric oxide (NO):** Through various pathways, oxidative stress can reduce the NO bioavailability in OSA, and this is the cause of endothelial dysfunction (described later).
3. **Increased lipids, proteins, and DNA oxidation**: The excessively generated ROS may lead to increased oxidation of biological compounds such as lipids, proteins, and DNA. This is often the cause of atherosclerosis (Bauters et al. 2016).
4. **Reduced antioxidant capacity:** An Antioxidant system (AO) can counteract the effects of oxidative stress. But this AO system is impaired in untreated OSA (Vicente et al. 2016). Lower levels of superoxide dismutase (SOD) have been described in OSA (DeMartino et al. 2016), a key antioxidant enzyme.

In COVID, oxidative stress occurs through the following pathway: Both the viral infection and RAS activation produce ROS leading to oxidative burst. After viral entry, toll-like receptors activate macrophages, and there is secretion of TNF alpha, which in turn activates NADPH oxidase. NADPH oxidase is the main enzyme that stimulates the production of ROS. The resulting ROS targets the virus. But NADPH being overused by inflammation also brings certain challenges to the AO system (Delgado-Rochea and Mestab 2020).

14.4.3 Endothelial Dysfunction

In COVID: The beginning of endothelial dysfunction in COVID is perhaps the trigger to complications. The main causes of endothelial dysfunction are reduced activity of the ACE2/Angiotensin (1–7) axis and enhanced activity of the ACE/Angiotensin II axis in oxidative stress. This leads to an imbalance between two systems of nitric oxide synthesis. It decreases the endothelial NO synthase (eNOS), which is anti-inflammatory but increases the inducible NO synthase (iNOS), which is pro-inflammatory and produces NO radicals. This imbalance between the AO-PO system results in the destruction of endothelial cells and a prothrombotic state. This causes blockage of small vessels and a hypoxic environment. In the hypoxic situation, there are ROS generation and hypoxia-inducible factor 1Alpha (HIF-1) activation (Delgado-Rochea and Mestab 2020). Consequently, it induces the expression of furic enzymes and further viral activation.

In OSA: Several studies have suggested impaired endothelial functions in OSA (Budhiraja et al. 2007; Ip et al. 2004). The three potential aetiologies include—hypoxemia, oxidative stress, and systemic inflammation. Endothelial injury results in the alteration of endothelial hormones responsible for maintaining vascular tone and preventing abnormal cell proliferation, increasing coagulopathy, and altered leukocyte trafficking. This also exposes subendothelial structures to diverse growth factors in the blood (Budhiraja et al. 2004). The resultant vasoconstriction, vascular smooth muscle proliferation, and hypercoagulability may lead to adverse cardiovascular consequences associated with OSA, such as hypertension, cardiovascular disorders, and stroke.

14.4.4 Immune Dysfunction, Cytokine Storm, and ARDS

In COVID-19 pneumonia, the aberrant release of pro-inflammatory cytokines leads to lung injury by damaging the epithelial and endothelial barriers, microvascular leakage, alveolar edema, and hypoxia. The antigen presenting cells (APCs), including macrophages, present COVID-19 antigens to T cells. T cells are activated to produce various cytokines via Th17 and others by a positive feedback loop (DeMartino et al. 2016). Also, during viral replication, the NF-k beta pathway is activated. The uncontrolled release of pro-inflammatory factors like IL-6, IL-8, IL-1Beta, GM-CSF, and chemokines such as CCL-2, CCL-5, and CCL-3, together with reactive oxygen species (ROS) cause ARDS leading to pulmonary fibrosis and death (Delgado-Rochea and Mestab 2020).

In OSA: The chances of ARDS and pulmonary complications are more in OSA. This was nicely depicted by one study which evaluated the chances of peri and postoperative pulmonary complications in surgical and orthopedic patients (Memtsoudis et al. 2011). They selected these groups to avoid or reduce the potential biases of medical diseases contributing to ARDS and other complications. The study was conducted from 1998 to 2007 and recruited 2,610,441 postoperative orthopedic patients and 3,441,262 surgical post-operative patients. Of those, 2.52% and 1.4% had OSA, respectively.

The following Table 14.1 shows the rate of postoperative complications among OSA and non-OSA patients.

So, from the above table (Memtsoudis et al. 2011), the researchers concluded that OSA is an independent risk factor for perioperative pulmonary complications. So, it is indirect evidence that OSA patients are at increased risk of pulmonary complications when confronted with any acute insults.

Table 14.1 The rate of postoperative complications among OSA and non-OSA patients

Pulmonary complications	Orthopedic with OSA (%)	Orthopedic without OSA (%)	Surgical with OSA (%)	Surgical without OSA (%)
Aspiration pneumonia	1.18	0.84	2.79	2.05
ARDS	1.06	0.45	3.79	2.44
Mechanical ventilation	3.99	0.79	10.8	5.94
Pulmonary thromboembolism	0.51	0.42	**0.45**	**0.49**

So, from the above discussions, it is evident that COVID and OSA follow the same pathways in different time frames. The already overwhelming pathways in OSA should respond more deadly during the acute insults.

The pathophysiological manifestations of OSA are due to poor sleep and obstructive respiratory events during sleep. So, sleep, per se, is also an important determinant of cellular events. Sleep deprivation from any cause, be it OSA, or another cause, will jeopardize the body milieu. This sleep deprivation can occur in OSA, shift work, or due to other causes.

14.5 Hypoxia Is the Major Determinant of Severity

One of the major determinants of the severity of COVID-19 disease is low oxygen saturation or hypoxia (Somers et al. 2020). The interaction between "acute happy hypoxia" of COVID-19 and "chronic happy hypoxia" of OSA is a subject of major debate. The commonly used parameter for defining sleep-disordered breathing (SDB) known as the event-based apnea-hypopnea index (AHI) may not fully capture the intricate underlying pathophysiological mechanisms of SDB. Observational studies have shown that metrics assessing nocturnal hypoxemia refers to a condition during sleep where the level of oxygen saturation (SaO_2) in the blood falls below 90%. This is measured as the percentage of total sleep time (TST) that is spent with oxygen saturation below this threshold (TST $<$ 90%).provide more accurate predictions of outcomes in SDB, including mortality, compared to the AHI (Pena Orbea et al. 2021; Oldenburg et al. 2016; Azarbarzin et al. 2019; Punjabi et al. 2009).

14.6 The Prevalence of OSA in COVID

A case-control study was conducted within the Cleveland Clinic Health System, encompassing Ohio and Florida, involving 5402 participants. The study encompassed all individuals who underwent COVID-19 testing within the period of March 8 to November 30, 2020, and had available sleep study records. By utilizing overlap propensity score weighting, the researchers evaluated sleep indices and SARS-CoV-2 positivity. They also assessed COVID-19 clinical outcomes using the institutional registry (Pena Orbea et al. 2021). The findings of the study indicated that there was no association between sleep-disordered breathing (SDB) and sleep-related hypoxia with an increased likelihood of SARS-CoV-2 positivity. While there was no association found between SARS-CoV-2 positivity and sleep-related hypoxia, it was observed that following SARS-CoV-2 infection, the presence of sleep-related hypoxia was observed, became a risk factor associated with adverse outcomes in COVID-19 (Pena Orbea et al. 2021).

In one of the largest case series of 5700 patients admitted in 12 different New York hospitals with a median age of 63 years (IQR 52–75), the commonest comorbidity was hypertension (56.6%), but they also stated that they had 41.7% obese patients and 2.9% patients with OSA (Richardson et al. 2020). A recent Finnish study evaluated the existence of OSA in COVID-19 patients (Feuth et al. 2020). They found that OSA was present in 29% of the patients hospitalized with COVID-19, being almost 10 times higher than its prevalence. Other small studies have shown that OSA was present in 21% of the patients in a case series in Seattle and 28.6% of the critically ill COVID-19 patients in Washington (Bhatraju et al. 2020; Arentz et al. 2020).

14.7 Covibesity

Another factor that can lead to an increased incidence of OSA during lockdown is an increased chance of getting obese due to various factors (Bhat and Chokroverty 2022). They are a sedentary lifestyle, overeating, decreased physical activity outdoors, and increased alcohol use. This type of obesity is given the name "Covibesity" or COVID-obesity.

14.8 COVID Pandemic Impact on CPAP Usage

Pandemic fear has also brought changes in compliance with various treatments. CPAP, or continuous positive airway pressure, represents the primary therapeutic approach of treatment of OSA. However, long-term compliance with CPAP is often poor in OSA patients. But a recent small study conducted among patients of a New York City sleep practice showed that a majority of patients (88%) continued to use CPAP during the early stages of the pandemic, when they were located in the epicenter, and among the 38% who expressed concern that their underlying OSA would make them susceptible to COVID-19, several expressed the sentiment that use of CPAP more often and for longer periods would be of clinical benefit, despite the absence of clear data in this regard (Thorpy et al. 2020).

One potential concern with CPAP use is the generation and spread of aerosols which can shed the virus, thus infecting the household members. International medical societies recommend continuing the use of CPAP in patients with OSA, albeit with measures taken to avoid the potential spread of the virus, such as sleeping in a separate room and diligent cleaning and disinfecting of equipment (Bhat and Chokroverty 2022).

14.9 Other Sleep Disorders During the COVID Pandemic

COVID-19 is believed to be associated with various factors contributing to sleep disturbances. A recent meta-analysis has examined this issue; the meta-analysis revealed significant prevalence rates of anxiety symptoms (ranging from 6.33% to 50.9%), depression symptoms (ranging from 14.6% to 48.3%), and post-traumatic stress disorder symptoms. (PTSD) (7% to 53.8%), psychological distress (34.43% to 38%), and stress (8.1% to 81.9%) in the general population across the globe during the COVID19 pandemic (Xiong et al. 2020). These changes in mental health gave rise to insomnia, disrupted sleep continuity, changes in the sleep-wake cycle, feelings of non-restorative sleep, and decreased sleep quality arising due to stress (Bhat and Chokroverty 2022). These sleep problems have been grouped under a new term, "COVID somnia" (Gupta and Pandi-Perumal 2020). But two major sleep disorders need special mention- insomnia and circadian rhythm sleep disorders. These two disorders are directly affected by the COVID pandemic and lockdown.

14.10 Insomnia During the COVID Pandemic

Insomnia refers to the condition characterized by difficulty initiating sleep or difficulty in maintaining sleep or early morning arousal, despite adequate opportunity to sleep, and always has daytime consequences of sleep deprivation. The mental health problems related to the COVID pandemic have been implicated in causing insomnia in a substantial population. In a Chinese study among nearly 12,000 adolescents and young adults, the prevalence of insomnia symptoms during part of the COVID-19 epidemic period was 23.2%, with female sex (Zhou et al. 2020). A large Spanish study conducted on over 15,000 subjects in the early phase of the lockdown showed high rates of difficulty initiating or maintaining sleep (23.9%), associated with higher age, female sex, reduction in income, having elderly dependents, alcohol use, depression, anxiety, and stress (Dal Santo et al. 2022). Increased use of social media and electronic gadgets also leads to poor sleep hygiene and insomnia. COVID-19 infection itself can cause various sleep problems like insomnia in the acute phase or in the post-acute phase called "long COVID." ICU admission in COVID is a major stressor and can give rise to long-term consequences of insomnia. Anxiety has already been stated to be associated with worse outcomes during ICU stay (Kompaniyets et al. 2021). In one study among 572 patients hospitalized for COVID-19, the prevalence of insomnia was 11% (Goldstein et al. 2021). In another large study conducted in Wuhan, China, 1733 patients discharged after hospitalization for COVID-19 were followed up (median follow-up time after symptom onset was 186 days). Fatigue or muscle weakness (seen in 63%) and sleep difficulties (seen in 26%) were the most common symptoms, with anxiety or depression reported among 23% of patients (Huang et al. 2021).

14.11 Circadian Rhythm Disorders (CRD) in the COVID Pandemic

Circadian rhythm disorders (CRD) are sleep disorders where our biological clock is not synchronous to our social clock. This causes sleep at inappropriate times. They are classified as delayed sleep phase syndrome (DSPS), advanced sleep phase syndrome (ASPS), shift work disorder, or jet lag. These disorders are direct effects of lockdown and home confinement. Conversely, increased time spent at home and decreased travel time have also provided increased opportunities to sleep. So, there was no unanimous trend related to circadian disorders. Large studies are lacking on this issue to provide a clear picture.

14.12 The COVID-19 Effects on the Sleep of Healthcare Workers

Healthcare workers faced tremendous mental and physical workloads during this pandemic. That took a toll on their sleep health. They faced extreme work pressure, fast adaptation to intense critical care situations, an unseen amount of severe critical patients, numerous deaths of patients, and risks of infection (Marvaldi et al. 2021). A recent meta-analysis from France on 70 cross-sectional studies across the globe involving 101,017 healthcare participants concluded the pooled prevalence of anxiety was 30% (95%CI 24.2–37), depression was 31% (95%CI 25.7–36.8), Post-traumatic stress disorder (PTSD) was 31.4% (95%CI −17.5 to −47.3). The pooled prevalence of sleep disorders was 44% (95%CI 24.56–64.97), and this was mainly measured for insomnia. Subgroup analysis identified female subjects were associated with lesser sleep problems (Marvaldi et al. 2021).

14.13 Impact of COVID Pandemic on Sleep in Children

Although SARS-CoV-2 infection is associated with asymptomatic or mild disease courses in children, the indirect accompaniments of the pandemic, such as lockdowns, travel restrictions, and closure of schools and recreational activities have also influenced the lifestyle and movement behaviors in children (Castagnoli et al. 2020; Mustafa and A Selim L. 2020). A recent systematic review and meta-analysis of ten studies focused on the issue of sleep disorders in children during the COVID pandemic (Sharma et al. 2021).

Before the pandemic, different studies have reported different prevalences of sleep problems in typically developing children across different age groups, with a range of 25–50% in preschoolers to 6–37% school children and 40% in adolescents (Blader et al. 1997; Owens et al. 2000; Vignau et al. 1997; Pagel et al. 2007). With sleep problems reported in nearly 25% of typically developing children otherwise, the pooled prevalence observed (more than two times) during the pandemic is alarming. However, the pooled prevalence was lower in preschoolers as compared with pre-pandemic prevalence (although not statistically significant) (Liu et al. 2021; Di Giorgio et al. 2021).

Preschool children behaved differently in the context to sleep during the pandemic when compared to older children and adolescents (Liu et al. 2021). There are many underlying factors that may be responsible for the contrasting outcome in preschoolers. Preschool children spend most of their time at home and usually do not have large social friend groups as in school. Besides, they are too puerile to understand or get affected by any viral disease or calamity. Their comfort zone with family at home is their sole need, which happened during the first 6 months of the COVID-19 pandemic. With work-from-home and home confinement regulations in place, family time increased.

However, school-going children and adolescents could understand the prevailing conditions in the world. They were impacted by the closure of schools and recreational activities and distancing from peers. All these factors directly or indirectly affect their sleep. Besides, the relaxation due to the closure of schools resulted in delayed bedtime and wake-up time during the initial phase of the pandemic and home confinement. The delayed wake-up time might have been useful for parents and caregivers who could cluster most of the household chores while the child was asleep.

Based on these observations, the present study concluded that the prevalence of sleep problems in children and adolescents during the COVID-19 pandemic is alarming. Preschool children tended to have relatively fewer sleep disturbances compared to pre-pandemic times. Sleep duration recommendations were not met in nearly half of healthy children during the pandemic.

14.14 Sleep Practice During the COVID Pandemic

Awareness of sleep disorders and sleep diagnostics is already very sparse in different parts of the globe. The pandemic caused a further downfall in that trend. Sleep services were not included in the essential healthcare services, and sleep laboratories were mostly closed during the lockdown. Both World Health Organization (WHO) and the Center for Disease Control (CDC) declared the potential for viral transmission via aerosol-generating procedures, like noninvasive ventilation. American Academy of Sleep Medicine (AASM) first published guidance on March 19, 2020, regarding mitigation strategies for the spread of COVID-19 disease in sleep practice. It recommends postponement and rescheduling of in-lab polysomnography and PAP administration except in emergencies. This recommendation resulted in the closure of many sleep laboratories and the reduction of Home Sleep Apnea Testing (HSAT). In many places, physical consultation for sleep disorders was also stopped, and few adopted online platforms and teleconsultations for such practices. Though there were no large-scale studies on the economic impact of such closure, many sleep laboratories suffered severe losses and permanent closures. Many laboratories adopted new technologies with portable devices with disposable consumables. But overall sleep diagnostics were at its worst status during the pandemic.

Conflict of Interest None.

References

Arentz M, Yim E, Klaff L et al (2020) Characteristics and outcomes of 21 critically ill patients with COVID-19 in Washington State. JAMA 323(16):1612–1614. https://doi.org/10.1001/jama.2020.4326

Azarbarzin A, Sands SA, Stone KL et al (2019) The hypoxic burden of sleep apnoea predicts cardiovascular disease-related mortality: the Osteoporotic Fractures in Men Study and the Sleep Heart Health Study. Eur Heart J 40(14):1149–1157. https://doi.org/10.1093/eurheartj/ehy624

Barceló A, Elorza MA, Barbé F, Santos C, Mayoralas LR, Agusti AG (2001) Angiotensin converting enzyme in patients with sleep apnoea syndrome: plasma activity and gene polymorphisms. Eur Respir J 17(4):728–732. https://doi.org/10.1183/09031936.01.17407280

Bauters F, Rietzschel ER, Hertegonne KB, Chirinos JA (2016) The link between obstructive sleep apnea and cardiovascular disease. Curr Atheroscler Rep 18:1. https://doi.org/10.1007/s11883-015-0556-z

Bhat S, Chokroverty S (2022) Sleep disorders and COVID-19. Sleep Med 91:253–261. https://doi.org/10.1016/j.sleep.2021.07.021

Bhatraju PK, Ghassemieh BJ, Nichols M et al (2020) Covid-19 in critically ill patients in the Seattle region—case series. N Engl J Med 382(21):2012–2022. https://doi.org/10.1056/NEJMoa2004500

Blader JC, Koplewicz HS, Abikoff H et al (1997) Sleep problems of elementary school children. A community surveys. Arch Pediatr Adolesc Med 151(5):473–480

Budhiraja R, Tuder RM, Hassoun PM (2004) Endothelial dysfunction in pulmonary hypertension. Circulation 109:159–165

Budhiraja R, Parthasarathy S, Quan SF (2007) Endothelial dysfunction in obstructive sleep apnea. J Clin Sleep Med 3(4):409–415

Castagnoli R, Votto M, Licari A et al (2020) Severe acute respiratory syndrome coronavirus 2 (SARS-CoV-2) infection in children and adolescents: a systematic review. JAMA Pediatr 174(9):882–889

Dal Santo F, Gonzalez-Blanco L, Rodríguez-Revuelta J et al (2022) Early impact of the COVID-19 outbreaks on sleep in a large Spanish sample. Behav Sleep Med 20:100–115

Delgado-Rochea L, Mestab F (2020) Oxidative stress as key player in severe acute respiratory syndrome coronavirus (SARS-CoV) infection. Arch Med Res 51:384–387. https://doi.org/10.1016/j.arcmed.2020.04.019

DeMartino T, Ghoul RE, Wang L, Bena J, Hazen SL, Tracy R, Patel SR, Auckley D, Mehra R (2016) Oxidative stress and inflammation differentially elevated in objective versus habitual subjective reduced sleep duration in obstructive sleep apnea. Sleep 39:1361–1369. https://doi.org/10.5665/sleep.5964

Di Giorgio E, Di Riso D, Mioni G et al (2021) The interplay between mothers' and children behavioral and psychological factors during COVID-1: an Italian study. Eur Child Adolesc Psychiatry 30:1401–1412. https://doi.org/10.1007/s00787-020-01631-3

Ekiz T, İnönü Köseoğlu H, Pazarlı AC (2020) Obstructive sleep apnea, renin-angiotensin system, and COVID-19: possible interactions. J Clin Sleep Med 16(8):1403–1404

Feuth T, Saaresranta T, Karlsson A et al (2020) Is sleep apnoea a risk factor for Covid-19? Findings from a retrospective cohort study. medRxiv. https://doi.org/10.1101/2020.05.14.20098319

Fraga-Silva RA, Sorg BS, Wankhede M et al (2010) ACE2 activation promotes antithrombotic activity. Mol Med 16(5–6):210–215. https://doi.org/10.2119/molmed.2009.00160

Goldstein CA, Rizvydeen M, Conroy DA et al (2021) The prevalence and impact of pre-existing sleep disorder diagnoses and objective sleep parameters in patients hospitalized for COVID-19. J Clin Sleep Med 126:1039–1050. https://doi.org/10.5664/jcsm.9132

Gupta R, Pandi-Perumal SR (2020) COVID-somnia: how the pandemic affects sleep/wake regulation and how to deal with it? Sleep Vigil 4:51–53. https://doi.org/10.1007/s41782-020-00118-0

Halder AK (2021) COVID-19 and obstructive sleep apnea. Indian J Sleep Med 16(3):86–89

Heinzer R, Vat S, Marques-Vidal P, Marti-Soler H, Andries D, Tobback N et al (2015) Prevalence of sleep-disordered breathing in the general population: the HypnoLaus study. Lancet Respir Med 3(4):310–318

Huang C, Huang L, Wang Y et al (2021) 6-month consequences of COVID-19 in patients discharged from hospital: a cohort study. Lancet 397(10270):220–232

Ip MS, Tse HF, Lam B, Tsang KW, Lam WK (2004) Endothelial function in obstructive sleep apnea and response to treatment. Am J Respir Crit Care Med 169:348–353

Jin ZN, Wei YX (2016) Meta-analysis of effects of obstructive sleep apnea on the renin-angiotensin-aldosterone system. J Geriatr Cardiol 13(4):333–343. https://doi.org/10.11909/j.issn.1671-5411.2016.03.020

Kompaniyets L, Pennington AF, Goodman AB, Rosenblum HG, Belay B, Ko JY et al (2021) Underlying medical conditions and severe illness among 540,667 adults hospitalized with COVID-19, March 2020–March 2021. Prev Chronic Dis 18:210123. https://doi.org/10.5888/pcd18.210123

Lee SD, Kang SH, Ju G, Han JW, Kim TH, Lee CS et al (2014) The prevalence of and risk factors for sleep-disordered breathing in an elderly Korean population. Respiration 87(5):372–378

Liu Z, Tang H, Jin Q et al (2021) Sleep of preschoolers during the coronavirus disease 2019 (COVID-19) outbreak. J Sleep Res 30:e13142

Marvaldi M, Mallet J, Dubertret C, Moro MR, Guessoum SB (2021) Anxiety, depression, trauma-related, and sleep disorders among healthcare workers during the COVID-19 pandemic: a systematic review and meta-analysis. Neurosci Biobehav Rev 126:252–264. https://doi.org/10.1016/j.neubiorev.2021.03.024

McSharry D, Malhotra A (2020) Potential influences of obstructive sleep apnea and obesity on COVID-19 severity. J Clin Sleep Med 16(9):1645. https://doi.org/10.5664/jcsm.8538

Memtsoudis S, Liu SS, Ma Y et al (2011) Perioperative pulmonary outcomes in patients with sleep apnea after noncardiac surgery. Anesth Analg 112(1):113–121. https://doi.org/10.1213/ANE.0b013e3182009abf

Mustafa NM, A Selim L. (2020) Characterisation of COVID-19 pandemic in paediatric age group: a systematic review and meta-analysis. J Clin Virol 128:104395

Oldenburg O, Wellmann B, Buchholz A et al (2016) Nocturnal hypoxaemia is associated with increased mortality instable heart failure patients. Eur Heart J 37(21):1695–1703. https://doi.org/10.1093/eurheartj/ehv624

Owens JA, Spirito A, McGuinn M et al (2000) Sleep habits and sleep disturbance in elementary school-aged children. J Dev Behav Pediatr 21(1):27–36

Pagel JF, Forister N, Kwiatkowki C (2007) Adolescent sleep disturbance and school performance: the confounding variable of socioeconomics. J Clin Sleep Med 3(1):19–23

Pena Orbea C, Wang L, Shah V et al (2021) Association of Sleep-Related Hypoxia with risk of COVID-19 hospitalizations and mortality in a large integrated health system. JAMA Netw Open 4(11):e2134241. https://doi.org/10.1001/jamanetworkopen.2021.34241

Punjabi NM, Caffo BS, Goodwin JL et al (2009) Sleep-disordered breathing, and mortality: a prospective cohort study. PLoS Med 6(8):e1000132. https://doi.org/10.1371/journal.Med.1000132

Reddy EV, Kadhiravan T, Mishra HK, Sreenivas V, Handa KK, Sinha S et al (2009) Prevalence and risk factors of obstructive sleep apnea among middle-aged urban Indians: a community-based study. Sleep Med 10(8):913–918

Richardson S, Hirsch JS, Narasimhan M et al (2020) Presenting characteristics, comorbidities, and outcomes among 5700 patients hospitalized with COVID-19 in the New York City Area. JAMA 323(20):2052–2059. https://doi.org/10.1001/jama.2020.6775

Sharma M, Aggarwal S, Madaan P, Saini L, Bhutani M (2021) Impact of COVID-19 pandemic on sleep in children and adolescents: a systematic review and meta-analysis. Sleep Med 84:259–267. https://doi.org/10.1016/j.sleep.2021.06.002

Somers VK, Kara T, Xie J (2020) Progressive hypoxia: a pivotal pathophysiologic mechanism of COVID-19 pneumonia. Mayo Clin Proc 95(11):2339–2342. https://doi.org/10.1016/j.mayocp.2020.09.015

Thorpy M, Figuera-Losada M, Ahmed I et al (2020) Management of sleep apnea in New York City during the COVID-19 pandemic. Sleep Med 74:86–90

Tufik S, Santos-Silva R, Taddei JA, Bittencourt LRA (2010) Obstructive sleep apnea syndrome in the Sao Paulo epidemiologic sleep study. Sleep Med 11(5):441–446

Vaduganathan M et al (2020) Renin–Angiotensin–Aldosterone system inhibitors in patients with Covid-19. N Engl J Med 382:1653–1659. https://doi.org/10.1056/NEJMsr2005760

Vicente E, Marin JM, Carrizo SJ, Osuna CS, González R, Marin-Oto M, Forner M, Vicente P, Cubero P, Gil AV et al (2016) Upper airway and systemic inflammation in obstructive sleep apnoea. Eur Respir J 48:1108–1117. https://doi.org/10.1183/13993003.00234-2016

Vignau J, Bailly D, Duhamel A et al (1997) Epidemiologic study of sleep quality and troubles in French secondary school adolescents. J Adolesc Health 21(5):343–350

Xiong J, Lipsitz O, Nasri F et al (2020) Impact of COVID-19 pandemic on mental health in the general population: a systematic review. J Affect Disord 277:55–64

Zhou SJ, Wang LL, Yang R et al (2020) Sleep problems among Chinese adolescents and young adults during the coronavirus-2019 pandemic. Sleep Med 74:39–47

Chapter 15
Sleep Disturbances in Hospitalized and Intensive Care Unit Patients

Pratyaksh P. Vaishnav, Ashutosh Suresh, Sreelakshmi Kooragayalu, and Shravan Kooragayalu

Abstract Sleep disturbance is a commonly overlooked complication of hospitalized and intensive care unit (ICU) patients. Sleep in critically ill patients is often characterized by frequent disturbances to sleep structure, sleep depth, and changes in circadian rhythm that occur for a multitude of reasons. In an acute care setting, the quality of sleep is generally neglected due to the ongoing care of the patient's illness. Medical personnel do not attach much importance to a patient's sleep quality as they have many other duties and need to provide holistic care. Sleep disturbance due to increased awakenings in these clinical settings can have similar consequences and adverse health effects as sleep deprivation. It can negatively impact multiple organ systems, impeding overall well-being and prolonging the length of stay of these patients. There are various sleep measurement methods in ICU patients, although they do not all measure the same dimensions, and some of them are subjective and a few objective measurements. Many factors disturb the sleep of patients in an ICU setting, some of which consist of environmental factors, patient care activities including pharmacological and mechanical interventions, and patient-related factors. Changes in sleep structure developing during an ICU stay may significantly contribute to sleep disorders that carry forward even after the hospitalization is over. Some of the measures aim at increasing patients' comfort, reducing light and noise intensity at night, and an appropriate organization and aggregation of nursing care interventions to prevent sleep interruptions have been proven effective in sleep interruptions in an ICU setting. Awareness of the medical personnel and hospital

---

P. P. Vaishnav
J.J.M Medical College, Davanagere, Karnataka, India

Vasavi Hospital, Bangalore, Karnataka, India

A. Suresh
J.J.M Medical College, Davanagere, Karnataka, India

University of Pittsburgh Medical Center—Western Maryland Health System, Cumberland, MD, USA

S. Kooragayalu · S. Kooragayalu (✉)
University of Pittsburgh Medical Center—Western Maryland Health System, Cumberland, MD, USA

A. S. BaHammam, M. Hunasikatti (eds.), *Sleep Apnea Frontiers*, Progress in Sleep Research, https://doi.org/10.1007/978-981-99-7901-1_15

employees along with interaction within the therapeutic team regarding the sleep-disturbing factors in the ICU setting and techniques for preventing it can significantly improve the quality of sleep of the patients. Here, in this chapter, we have highlighted the factors causing sleep disturbances, their physiological effects on different systems of the body and a few possible methods to mitigate these disturbances.

Keywords Sleep disturbance · Hospitalized patients · Intensive care unit (ICU) · Sleep structure · Sleep deprivation

15.1 Introduction

Sleep is a fundamental physiological process that plays critical roles in various aspects of well-being and functioning. It serves as a crucial mechanism for recovery, conserving energy, and ensuring survival. Furthermore, sleep is closely linked to essential functions including neural development, cognitive processes like learning and memory, emotional regulation, cardiovascular and metabolic health, and the clearance of cellular toxins (Mukherjee et al. 2015).

In an intensive care unit setting, patients are artificially put to sleep by medications. What makes physiological sleep different from similar conditions, such as a coma, hibernation, or anesthesia, is the fast return to wakefulness after a strong stimulus (Lewandowska et al. 2020). Physiological sleep and stable sleep/wake cycles are linked to better physical and mental health throughout life. Sleep disturbance is defined as the insufficient quantity or quality of sleep with subsequent daytime impairment and significantly reduced general overall well-being (Lewandowska et al. 2020).

15.2 Physiological Sleep and Circadian Rhythm

The average duration of nocturnal rest recommended for a healthy adult is around 7–8 h. The biological requirement for sleep is, however, hard to estimate. It varies individually and depends on genetic factors, age, and gender. Entire nocturnal sleep can be divided into cycles. Each cycle lasts for 90–110 min (Medrzycka-Dabrowska et al. 2018). Sleep structure is defined by criteria developed by the American Academy of Sleep Medicine. Figure 15.1 briefly describes the architecture of the normal sleep cycle. Each night cycle consists of two phases, namely REM (rapid eye movement) and NREM (non-rapid eye movement). The REM phase is characterized by rapid eyeball movement, muscle atonia, and dreams. The NREM phase is divided into three additional stages: N1—the lightest sleep, the transition from wakefulness to sleep; N2—the moment of falling asleep, characterized by light sleep; N3—deep sleep, the hardest to interrupt (Lewandowska et al. 2020). An additional factor, which normalizes nocturnal rest, is the circadian rhythm. This is mainly influenced

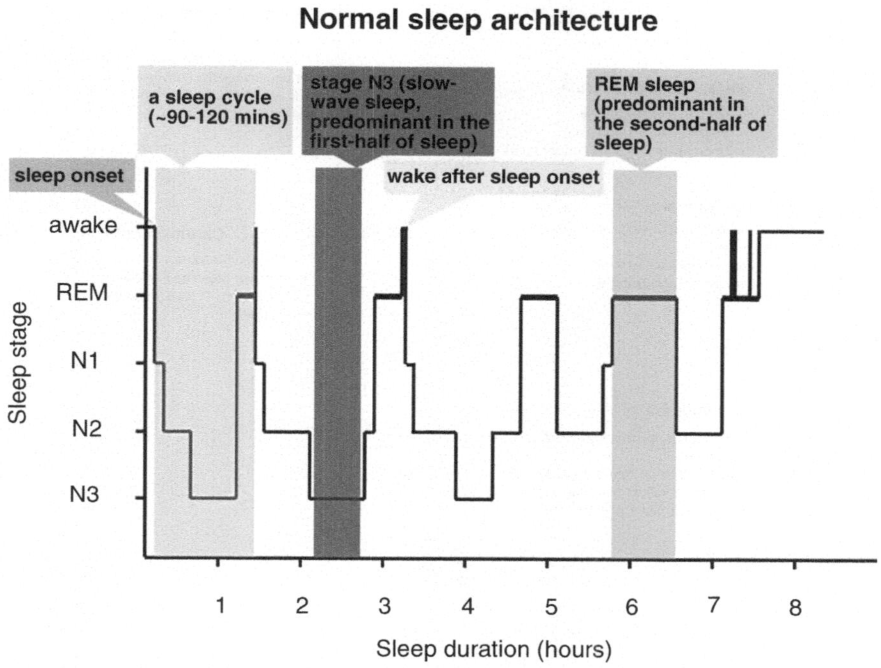

Fig. 15.1 Normal Sleep architecture (Tan et al. 2019)

by light, which indicates the time for rest. Light stimuli that fall on the retina influence the functioning of the anterior subthalamus, i.e., suprachiasmatic nuclei. This structure is responsible for synchronizing all circadian systems of the body (Medrzycka-Dabrowska et al. 2018).

15.3 Sleep in ICU

The intensive care unit (ICU) is widely regarded as one of the most critical and crucial healthcare settings. In here medical staff save lives by treating the most severely ill patients. It is a complex atmosphere and is based on continuous observation, monitoring, and intervention involved in patient care. Given the overall workload of ICU medical staff, the physical and technical aspects of care become the priority, overshadowing the psychological aspects of care and the needs of the patient. Patients in an ICU are exposed to multiple stimuli including continuous noise made by the monitoring devices, light emitted by various devices, and frequent medical interventions. Especially during the nighttime, these disturbances can interrupt and alter patients' sleep and have a negative impact. Due to the certain specificities of this unit, medical personnel is not able to eliminate the factors

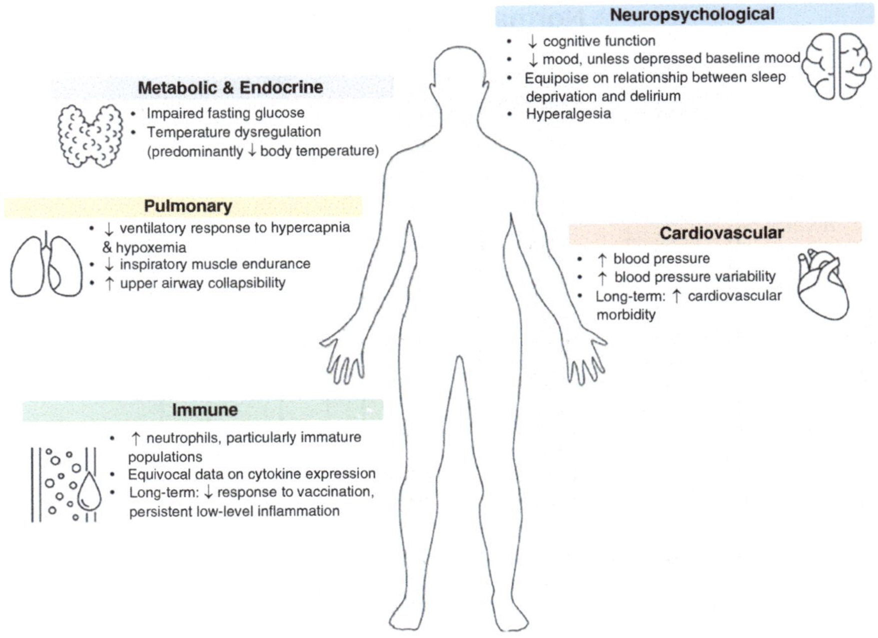

Fig. 15.2 Effects of Sleep disruption on organ systems (Chang et al. 2020)

responsible for poor sleep quality and lack of physiological rest (Lewandowska et al. 2020).

Although patients may achieve an acceptable amount of sleep during 24 h while receiving care in the ICU, it is noted that this sleep is usually fragmented and has a significantly abnormal architecture. Patients have an increased amount or percentage of stage 1 sleep which typically suggests sleep fragmentation due to a sleep disorder (Pulak and Jensen 2016).

An increased number of awakenings may have similar consequences and adverse health effects as sleep deprivation. These include weakening of the immunological system leading to an increased risk of infections, reduced respiratory parameters, increased perception of pain, reduced glucose tolerance, increased sympathetic activity, impaired wound healing, and increased protein catabolism. The above-mentioned factors were identified during studies performed on healthy people. In a hospital or an ICU setting sleep disturbances in patients therefore can lead to more pronounced consequences of sleep deprivation, which in turn can hamper the recovery process (Medrzycka-Dabrowska et al. 2018). Sleep deprivation impairs cognition, ranging from apathy and confusion to delirium, all of which may lead to increased morbidity and mortality (Lewandowska et al. 2020).

Below is a brief description of how sleep disturbance in any hospitalized or ICU patient can affect different vital organ systems (Fig. 15.2):

15.4 Neurophysiological Effects

Drouot et al. combined EEG spectral analysis and visual quantitative EEG analysis during sleep and wakefulness and identified subclinical alterations in sleep/wake states, which may affect one-third of conscious non-sedated ICU patients with respiratory failure (Drouot et al. 2012). These abnormalities may reflect a subclinical form of brain dysfunction that may be associated with an increased risk of delirium (Pulak and Jensen 2016).

An association between delirium and sleep disturbance in the ICU is suggested, although medications used to control delirium may be contributing to deleterious effects on sleep and overall psychiatric outcomes (Pulak and Jensen 2016).

Studies with sleep-promoting interventions have been shown to reduce the number and duration of delirium episodes among general medical wards, medical ICU, and surgical ICU patients (Chang et al. 2020).

Other studies revealed that during their stay in a hospital or an ICU, patients could remember frequent interruptions, pain, anxiety, and fear, which all reduced their ability to sleep (Medrzycka-Dabrowska et al. 2018). One study, which included 464 patients, demonstrated that 51% of them experienced dreams and nightmares, whereas 14% of these subjects reported nightmares negatively influencing their quality of life even 6 months after their ICU stay (Granja et al. 2005).

15.5 Cardiovascular System

Elevated blood pressure is a well-known consequence of short-term sleep deprivation (Chang et al. 2020). The hypertensive effect of sleep deprivation has been attributed to an increase in sympathetic activity (Zhong et al. 2005), a reduction in maximum endothelium-dependent vasodilation, and an upward resetting of the baroreflex set point (Chang et al. 2020). In addition to its hypertensive effect, sleep deprivation is also correlated with increased blood pressure variability (Zhong et al. 2005).

The concentrations of C-reactive protein (CRP), a marker of inflammation that has been shown to have predictive value on future myocardial infarction and stroke in apparently healthy individuals, are higher after experimental sleep deprivation (Chang et al. 2020).

15.6 Respiratory System

Most studies have found a reduced ventilatory response to hypercapnia and hypoxemia after 24 h of sleep deprivation (Chang et al. 2020), although this finding has been disputed by a newer study performed with stricter control of subject behavior and environment (Spengler and Shea 2000). Short-term sleep deprivation has also been shown to lessen inspiratory muscle endurance (Chang et al. 2020).

Experimental sleep deprivation has also been found to increase upper airway collapsibility, which can cause or contribute to preexisting obstructive sleep apnea (OSA) and lead to challenges in successful extubation (Chang et al. 2020).

Studies on patients with OSA have also demonstrated that cerebral blood flow (CBF) decreases during episodes of apneas, and intracranial pressure increases in linear proportion to the length of an apneic event. The impact of sleep deprivation on pulmonary function is relevant in the Neuro ICU given the proportion of patients who require ventilatory support and the prevalence of sleep-disordered breathing among stroke and transient ischemic attack patients (Chang et al. 2020).

The reduction of CBF in OSA may be attributed to damage to brain areas essential for vascular regulation (including the insular, cingulate, and medial frontal cortices; raphe and ventrolateral medulla) and an unexplained blunting of the typical vasodilator response to increased PaCO2 (Yadav et al. 2013). Such results highlight the possible complications of sleep deprivation for Neuro ICU patients with ischemic brain injury but may not be generalizable (Chang et al. 2020).

15.7 Effects of Sleep Deprivation on Immune Function

Sleep plays a vital role in promoting adaptive immune responses, which ultimately contribute to the development and enhancement of the immune system's ability to recognize and respond to pathogens and other threats. This impact is rooted in the reciprocal interaction between the central nervous system (CNS) and the immune system, along with the direct neural connections between the autonomic nervous system and the immune system.

Below are some different ways in how sleep deprivation can potentially affect healing, both in terms of tissue repair and cellular immune function (Pulak and Jensen 2016).

- Interleukin 1 (IL-1) and tumor necrosis factor have been shown to regulate NREM sleep; therefore, loss of sleep may result in impairment of host defenses against infection.
- Significant reduction in the number and phagocytic activity of lymphocytes was found to be decreased by 13% as compared with non-sleep deprived individuals.
- Benedict et al. (Beauchemin and Hays 1998) found that Interleukin-7 levels were increased during normal sleep.
- IL-7 facilitates the transition of CD8þ effector to memory T-cells and lengthens the survival of the T-cell memory cells.
- Natural killer T-cell activity was found to be negatively correlated with the severity of insomnia.
- Sleep restriction can also result in reduced antioxidant activity (catalase and glutathione), decreased spleen weight, alterations in leukocyte and lymphocyte

counts, and the production of serum antibodies without the presence of an antigen.

15.8 Hormonal System and Metabolism

Lack of sleep leads to significant alterations in homeostatic mechanisms and has a profound impact on neuroendocrine stress systems. Wakefulness enhances catabolism, while sleep shifts metabolism in favor of anabolism. Urinary nitrogen secretion increases markedly with sleep reversal (change in diurnal activity) and sleep deprivation (Pulak and Jensen 2016).

Spiegel et al. (Caldwell et al. 2002) found that sleep-deprived patients showed glucose intolerance, insulin resistance, lower thyrotropin, and elevated evening cortisol levels (Spiegel et al. 1999). Sleep-deprived patients are found to have higher oxygen consumption, carbon dioxide production, and catecholamine levels, likely as a stress response (Pulak and Jensen 2016).

15.9 Factors Affecting Sleep

Many factors disturb the sleep of patients in an ICU setting with specific environments and invasive therapeutic interventions. The inability to obtain physiological sleep depends on the patient's illness, varying severity of the illness, and previous sleep experience. Below we describe the factors affecting sleep in an ICU environment under the classification of environmental factors, patient care activities including pharmacological interventions, mechanical interventions, and patient-related factors.

15.9.1 Environmental Factors

ICU could be a place of anxiety, fear, and apprehension for any patient. Patients rarely have time to prepare for acute change mentally and physically in such a fast-paced critical environment. These factors are further exacerbated by sedative medications, the inability to communicate or move, leading to a state of low morale and sleep disturbance. ICU patients require frequent monitoring and repeated interventions for diagnostic tests, vital signs, blood collection, and medication administration. This leads to a constant awakening, heightened noise, and disturbing light levels in an ICU setting, thus all contributing to disrupted sleep (Pulak and Jensen 2016).

15.9.1.1 Noise

The World Health Organization's (WHO) external review policy on community noise has produced guidelines that identify specific noise levels within hospital environments. The recommendations include noise levels not exceeding 35 dB (A) during the night and 40 dB(A) during the day. In general, this is far exceeded by the average noise levels in an ICU. Recent studies show the average noise level in ICU to be 55–66 dB(A) with peaks reaching as high as 85 dB(A). These noise levels are considered enough to stimulate cardiovascular and endocrine systems disrupting sleep as a result of noise-induced stress modulation (Pulak and Jensen 2016). [For reference, the noise levels of an average vacuum cleaner are about 70 db] (Tan et al. 2019).

Furthermore, noise can diminish the magnitude of overnight melatonin release in healthy humans (Huang et al. 2015). Thus, it could be speculated that noise in the hospital may lead to the development of sleep disturbances and circadian disruption in patients. Supporting this hypothesis, two studies on the general medicine and oncology floors found that 27–43% of patients identified noise as a major cause of poor sleep (Grossman et al. 2017; Yoder et al. 2012).

A systematic review by K. Honarmand showed that among ICU-related factors, patients reported noise as being one of the most disruptive factors for sleep in the ICU (Honarmand et al. 2020).

A study by S. Bihari et al. concluded that the most disruptive source of ICU noise is bedside phones and ICU mechanical alarms. However, the study did not support the hypothesis that ICU noise is the most disruptive stimulus (Bihari et al. 2012).

Freedman et al. used PSG and time-synchronized recording of environmental noise, directly linked to noise arousals from sleep, and determined that noise was responsible for only 15% of all arousals and awakenings (Freedman et al. 2001).

Healthy individuals slept relatively well in the potentially disruptive environment of the ICU and although noise accounted for a significant proportion of sleep disruption, its extent was not pathologic. High-peak noise may be more disruptive than general background noise; however, even peaks were not found that problematic. Therefore, the relative contribution of noise and other components of the ICU environment to the pathogenesis of ICU sleep disruption is largely unknown and remains a complex issue (Pulak and Jensen 2016).

15.9.1.2 Light

The light–dark cycle is probably the most powerful entraining factor in the human sleep–wake cycle and ultimately the development of restful surroundings (Pulak and Jensen 2016). Unnatural light characteristics that are often found in the hospital may contribute to impaired sleep and circadian disruption. Artificial blue-enriched light during evening hours delays the production of melatonin and impairs sleep quality (Tan et al. 2019). Medical inpatients are often subjected to dim light conditions during the daytime and bright light during the evening and night hours (Chang et al. 2020). For example, the maximal nocturnal luminance in ICU can reach as high as 1000–3000 lux during the night (Meyer et al. 1994). A potential reason why ICU patients lack the typical nocturnal peak of melatonin secretion and show a delayed circadian rhythm (Tan et al. 2019).

Dunn et al. recorded ICU activity that was occurring or not occurring, while various light sources were illuminated. The activity associated with the greatest amount of light exposure was obtaining samples for laboratory tests, while the second activity most often recorded while the light was on was "none" (Dunn et al. 2010). This suggested a lack of simple vigilance on the part of healthcare providers (Pulak and Jensen 2016).

Natural light exposure is often unevenly distributed in different rooms of the hospital. This may hamper the ability of patients sleeping further from the window to be in sync with the natural light/dark cycle, with possible undesirable implications of sleep disturbance (Tan et al. 2019). A study involving 1057 patients from various medical floors pointed to the role of natural daylight exposure on patients' length of stay. Patients located in southwest-facing rooms (mean 220 lux between 8 am and 6 pm) had a 29% shorter average length of stay, compared to patients in the northwest-facing rooms (mean 101 lux) (Choi et al. 2012). Furthermore, brighter compared to darker hospital rooms correlated with reduced duration of hospitalization and decreased mortality rates. in patients with myocardial infarction ($n = 628$) (Beauchemin and Hays 1998). Such studies implicate that circadian disruption due to weak light exposure may adversely affect patients' health and recovery (Tan et al. 2019).

15.9.2 Patient Care Activities

The need for clinical interventions increases exponentially in ICU as many different assessments are scheduled on an hourly basis such as vital sign monitoring, fluid balance assessments, neurological assessments, and pharmacological administration. These interventions in combination with basic care needs establish an environment in which sleep is easily deprioritized and disrupted (Delaney et al. 2015).

In Lee et al.'s study, 50% of patients reported being woken 2–5 times throughout the night caused difficulties falling back to sleep once woken (Lee et al. 2007). The frequency of disruptions for mechanically ventilated patients was found to be substantially higher, with Le et al. reporting that patients experienced 20–60 disruptions per hour (Le et al. 2012).

However, a systematic review by K. Honarmand et al. showed that care interventions were not associated with sleep disruption in the majority studies (Honarmand et al. 2020). Approximately 20% of patient care activities resulted in arousal or awakening, accounting for only 7% of observed sleep disruption. Therefore, patient care activities, although frequent, may not be a predominant source of sleep disruption in ICU patients (Pulak and Jensen 2016).

15.9.2.1 Pharmacological Intervention

Medications can affect sleep in various ways. They may contribute to disturbed sleep and adversely affect normal sleep physiology and architecture. Medications may affect the CNS directly by penetrating the blood–brain barrier or indirectly by affecting a medical or psychiatric illness that results in altered sleep. Some may disrupt sleep by their effect on preexisting sleep disorders and others have an equally disruptive effect when withdrawn abruptly (Pulak and Jensen 2016).

Sedatives

Sedation is a mode in which patients are put into a state of sleep that is not physiological, not circadian, or fully reversible with external stimuli. The most common ICU sedatives are those that interact with g-aminobutyric acid (GABA) receptors in the CNS (Pulak and Jensen 2016). At low doses, benzodiazepines and propofol suppress slow wave sleep (SWS) and may decrease REM sleep. They shorten sleep latency, decrease arousal, and increase NREM stage 2 sleep. Higher doses are associated with a characteristic slowing of the EEG and both can lead to a burst suppression pattern, thus the EEG shows more of a comatose pattern. With repeated use, benzodiazepines can completely abolish NREM stage 4 sleep (Pulak and Jensen 2016).

Dexmedetomidine, a selective alpha 2-adrenoreceptor agonist, is increasingly utilized in the Neuro ICU due to its analgesic, sedative, and anxiolytic effects while maintaining patient arousability and minimizing respiratory depression (Chang et al. 2020). [The use of dexmedetomidine has effects closer to the naturally occurring sleep pathway in the CNS as it inhibits Norepinephrine release by locus ceruleus. Dexmedetomidine preserves a natural sleep pattern and induces cooperative sedation] (Pulak and Jensen 2016). While dexmedetomidine is more expensive, several studies have found it to be more cost-effective than traditional sedatives as it is associated with reduced ICU stay costs and shorter times to extubation (Chang et al. 2020).

Analgesics

Opioids form a major part of pain management in critically ill patients. They affect the natural sleep pattern via the Ponto-thalamic arousal pathway. Even single doses of opioids potently suppress SWS (Pulak and Jensen 2016). Rapid eye movement (REM) sleep suppression by opioids is a dose-dependent phenomenon (Pulak and Jensen 2016). Opioids increase NREM stage 2 sleep and increase wakefulness. However, if pain is a predominant cause of disturbed sleep, the overall effects of opioids would likely be to improve sleep (Caldwell et al. 2002).

Nonsteroidal anti-inflammatory medications can adversely affect sleep by decreasing sleep efficiency and increasing awakenings. These effects may be because of inhibition of prostaglandin synthesis, decreased melatonin secretion, attenuation in nocturnal body temperature, or gastric irritation (Murphy et al. 1994).

Cardiovascular Medications

Cardiovascular medications are widely used medications in an ICU environment (Pulak and Jensen 2016). Beta-blockers are known to cause nightmares, insomnia, and REM sleep suppression. These CNS effects are mainly due to their lipid solubility allowing them to cross the blood-brain barrier (Pulak and Jensen 2016). Adrenergic receptor agonists like norepinephrine, epinephrine, and dopamine are associated with insomnia and suppression of REM sleep and SWS (Bourne and Mills 2004). Amiodarone, a highly effective antiarrhythmic has also been known to cause neurological side effects in up to 40% of patients. These include insomnia and nightmares (Pulak and Jensen 2016). Diuretics disrupt sleep due to increased frequency of urination or requirement of painful urinary catheters (Chang et al. 2020).

Respiratory Medications

Patients on mechanical ventilation and those who have acute or chronic respiratory conditions with symptoms are likely to receive inhaled b-adrenergic receptor agonists while in the ICU. Central nervous system stimulation with associated restlessness, agitation, and insomnia is a well-known adverse effect. The overall effect on sleep may be positive, only if these medications alleviate dyspnea and desaturations (Pulak and Jensen 2016).

Corticosteroids

Corticosteroids are known to cause REM sleep suppression and increased nocturnal awakenings. Steroid psychosis or hypermania can cause insomnia (Wolkowitz et al. 1990).

15.9.2.2 Mechanical Interventions

Radiological and Laboratory Testing

In any ICU setting patients frequently undergo multiple testing, including blood draws, chest X-rays, echocardiogram, and sometimes patients need to be transported out of the unit for investigations including a computed tomography scan (CT scan) of various organs. Patients also undergo necessary procedures which are usually performed at the bedside if feasible given the critical condition of the patients. All these above-mentioned interventions can cause significant disturbances in patients' sleep.

Grooming of the Patient

Hospitalized patients need to be assisted with activities of daily living including bathing, feeding, and grooming and most of these interventions are performed during the hospitalization by nurses or nursing assistants. For some of the patients these activities could come in the way of their sleep causing sleep disturbances.

Mechanical Ventilation

Both invasive and noninvasive ventilation (NIV) modes of mechanical ventilation can affect a patient's sleep. Regardless of certain limitations arising from a high variance in underlying diseases, patient groups, and types of ventilation (endotracheal tube, tracheostomy, mask), there are many hints that the mode of ventilation will also influence sleep quality and quantity (Chang et al. 2020).

Mechanical ventilation can disrupt sleep due to a mismatch between patient demand and ventilator supply (inadequate support leading to dyspnea or excessive support leading to apnea) (Chang et al. 2020). Major causes of sleep disturbance associated with mechanical ventilation include patient–ventilator asynchronies, central apneas due to overventilation, and inadequate ventilatory support due to improper settings or air leaks leading to increased respiratory effort (Chang et al. 2020).

Studies by Parthasarathy and Tobin (2002) and Toublanc et al. (2007) found that pressure support ventilation (PSV) led to episodic hyperventilation with resulting decreased partial pressure of carbon dioxide, causing apneic events in comparison to assist control ventilation (ACV). However, work by Cabello and colleagues showed no differences in sleep architecture between non-sedated ICU patients on clinically adjusted PSV, automatically adjusted PSV, or ACV. The authors, therefore, suggested that it is not PSV itself that causes sleep disruption but the specific ventilator settings: They used lower levels of pressure support, standard minute ventilation, and tidal volume across all modes to better mimic physiological conditions (Cabello et al. 2008).

Other studies looking at proportional assist ventilation in which pressure is applied in proportion to the patient's inspiratory effort and neurally adjusted ventilatory assist in which pressure is applied in proportion to the electrical activity of the patient's diaphragm also find improved sleep quality. Thus, evidence from multiple studies indicates that patient–ventilator mismatch, rather than any given ventilator mode, is most responsible for sleep disruption (Chang et al. 2020).

Noninvasive Ventilation

Patients with hypercapnic respiratory failure requiring >24 h of noninvasive ventilation (NIV) exhibit marked sleep disturbances as assessed by PSG. Sleep alterations including circadian sleep cycle disruption and shorter REM sleep time were associated with subsequent delirium and late NIV failure (Pulak and Jensen 2016).

15.9.3 Patient Specific Factors

15.9.3.1 Pain

Pain could be either related to illness severity or on its own is associated with poor sleep and increased arousal due to discomfort. Pain causing loss of sleep may trigger an unfortunate feedback loop in which the lack of sleep leads to hyperalgesia, which in turn inhibits adequate rest (Chang et al. 2020).

The most common patient-reported physiological factor is pain and breathing difficulty in those reporting poor sleep as stated in a systematic review of patient-reported studies by K. Honnarmand et al. (2020).

Bed Rest and Physical Activity

Daytime physical activity improves nighttime sleep and supports the maintenance of a healthy circadian rhythm. Therefore, reduced physical activity due to limited opportunities to engage in exercise or outdoor activities might result in impaired sleep and circadian disruption during hospitalization. This in turn may decrease daytime physical activity due to a lack of motivation (Tan et al. 2019). Another common behavior among inpatients is napping. A daytime nap may compensate in part for less sleep at night and promote recovery from acute illness. However, excessive napping might be a marker of other pathology, including undiagnosed sleep disorders (Tan et al. 2019).

15.9.3.2 Anxiety and Fear

The ability to sleep is most affected by fear-related factors. A strong sense of powerlessness and helplessness related to their critical clinical condition leads to sleeping difficulties. Some of the patients are afraid to sleep due to recurring nightmares, which leads to constant tiredness and anxiety. The death of a fellow patient staying in the same ward also contributes to feelings of insecurity, causing patients to experience sleeping difficulties. In an ICU setting where death is not uncommon, patients were too frightened to go to sleep. Another factor that contributed to lack of sleep was loneliness, longing for family, and lack of communication with them (Lewandowska et al. 2020).

15.10 Sleep Measurement in Critically Ill Patients

Sleep is a multidimensional construct, composed of dimensions such as total sleep time, percent of sleep stages, frequency of awakenings or arousals, expectations, global perceptions, sleep movements, tiredness upon awakening, daytime energy levels, and functional impairments. Although there are various methods of measurement of sleep in ICU patients, they do not all measure the same dimensions, and we do not expect them to demonstrate 100% agreement (Medrzycka-Dabrowska et al. 2018). Below is a brief description of the different types of measurements that exist.

15.10.1 Polysomnography

Polysomnography is the gold standard test and a comprehensive method for sleep assessment. Such an examination includes electroencephalography (EEG), electrooculography (EOG), subcutaneous electromyography (EMG), electrocardiography (ECG), monitoring of thorax movements, and pulse oximetry. In an ICU setting, the course of polysomnographic examination is additionally hampered by the presence of several sensors monitoring basic vital functions, which are attached to the patient's body (Medrzycka-Dabrowska et al. 2018).

15.10.2 Bispectral Index

The bispectral index (BIS) is used to minimize the amount of equipment required to perform EEG. BIS is calculated by the multiple analysis of the native EEG waveform, including power spectroscopy analysis, bispectral analysis, and timing analysis of the suppression to stimulus index. A BIS value ranges from 0 (no EEG signal) to one hundred, the higher numbers indicating awakeness, and a score one hundred indicates a clinically fully awake patient. BIS may potentially assess the depth of sleep, yet it is not a standard method for assessing sleep in the ICU setting (Kamdar et al. 2012a).

15.10.3 Actigraphy

Actigraphy is one of the methods for sleep assessment commonly referred to in the literature. It is a continuous method for registering a patient's activity during the day and night using a special device called an actigraph (usually lasting several days). Technically, the device comes in the form of a watch. It is placed on the patient's wrist, which enables researchers to evaluate the patient's wakefulness based on their movements. Once the subject is asleep the device detects no movement at all (Pulak and Jensen 2016). Actigraphy allows the calculation of total and average sleep duration, sleep onset latency (the length of time that it takes to fall asleep), the number of awakenings during sleep, the number of naps during the day, and the amount of active and non-active time during the day (Medrzycka-Dabrowska et al. 2018).

15.10.4 Patient Assessment

Subjective methods make it possible to assess sleep in ICU patients. One such method is the RCSQ (Richards-Campbell Sleep Questionnaire). This five-item scale evaluates the patient's perception of their depth of sleep, sleep onset latency, sleep duration, sleep onset latency after awakening, and overall sleep quality. Patients who can maintain logical communication mark their score on a scale of 0–10 (Medrzycka-Dabrowska et al. 2018).

15.10.5 Nurse Assessment

This is a subjective method where nurses may assess a patient's sleep by direct observation, as well as by using scales and questionnaires. Recent studies proved that direct observation by nurses may be insufficiently accurate for an evaluation of the amount of sleep and for identifying sleep disorders (Bourne et al. 2007). Results of a study published in 2012, which compared a patient's self-assessed sleep and an assessment performed by the nurse taking care of the patient during a particular shift using the RCSQ questionnaire, showed that nurses tend to overestimate the score (Kamdar et al. 2012b). Due to the nature of the constant care that nurses provide to patients, they should take sleep and rest into consideration in the process of taking care of the patient (Medrzycka-Dabrowska et al. 2018) (Table 15.1).

Table 15.1 Methods of sleep assessment (Medrzycka-Dabrowska et al. 2018)

Sleep assessment method	Nature and reliability	Advantages	Disadvantages	Clinical application
Polysomnography	Gold standard Inter-rater reliability in critical care, kappa = 0.79–0.83	Monitors the amount and quality of sleep. Distinguishes particular sleep stages	Continuous attendance of a technician is needed during monitoring to obtain adequate results. Partially subjective assessment during the first stage of sleep. The amount of equipment hampers sleep in non-sedated patients. Cost-expensive in the ICU setting. A patient's critical condition may result in EEG disturbances	Not performed in routine clinical sleep assessment
BIS	All patients with BIS values below 80 were asleep	BIS sensor easily applied. The entire equipment to measure BIS is easy to use. Inexpensive examination. Outcome is readily available	BIS sensors may disturb the patient. Muscle movements may disturb BIS measurement. Once the measurement is complete, it is necessary to upload the data to a personal computer for analysis	Not performed in routine clinical sleep assessment. Validation of the examination and algorithm development are required
Actigraphy	Correlation with polysomnography for TST 0.72–0.98 (Chang et al. 2020; Grossman et al. 2017). Not validated versus polysomnography in the ICU setting	Noninvasive. May be performed by non-technicians and non-specialists. Inexpensive once the device and accessories are purchased	Neuromuscular fatigue increases the risk of overestimating sleep quality. Nursing staff may displace the sensor, in the form of a watch, during nursing care interventions	Used to determine a circadian rhythm

(continued)

Table 15.1 (continued)

Sleep assessment method	Nature and reliability	Advantages	Disadvantages	Clinical application
			periods of physical inactivity—the chronic state in ICU patients, may be scored by the sensor as sleep	
Patient assessment	Sleep in the Intensive Care Unit Questionnaire—not validated versus polysomnography (Yoder et al. 2012) Richards-Campbell Sleep Questionnaire reliability (Cronbach's alpha = 0.90) correlation versus polysomnographic sleep efficiency index in ICU patients = 0.58	If capable, a patient can compare the baseline sleep value with the one experienced. Relatively quick	Cannot be used in patients with cognitive dysfunction. Memory problems may reduce its accuracy. Cannot be performed in sedated patients. Circadian rhythm disturbances may adversely affect nocturnal sleep perception	RCSQ scale does not apply to sedated patients, patients with cognitive dysfunction and with delirium
Nurse assessment	Direct observation at 15-min intervals, assessment of sleep state compared to polysomnography correct for 81.9% of the time (Honarmand et al. 2020). RCSQ—reliability versus patient's assessment (Cronbach's alpha = 0.83–0.95)	Easy to perform during routine nursing care	Overestimated TST. Frequent patient assessment is required. Risk of data loss caused by other direct and indirect nursing care activities. It is not possible to present an accurate report of the patient's total sleep quality	Easy and targeted tool for sleep assessment on ICU patients who meet the requirements for such an evaluation

15.11 Strategies to Promote Sleep in the ICU

It has been proven that increasing awareness of this issue and interaction within the therapeutic team may minimize sleep disorders. The main measures should aim at increasing the comfort of patients, reducing light and noise intensity at night, and an appropriate organization and aggregation of nursing care interventions to prevent sleep interruptions. An adequate mode of mechanical ventilation is also important for the comfort of patients. Regular assessments of pain intensity and appropriate methods of pain control play a key role as well (Medrzycka-Dabrowska et al. 2018).

15.11.1 Vital Sign Monitoring

Vital sign monitoring is an essential part of hospital routines. Yet, nocturnal measurements (normally repeated 2–3 times overnight) could impair patients' sleep quality.

An intervention to the said problem is to introduce wireless technologies that transmit patient vital signs information to a central location (for example heart rate, body temperature, blood glucose level, and blood pressure) (Tan et al. 2019). Some portable, continuous vital sign monitoring systems have already been piloted in the clinical setting and might be promising for promoting sleep while safeguarding early recognition of possible deterioration (Weenk et al. 2017).

15.11.2 Physical Activity

Reduced physical activity may result in impaired sleep. Therefore, interventions such as encouraging patients to take a walk outside, serving meals in a separate room, or offering bed exercises, all of which are likely to aid sleep and support the maintenance of a normal circadian rhythm. Additionally, such interventions increase exposure to natural light with additional direct positive effects on sleep and circadian rhythm (Tan et al. 2019).

15.11.3 Pain

Pain and sleep interact in a bidirectional manner (Bjurstrom and Irwin 2016) and could therefore cause a vicious cycle. Chronotherapy, referring to the appropriate timing of medication to relieve illness symptoms at times when they are most disruptive (for example: by targeting pain during nocturnal sleep) should be considered (Tan et al. 2019).

15.11.4 Ambient Temperature

In an article that studied the impact of temperature, researchers exposed 40 patients to different nighttime temperatures over three nights and found increased total sleep time and greater sleep efficiency on nights with lower temperatures (DuBose and Hadi 2016).

15.11.5 Melatonin Therapy

Reduction in plasma melatonin levels, low melatonin secretion, and lack of circadian increase in melatonin has been found in critical care patients undergoing mechanical ventilation. In contrast to other sedatives and hypnotics, melatonin causes minimal side effects even in extremely high doses (700 mg/day). Tolerance to melatonin does not seem to develop during treatment as it does with other sedatives, so there is no need for increasing doses for sleep induction (Shilo et al. 2000). Melatonin therapy is effective in the resetting of sleep–wake cycles and the entrainment of circadian rhythms (Pulak and Jensen 2016).

15.11.6 Noise Reduction

Noise represents an easily modifiable factor that may impact patients' sleep patterns in an ICU setting. Patient satisfaction is said to go up when the unit is less noisy, and sleep is improved. The introduction of guidelines to decrease noise levels includes relocating all acoustic alarms to outside the patient's rooms; adapting all mechanical alarms (pumps and ventilators) to the nighttime environment; closing the door or curtain if this can be safely done; and discouraging noisy tasks at night such as garbage emptying and linen collection (Pulak and Jensen 2016).

Basic noise reduction interventions such as earplugs and white noise machines have been shown to promote longer and better-quality sleep. Newer ICUs can consider noise reduction in their structural designs. Quiet time to allow naps in the afternoon and evening hours is a simple policy that may increase patient satisfaction with sleep, especially as patient satisfaction is becoming an important benchmark of care (Chang et al. 2020).

15.11.7 Light

Promoting sleep–wake rhythms through various day/night environmental interventions could improve the patients' naturally occurring circadian rhythms and was associated with a reduction of the prevalence of delirium. Such interventions comprise sufficient daylight, encouragement of physical and cognitive activities during the day, dimming the lights, closing curtains, providing eye masks and/or earplugs, and reducing nursing activities during the night (Nilius et al. 2021).

Daylight plays an important role in sleep regulation. It suppresses melatonin, increasing alertness and entraining the circadian rhythm to support better sleep at night. Elderly patients exposed to bright daytime light for a week slept longer and awoke later. The use of bright light in the early evening was also found to be effective in increasing total sleep time and reducing awakenings for elderly patients (DuBose and Hadi 2016).

15.12 Conclusion

Sleep is an essential component of daily life. It plays a more important role in people with ailments as it aids in recovery and promotes general health. The ICU is a high-intensity environment where the technical and analytical aspects of healthcare could be given importance over the psychological aspects and comfort of the patient. While this approach has its benefits, through this chapter we have highlighted the importance of sleep and the effects of sleep disruption. Healthcare workers have to be mindful of both the impacts of sleep and sleep disorders and take an active approach to promote sleep among patients in hospital and ICU settings. This is of particular importance because recent studies have shown an overall significant reduction in the length of stay for hospitalized and ICU patients. Each patient has several factors that impact sleep in the ICU, at the same time each factor is responsible only for a small percentage of the sleep disruption. Thus, a more complete action plan with multiple approaches to promote sleep should be implemented. Finally, more information on the approaches to promote sleep in ICU patients must be discovered through robust studies to tackle this problem.

References

Beauchemin KM, Hays P (1998) Dying in the dark: sunshine, gender, and outcomes in myocardial infarction. J R Soc Med 91(7):352–354. https://doi.org/10.1177/014107689809100703

Bihari S, Doug McEvoy R, Matheson E, Kim S, Woodman RJ, Bersten AD (2012) Factors affecting sleep quality of patients in intensive care unit. J Clin Sleep Med 8(3):301–307. https://doi.org/10.5664/jcsm.1920

Bjurstrom MF, Irwin MR (2016) Polysomnographic characteristics in nonmalignant chronic pain populations: a review of controlled studies. Sleep Med Rev 26:74–86. https://doi.org/10.1016/j.smrv.2015.03.004

Bourne RS, Mills GH (2004) Sleep disruption in critically ill patients—pharmacological considerations. Anaesthesia 59(4):374–384. https://doi.org/10.1111/j.1365-2044.2004.03664.x

Bourne RS, Minelli C, Mills GH, Kandler R (2007) Clinical review: sleep measurement in critical care patients: research and clinical implications. Critical care (London, England) 11(4):226. https://doi.org/10.1186/cc5966

Cabello B, Thille AW, Drouot X, Galia F, Mancebo J, d'Ortho MP, Brochard L (2008) Sleep quality in mechanically ventilated patients: comparison of three ventilatory modes. Crit Care Med 36(6):1749–1755. https://doi.org/10.1097/CCM.0b013e3181743f41

Caldwell JR, Rapoport RJ, Davis JC, Offenberg HL, Marker HW, Roth SH, Yuan W, Eliot L, Babul N, Lynch PM (2002) Efficacy and safety of a once-daily morphine formulation in chronic, moderate-to-severe osteoarthritis pain: results from a randomized, placebo-controlled, double-blind trial and an open-label extension trial. J Pain Symptom Manag 23(4):278–291. https://doi.org/10.1016/s0885-3924(02)00383-4

Chang VA, Owens RL, LaBuzetta JN (2020) Impact of sleep deprivation in the neurological intensive care unit: a narrative review. Neurocrit Care 32(2):596–608. https://doi.org/10.1007/s12028-019-00795-4

Choi J, Beltrán LO, Kim H (2012) Impacts of indoor daylight environments on patient average length of stay (ALOS) in a healthcare facility. Build Environ 50:65–75

Delaney LJ, Van Haren F, Lopez V (2015) Sleeping on a problem: the impact of sleep disturbance on intensive care patients—a clinical review. Ann Intensive Care 5:3. https://doi.org/10.1186/s13613-015-0043-2

Drouot X, Roche-Campo F, Thille AW, Cabello B, Galia F, Margarit L, d'Ortho MP, Brochard L (2012) A new classification for sleep analysis in critically ill patients. Sleep Med 13(1):7–14. https://doi.org/10.1016/j.sleep.2011.07.012

DuBose JR, Hadi K (2016) Improving inpatient environments to support patient sleep. Int J Qual Health Care 28(5):540–553. https://doi.org/10.1093/intqhc/mzw079

Dunn H, Anderson MA, Hill PD (2010) Nighttime lighting in intensive care units. Crit Care Nurse 30(3):31–37. https://doi.org/10.4037/ccn2010342

Freedman NS, Gazendam J, Levan L, Pack AI, Schwab RJ (2001) Abnormal sleep/wake cycles and the effect of environmental noise on sleep disruption in the intensive care unit. Am J Respir Crit Care Med 163(2):451–457. https://doi.org/10.1164/ajrccm.163.2.9912128

Granja C, Lopes A, Moreira S, Dias C, Costa-Pereira A, Carneiro A, JMIP Study Group (2005) Patients' recollections of experiences in the intensive care unit may affect their quality of life. Critical care (London, England) 9(2):R96–R109. https://doi.org/10.1186/cc3026

Grossman MN, Anderson SL, Worku A, Marsack W, Desai N, Tuvilleja A, Ramos J, Francisco MA, Lafond C, Balachandran JS, Mokhlesi B, Farnan JM, Meltzer DO, Arora VM (2017) Awakenings? Patient and hospital staff perceptions of nighttime disruptions and their effect on patient sleep. J Clin Sleep Med 13(2):301–306. https://doi.org/10.5664/jcsm.6468

Honarmand K, Rafay H, Le J, Mohan S, Rochwerg B, Devlin JW, Skrobik Y, Weinhouse GL, Drouot X, Watson PL, McKinley S, Bosma KJ (2020) A systematic review of risk factors for sleep disruption in critically ill adults. Crit Care Med 48(7):1066–1074. https://doi.org/10.1097/CCM.0000000000004405

Huang HW, Zheng BL, Jiang L, Lin ZT, Zhang GB, Shen L, Xi XM (2015) Effect of oral melatonin and wearing earplugs and eye masks on nocturnal sleep-in healthy subjects in a simulated intensive care unit environment: which might be a more promising strategy for ICU sleep deprivation? Crit Care (London, England) 19(1):124. https://doi.org/10.1186/s13054-015-0842-8

Kamdar BB, Needham DM, Collop NA (2012a) Sleep deprivation in critical illness: its role in physical and psychological recovery. J Intensive Care Med 27(2):97–111. https://doi.org/10.1177/0885066610394322

Kamdar BB, Shah PA, King LM, Kho ME, Zhou X, Colantuoni E, Collop NA, Needham DM (2012b) Patient-nurse interrater reliability and agreement of the Richards-Campbell sleep questionnaire. Am J Crit Care 21(4):261–269. https://doi.org/10.4037/ajcc2012111

Le A, Friese RS, Hsu CH, Wynne JL, Rhee P, O'Keeffe T (2012) Sleep disruptions and nocturnal nursing interactions in the intensive care unit. J Surg Res 177(2):310–314. https://doi.org/10.1016/j.jss.2012.05.038

Lee CY, Low LP, Twinn S (2007) Older men's experiences of sleep in the hospital. J Clin Nurs 16(2):336–343. https://doi.org/10.1111/j.1365-2702.2005.01491.x
Lewandowska K, Mędrzycka-Dąbrowska W, Pilch D, Wach K, Fortunato A, Krupa S, Ozga D (2020) Sleep deprivation from the perspective of a patient hospitalized in the intensive care unit-qualitative study. Healthcare (Basel, Switzerland) 8(3):351. https://doi.org/10.3390/healthcare8030351
Medrzycka-Dabrowska W, Lewandowska K, Kwiecień-Jaguś K, Czyż-Szypenbajl K (2018) Sleep deprivation in Intensive Care Unit—systematic review. Open Med 13(1):384–393. https://doi.org/10.1515/med-2018-0057
Meyer TJ, Eveloff SE, Bauer MS, Schwartz WA, Hill NS, Millman RP (1994) Adverse environmental conditions in the respiratory and medical ICU settings. Chest 105(4):1211–1216. https://doi.org/10.1378/chest.105.4.1211
Mukherjee S, Patel SR, Kales SN, Ayas NT, Strohl KP, Gozal D, Malhotra A, American Thoracic Society Ad hoc Committee on Healthy Sleep (2015) An official American Thoracic Society statement: the importance of healthy sleep. Recommendations and future priorities. Am J Respir Crit Care Med 191(12):1450–1458. https://doi.org/10.1164/rccm.201504-0767ST
Murphy PJ, Badia P, Myers BL, Boecker MR, Wright KP Jr (1994) Nonsteroidal anti-inflammatory drugs affect normal sleep patterns in humans. Physiol Behav 55(6):1063–1066. https://doi.org/10.1016/0031-9384(94)90388-3
Nilius G, Richter M, Schroeder M (2021) Updated perspectives on the management of sleep disorders in the intensive care unit. Nat Sci Sleep 13:751–762. https://doi.org/10.2147/NSS.S284846
Parthasarathy S, Tobin MJ (2002) Effect of ventilator mode on sleep quality in critically ill patients. Am J Respir Crit Care Med 166(11):1423–1429. https://doi.org/10.1164/rccm.200209-999OC
Pulak LM, Jensen L (2016) Sleep in the intensive care unit: a review. J Intensive Care Med 31(1): 14–23. https://doi.org/10.1177/0885066614538749
Shilo L, Dagan Y, Smorjik Y, Weinberg U, Dolev S, Komptel B, Shenkman L (2000) Effect of melatonin on sleep quality of COPD intensive care patients: a pilot study. Chronobiol Int 17(1): 71–76. https://doi.org/10.1081/cbi-10010103
Spengler CM, Shea SA (2000) Sleep deprivation per se does not decrease the hypercapnic ventilatory response in humans. Am J Respir Crit Care Med 161(4 Pt 1):1124–1128. https://doi.org/10.1164/ajrccm.161.4.9906026
Spiegel K, Leproult R, Van Cauter E (1999) Impact of sleep debt on metabolic and endocrine function. Lancet (London, England) 354(9188):1435–1439. https://doi.org/10.1016/S0140-6736(99)01376-8
Tan X, van Egmond L, Partinen M, Lange T, Benedict C (2019) A narrative review of interventions for improving sleep and reducing circadian disruption in medical inpatients. Sleep Med 59:42–50. https://doi.org/10.1016/j.sleep.2018.08.007
Toublanc B, Rose D, Glérant JC, Francois G, Mayeux I, Rodenstein D, Jounieaux V (2007) Assist-control ventilation vs. low levels of pressure support ventilation on sleep quality in intubated ICU patients. Intensive Care Med 33(7):1148–1154. https://doi.org/10.1007/s00134-007-0659-2
Weenk M, van Goor H, Frietman B, Engelen LJ, van Laarhoven CJ, Smit J, Bredie SJ, van de Belt TH (2017) Continuous monitoring of vital signs using wearable devices on the general Ward: pilot study. JMIR Mhealth Uhealth 5(7):e91. https://doi.org/10.2196/mhealth.7208
Wolkowitz OM, Rubinow D, Doran AR, Breier A, Berrettini WH, Kling MA, Pickar D (1990) Prednisone effects on neurochemistry and behavior. Preliminary findings. Arch Gen Psychiatry 47(10):963–968. https://doi.org/10.1001/archpsyc.1990.01810220079010
Yadav SK, Kumar R, Macey PM, Richardson HL, Wang DJ, Woo MA, Harper RM (2013) Regional cerebral blood flow alterations in obstructive sleep apnea. Neurosci Lett 555:159–164. https://doi.org/10.1016/j.neulet.2013.09.033

Yoder JC, Staisiunas PG, Meltzer DO, Knutson KL, Arora VM (2012) Noise and sleep among adult medical inpatients: far from a quiet night. Arch Intern Med 172(1):68–70. https://doi.org/10.1001/archinternmed.2011.603

Zhong X, Hilton HJ, Gates GJ, Jelic S, Stern Y, Bartels MN, Demeersman RE, Basner RC (2005) Increased sympathetic and decreased parasympathetic cardiovascular modulation in normal humans with acute sleep deprivation. J Appl Physiol (Bethesda, Md.: 1985) 98(6): 2024–2032. https://doi.org/10.1152/japplphysiol.00620.2004

Chapter 16
Recent Advances on Sleep During Pregnancy and Postpartum

Ana Rita Brito and Miguel Meira e Cruz

Abstract Sleep differences between genders account for several biological factors impacted by a combination of environmental, social, and cultural influences. Anatomical, physiological, and psychological aspects affect sleep during pregnancy and often persist postpartum. For instance, sleep difficulties are often a common complaint in this vulnerable period of women's life. While indirect aspects such as back pain, gastroesophageal reflux, increased micturition, and night-time difficulties on repositioning, are frequent complaints among pregnant, there is a significant increase in the prevalence of particular sleep disorders like obstructive sleep apnea and restless leg syndrome. The overlap between symptoms related to pregnancy itself and concomitant affective disorders can seriously affect the mother's health, also impacting delivery-associated outcomes. As it is well known the putative role of inadequate or insufficient sleep in the cardiometabolic, neurohumoral, and psych-affective domains of mother health is associated with increased rates of illness and death in this population, recognition of this clinical relationship and adequate management of sleep health during pregnancy and postpartum should be a goal to be included as the standard of care. Therefore, in this chapter, the authors summarized the main interactions between sleep, complaints, sleep disorders, and health-related outcomes in pregnancy and postpartum with a focus on the new challenging strategies that may be implemented to guide a successful therapeutic approach.

Keywords Sleep differences · Gender · Pregnancy · Sleep disorders · Postpartum

A. R. Brito
Evelima Children's Hospital, Guy's and St Thomas NHS Foundation Trust, London, UK

M. Meira e Cruz (✉)
Cardiovascular Center, University of Lisbon, Lisbon School of Medicine, Lisbon, Portugal

European Sleep Center, Lisbon, Portugal
e-mail: mcruz@medicina.ulisboa.pt

A. S. BaHammam, M. Hunasikatti (eds.), *Sleep Apnea Frontiers*, Progress in Sleep Research, https://doi.org/10.1007/978-981-99-7901-1_16

16.1 Introduction

Sleep differs between men and women. There are significant gender differences in many sleep-regulating hormones concentration and specific female conditions, menstruation, pregnancy, and menopause induce changes in sleep (Porkka-Heiskanen et al. 2014).

Pregnancy induces sleep modulations as sleep efficiency already decreases during pregnancy from the first trimester (Porkka-Heiskanen et al. 2014) ensuring adequate sleep during pregnancy is crucial for well the-being of both the mother and the baby. Nevertheless women, frequently encounter sleep disruptions and a decline in sleep quality throughout the course of pregnancy and after giving birth. Circadian rhythms are closely linked to sleep problems and mood disorders (Gallaher et al. 2018).

Problems with sleep during this vulnerable period are reflected in the mother, with a higher risk for pregnancy-related comorbidities, mood dysregulation, and potential increased complications with delivery, but also in the fetus development (Porkka-Heiskanen et al. 2014). Sleep also has a significant impact on memory, cognitive abilities, hunger emotional state, and the ability to make decisions—all of which are crucial when getting ready to welcome a new baby into your household. For the mother, especially in the period after delivery with a newborn in hand, the nights without sleep result in tiredness and drowsiness during the day. For that reason, sleep education and support during pregnancy have been recently advocated.

Postpartum emotional distress is very common, with 10–20% of postpartum women reporting depressive or anxiety disorders. Sleep is a modifiable risk factor for emotional distress that has a pivotal role in postpartum adjustment (Dørheim et al. 2009).

16.2 Sleep in Women

Differences in hormone secretion profile between men and women affect sleep. Women report consistently worse sleep quality than men in all age groups. Sex steroids affect brain functions such as cognitive performance, mood, and sleep. In humans, estrogen decreases awakenings and promotes REM sleep by increasing time in REM and decreasing REM sleep latency (Manber and Armitage 1999).

In addition to sex steroids, hormones such as prolactin, growth hormone, cortisol, and melatonin have gender-dependent effects on sleep (Steiger 2002). Young women have higher melatonin rhythm amplitude but lower temperature rhythm amplitude than men, and sleep in women occurs later concerning biological time than in men. This might suggest that this delayed biological time of sleep can contribute to higher rates of Insomnia in women compared to men (Cain et al. 2010).

Regarding sleep disturbance, the most common in women is insomnia, with a prevalence of approximately 10–20%, and sleep disturbance is a risk factor for increased morbidity and mortality (Buysse 2013). Women are 1.3–1.6 times more likely to report sleep disturbances than men but the reasons for this are not very clear yet (Hajak 2001).

16.3 Sleep Architecture During Pregnancy and Postpartum

Poor sleep is widely recognized as a common complaint in pregnancy, and yet the data around the exact nature of these complaints and their implications for fetal and maternal health are incomplete. Also, there is uncertainty around how to best recognize and address these significant health issues.

A recent literature review suggests sleep architecture changes begin as early as the first trimester. There is evidence that primigravid sleep structure never returns to pre-pregnancy levels after birth. In addition, cesarean delivery frequency, early labor, labor length, depression, gestational hypertension, and gestational diabetes all appear to be influenced by sleep changes (Sweet et al. 2020).

A systematic review done by Gallaher and collaborators provided a summary of studies of circadian rhythms and associated sleep problems and maternal distress, among postpartum women. Circadian rhythm disturbances were strongly correlated with depression, social factors, and the mother's exposure to light postpartum (Gallaher et al. 2018).

The few well-controlled studies show alterations in sleep patterns consistent with subjective reports in assessments. The studies show that sleep efficiency already decreases during pregnancy, from the first trimester and especially in first-time mothers. However, some studies revealed that the total sleep time increases in late pregnancy or shows a U-shape with an increase in the first, a decrease in the second, and an increase in the third trimester (Lee 2000). Sleep onset and sleep maintenance are prolonged in pregnant women, with sleep latency being, on average, 10 min longer during pregnancy. The number of awakenings and the wake time after sleep onset increase significantly compared to non-pregnant (Lee 2000).

As pregnancy progresses, self-reported sleep quality worsens, with decreases in reported sleep duration and sleep efficiency (Hedman et al. 2002). In addition, over 40% of pregnant women suffer from insomnia in the first trimester, and this increases to over 60% by the third trimester (Román-Gálvez et al. 2018).

There is acute sleep loss seen on the nights preceding and following the birth, regardless of the time of day of the birth or manner of delivery (Beebe and Lee 2007). Following the birth, sleep quality continues to deteriorate, with worsened sleep efficiency (Kang et al. 2002) and further sleep loss at night, although some studies show relatively stable total sleep duration with the inclusion of scattered daytime naps (Bei et al. 2012). By the end of the first 3 months postpartum, the mothers' sleep has generally begun to improve but does not necessarily return to pre-pregnancy levels (Lee 2000; Kang et al. 2002).

Sleep efficiency is lower during postpartum (weeks 9–12) compared to women in late pregnancy, but it improves by 3–5 months and the total sleep time also decreases in postpartum. Increased wakefulness after sleep onset persists for 10 weeks after delivery but reduces by 3–5 months postpartum.

REM sleep duration decreases during the postpartum period and REM latency is shorter during the first month after delivery compared to the third trimester (Lee 2000). Postpartum breast feeding women seem to present increased slow wave sleep and lower amounts of light NREM sleep compared to bottle-feeding women or to age-matched non-pregnant women. The awakening threshold becomes lower during postpartum compared to late pregnant women too. Polysomnographic data from the third trimester has also shown increased wake after sleep onset and decreased rapid eye movement (REM) sleep compared to non-pregnant women (Hertz et al. 1992).

16.4 Common Sleep Problems in Pregnancy and the Postpartum Period and Its Consequences

Pregnancy-related sleep disturbances are already reported in the first trimester, with a prevalence of 10–15% (Baratte-Beebe and Lee 1999). Daytime sleepiness and fatigue are very common and may even be the first sign of pregnancy.

Women usually report sleep maintenance insomnia and nocturia as the main sleep problems, being this last one the most important cause of awakenings in more than 50% of night-time awakenings for 85–95% of pregnant women (Hedman et al. 2002; Baratte-Beebe and Lee 1999; Ward 2010). Nausea and vomiting, leg cramps, join pain, back pain, uncomfortable or unusual sleeping position, nasal congestion, thermoregulation problems, anxiety, and worries and nightmares are other problems that arose during pregnant women sleep (Baratte-Beebe and Lee 1999).

Although sleep challenges and their daytime consequences may be relieved during the second trimester, 20% of pregnant women report sleep difficulties (Lee 2000). Fetal movements and heartburn due to gastro-esophageal reflux may cause arousal and deteriorate sleep quality as pregnancy progresses. Sleep disturbances continue during the third trimester, remaining maintenance insomnia the most frequent form of insomnia as, not only the number but also the length of nocturnal awakenings increase, leading to decreased sleep efficiency (Hedman et al. 2002).

Sleep movement disorders, such as periodic limb movements and restless legs syndrome, are also frequent and interrupt sleep (Ward 2010; Manconi et al. 2012). Sleep-disordered breathing (SDB) and mood symptoms, especially depression and anxiety around delivery and motherhood may also be reasons for sleep problems (Ward 2010; Goyal et al. 2007; Skouteris et al. 2008).

A cross-sectional, population-based large study was conducted with 2830 women after they became mothers. The Pittsburgh Sleep Quality Index (PSQI) was utilized assess to sleep patterns, and depressive symptoms using the Edinburgh Postnatal Depression Scale (EPDS). The prevalence of sleep problems (PSQI > 5) was 57.7%, and the prevalence of depression (EPDS ≥ 10) was 16.5%.

Depression, previous sleep problems, being primiparous, not exclusively breastfeeding, or having a younger or male infant were factors associated with poor postpartum sleep quality. Poor sleep was also associated with depression when adjusted for other significant risk factors for depression, such as poor partner relationships, previous depression, depression during pregnancy, and stressful life events. The factors of sleep that had the strongest correlation with depression were sleep disruptions and the individual's perception of sleep quality (Dørheim et al. 2009).

During the postpartum period, sleep disruptions continue at least for the first 3 months after delivery and are more marked in first-time mothers, as the sleep efficiency drops from 90 to 77% from the third trimester to postpartum. The connection between sleep disruption and anxiety as well as depression has been well documented during pregnancy (Sedov and Tomfohr-Madsen 2021).

Maghami et al. found a significant relationship between sleep disturbances during pregnancy and postpartum depression (PPD). Women with sleep disorders are at an increased risk of developing PPD, which shows the importance of screening pregnant mothers for sleep disturbances. Furthermore, they discovered that as pregnancy progressed, the risk of postpartum depression (PPD) increased (Maghami et al. 2021).

Stress-related sleep disturbance was reported by 33.2% of women in a cross-sectional study that included a total of 2051 pregnant women in Peru. From this group of women, 24.9% had antepartum depression, 32.2% had generalized anxiety disorder, 30.9% had PTSD, and 27.6% were assessed as having a high risk of psychosis. It was found that women with stress-related sleep disturbances were more likely to experience antepartum depression (OR = 2.74; 95%CI: 2.22–3.38), generalized anxiety disorder (OR = 2.48; 95%CI: 2.04–3.02), PTSD (OR = 2.36; 95%CI: 1.93–2.88), and high risk for psychosis (OR = 2.07; 95%CI: 1.69–2.54) as compared to women without stress-related sleep disturbances (Sanchez et al. 2020).

Furthermore, they discovered that as pregnancy progressed, the risk of postpartum depression (PPD) increased (aOR: 2.77; 95% CI: 1.81–4.24), gestational diabetes mellitus (aOR: 1.96; 95% CI: 1.62–2.38), a cesarean delivery (aOR: 1.99; 95% CI: 1.70–2.33), depression (aOR: 3.98; 95% CI: 2.74–5.77), and preterm birth (aOR: 1.95; 95% CI: 1.55–2.45). The aim of this systematic review and meta-analysis was to further assess the overall relationship between four types of sleep disturbances (snoring, obstructive sleep apnea, extreme sleep duration, and poor sleep quality) and adverse perinatal outcomes (Yang et al. 2022).

16.4.1 Insomnia

Insomnia is defined as dissatisfaction with sleep quantity or quality, accompanied by associated distress or impairment during the daytime, that occurs at least three times a week for at least 3 months and is not attributable to another condition or a substance. This disorder is commonly experienced in adults, with prevalence rates ranging from 6 to 10% (American Psychiatric Association, 2013), insomnia complaints are more prevalent in samples of pregnant women (Mindell and Jacobson 2000; Mindell et al. 2015).

During late pregnancy, sleep complaints are common and multifactorial. Not all women with sleep complaints meet the diagnostic criteria for Insomnia. A study with 159 first-time mothers, with singleton pregnancies and without severe physical/ psychiatric conditions, reported that 37.1% of the women scored >7 on the Insomnia Severity Index, and 18.9% met DSM-5 criteria for Insomnia but without the 3 months duration criteria (Lee et al. 2018).

A meta-analysis about Insomnia symptoms during pregnancy included the analysis of 24 studies with a total of 15,564 participants. This analysis revealed that the overall prevalence of insomnia symptoms across samples was 38.2%, which is higher than expected in comparison to the non-pregnant population (Sedov et al. 2021). Large survey data indicate the prevalence of insomnia symptoms in women to be between 8.6% (Uhlig et al. 2014) and 15.6% (Morin et al. 2011). Therefore, women appear to be reporting more insomnia symptoms during pregnancy. Additionally, as pregnancy progresses women obtain less total sleep time experience increased wakefulness after initially falling asleep, have a longer duration of lighter sleep stages, have reduced amounts of deep sleep and REM sleep and awake more frequently relative to earlier in pregnancy (Izci-Balserak et al. 2018). Moreover, subjective dissatisfaction with sleep is commonly reported by pregnant women as almost half report a poor subjective sleep quality (Sedov et al. 2021).

A recent review summarized the existing data on the prevalence of insomnia symptoms during pregnancy and explored potential moderators, including trimester, gestational age, maternal age, symptoms of anxiety, and symptoms of depression.

It seems that women with more dysfunctional beliefs about sleep may catastrophize the consequences of waking and engage in maladaptive behaviors which might perpetuate insomnia (Morin et al. 2002; Woodley and Smith 2006). It is important to explore whether there are groups of women for whom insomnia symptoms are temporary and resolve at the end of pregnancy, as opposed to persisting into the postpartum period. One trajectory analysis of sleep quality from pregnancy to 6 months postpartum found that elevated symptom profiles from the third trimester of pregnancy continued into the postpartum period in a subset of women, which would suggest the need for prevention and treatment efforts to be implemented during pregnancy (Tomfohr et al. 2015).

Natural biological changes (e.g., reduced bladder capacity or physical pain) may act as a stressor that influences pregnant women to develop insomnia. For example, women who are vulnerable to insomnia may have more difficulty settling back to sleep after waking in the middle of the night due to pregnancy-related stressors. The prevalence of insomnia was higher in the third trimester (39.7%) than in both the second (27.2%) and first trimesters (25.3%) (Sedov et al. 2021).

A more recent investigation of insomnia trajectories indicated that insomnia symptoms are relatively stable in the early part of pregnancy but temporarily increase during the third trimester for some women (Sedov and Tomfohr-Madsen 2021). In pregnancy specifically, the third trimester and maternal age of 20 years or higher have also been associated with why women older than 20 years of age are at higher risk of insomnia are somewhat unclear; however, the third trimester is likely to be associated with higher insomnia risk because of the increased physical demands of pregnancy during this time (Mindell and Jacobson 2000).

Insomnia and sleep disturbances have been linked to increased odds of cesarean delivery and preterm birth (Okun et al. 2012). Sleep complaints, even in the absence of Insomnia, are associated with less adaptive sleep-related cognitions and poorer mental health. These findings contradict the notion that during the perinatal transition, sleep complaints are "expected" and do not merit clinical attention unless Insomnia is present. Women with sleep complaints who fall short on Insomnia criteria may be vulnerable to future insomnia and mood disorders, which need to be addressed (Lee et al. 2018).

16.4.2 Sleep Movement Disorders (RLS)

Sleep movement disorders including, restless legs syndrome (RLS), are frequently sleep observed disorders that can occur during pregnancy. Research has indicated that the prevalence of RLS among pregnant women varies between 10 and 34%. Fortunately, LSR of symptoms, typically resolve completely shortly after giving birth; however, in some patients, they may continue during the postpartum period. RLS is associated with several complications in pregnancy including pre-eclampsia and increased incidence of cesareans. Despite multiple hypotheses proposed to explain this association, the complete pathogenesis of restless legs syndrome (RLS) during pregnancy remains unclear. Current understanding suggests that factors such as a strong family history, low serum iron and ferritin levels, and high estrogen levels during pregnancy may play important roles. Vitamin D deficiency and disturbances in calcium metabolism may also contribute. Managing RLS medically during pregnancy is challenging because of the potential hazards it poses to both the mother and the developing fetus. However, in some cases, the severity of the condition may require treatment (Gupta et al. 2016).

A 2017 meta-analysis estimated the prevalence of RLS during pregnancy to be as high as 21%, compared to 2–10% in the general population, making it the most common gestational movement disorder and a significant contributor to reduced sleep quality and duration across all trimesters (Chen et al. 2018; Allen et al. 2005; Suzuki et al. 2003). A systematic review conducted in 2019 revealed that two very large studies found a statistically significant correlation between gestational RLS and pregnancy-induced hypertension (Darvishi et al. 2020; Liu et al. 2016; Ma et al. 2015). Three other smaller studies remained inconclusive regarding this correlation (Esposito et al. 2017; Sharma et al. 2016; Oyieng'o et al. 2016). This review also pointed toward a potentially strong relationship between peripartum depression and RLS during pregnancy, and a possible association between pre-eclampsia and RLS (Liu et al. 2016; Sarberg et al. 2016; Wesström et al. 2014).

Only a minority of women with RLS during pregnancy (6.5–9.9%) were already affected by the syndrome prior to their pregnancy (Manconi et al. 2012). Their symptoms usually worsened during the gestational period, with increasing prevalence and severity from the first (15.6%) to the last (38.8%) trimester, mildly decreasing during the last month of pregnancy (Neyal et al. 2015). Several longitudinal studies have demonstrated the disappearance of symptoms at the time of delivery, maintained thereafter throughout the puerperium. However, some authors found a high (34.8%) percentage of women were still symptomatic after delivery. Residual postpartum RLS correlated with lower ferritin, multiparity, and symptom onset during the second trimester (Neyal et al. 2015).

16.4.3 Sleep Disorder Breathing (SDB)

As pregnancy might be a vulnerable period, even a very low apnea–hypopnea index (AHI) may have unfavorable effects, especially on fetal outcomes. Physiological and hormonal changes during pregnancy affect. Sleep and breathing may either protect or even predispose for sleep disorder breathing (SDB) as there is an increased minute ventilation, preference for the lateral sleeping position during the third trimester and a decrease in REM sleep.

Gestational weight gain, decreased lung function reserve capacity, and increased arousal, increase the like hood of periodic breathing and obstructive sleep apnea (OSA) (Porkka-Heiskanen et al. 2014).

Progesterone acts as a respiratory stimulant and may protect from SDB during pregnancy despite gestational weight gain. This hormone also increases phrenic nerve activity and when negative upper airway pressure and the progesterone-induced respiratory drive increase during pregnancy, the propensity to upper airway collapse and periodic breathing also increases. Estrogen upregulates progesterone receptors and contributes to mucosal edema and mucus hypersecretion, which leads to the narrowing of the upper airway and increased resistance (Porkka-Heiskanen et al. 2014). Snoring is also common during pregnancy, ranging from 10 to 45% and up to 75% in cases of pre-eclampsia. Both SDB and snoring exacerbate in late pregnancy and alleviate or resolve completely within 3 months postpartum (Edwards and Sullivan 2008). Increased sympathetic activity during the third trimester contributes to the increased prevalence of SDB, which further increases tone and predisposes to pre-eclampsia during the third trimester (Edwards and Sullivan 2008). Snoring also seems to be associated with increased risk of gestational diabetes, in overweight pregnant women.

In the non-pregnant population, sleep-disordered breathing (SDB) confers a threefold increase in the risk of hypertension independent of other risk factors (Chen and He 2007). Hypertensive disorders of pregnancy (HDP) are the most common and serious medical disorders in pregnancy, affecting 3–8% of pregnant women, and responsible for over 60,000 maternal deaths and far greater numbers of perinatal deaths worldwide every year. Gestational hypertension (GH) is characterized by the new onset of hypertension after 20 weeks gestation, whereas pre-eclampsia (PE) is diagnosed when GH is accompanied by multisystem organ dysfunction (Redman and Sargent 2005; Sibai et al. 2005).

SDB during pregnancy appears to be linked to adverse outcomes for the mother and baby (Bourjeily et al. 2011; Fung et al. 2012) and was significantly associated with GH and pre-eclampsia in a recent large prospective cohort study (Facco et al. 2017). On the other hand, obesity is a risk factor for both increased blood pressure during pregnancy and PE (Wilson et al. 2018), it is also a well-recognized risk factor for SDB (Young et al. 2005) and a determinant of developing SDB in pregnancy (Pien et al. 2014).

A recent study explored the relationship between HDP and SDB (assessed with full polysomnography), using a closely matched design to control for the effect of obesity. Research hypothesized that pregnant women with HDP (both GH and PE) will have a higher rate of SDB compared with BMI- and gestation-matched normotensive pregnant women (Wilson et al. 2018). This single-center, cross-sectional study compared the prevalence of SDB in late pregnancy among women with HDP to women with uncomplicated pregnancies. The frequency of sleep-disordered breathing in the cases was 52.5% compared with 37.5% in the control group (Wilson et al. 2018).

More severe sleep-disordered breathing was more than twice as common in women with gestational hypertension or pre-eclampsia (35% versus 15%, $P = 0.039$). While more than half of women with a hypertensive disorder of pregnancy meet the clinical criteria for sleep-disordered breathing, it is also very common in normotensive women of similar body mass index.

Obstructive sleep apnea (OSA) occurs in one-fifth of women in late pregnancy and the severity of OSA is independently associated with symptoms of depression, while the presence of OSA is associated with symptoms of anxiety in pregnancy. Data from 126 women were analyzed for the presence of OSA (apnea–hypopnea index, AHI, ≥ 5) in an observational study where OSA was observed in 28 (22.2%) women. Symptoms of depression (defined as an EPDS score of ≥ 13) were observed in 11 (8.7%) women and symptoms of anxiety were observed in 16 (12.7%) women. Compared to women without depressive symptoms, those with depressive symptoms had higher AHI ($p = 0.03$). Women with symptoms of anxiety were more likely to have sleep apnea ($AHI \geq 5$; $p = 0.007$). Multivariate analyses showed that depressive symptoms were related to AHI and antidepressant use ($p < 0.001$) and symptoms of anxiety were related to the presence of sleep apnea ($AHI \geq 5$) and antidepressant use ($p < 0.001$) (Redhead et al. 2017).

16.4.4 Nocturnal Gastroesophageal Reflux Disease (Night-Time GERD)

I apologize for the error. "idAc" was a typo and should have been "Acid." Please disregard it. Here is the corrected response:

Acid reflux, also called gastroesophageal reflux, occurs when stomach acid flows back into the esophagus. While occasional reflux episodes are normal, frequent occurrences can result in the development of a medical condition called gastroesophageal reflux disease (GERD). Individuals with GERD may experience intensified symptoms, such as heartburn, particularly when they are sleeping or trying to sleep. In addition to heartburn, if stomach acid reaches the throat and larynx, a person may wake up coughing, choking, or experiencing chest pain.

Pregnant women often experience GERD, but their symptoms usually cease shortly after giving birth. During pregnancy, hormonal changes cause the valve responsible for keeping stomach acid in place to relax. As a result, existing acid reflux can worsen during this time. Furthermore, the increment of abdominal pressure during pregnancy can also contribute to this. Sleeping difficulties are frequently attributed to GERD. In a recent survey focusing on individuals experiencing frequent heartburn, it was found that approximately 60% reported GERD negatively impacting their sleep, while over 30% stated that it adversely affected their daytime performance (Vela et al. 2014; Shaker et al. 2003). Flare-ups of GERD symptoms after lying down can cause difficulties with sleep initiation and can lead to interruptions in sleep due to heartburn, chest pain, and coughing during the night. Research conducted in sleep clinics involving individuals with GERD has indicated that these symptoms are associated with reduced sleep quality (Kim et al. 2018).

16.5 New Challenges

The COVID-19 pandemic, which occurred in 2019, necessitated stringent containment measures that had varying effects on the daily lives of people worldwide. This brought new challenges to pregnancy and postpartum period to women, their families, and healthcare providers.

A cross-sectional survey was conducted with 305 postpartum women, from May to July 2020 to study postpartum women's mental health in Argentina during mandatory isolation. The study included five validated questionnaires to assess postpartum depression (Postpartum Depression Screening Scale-Short Form), insomnia (Insomnia Severity Index), memory complaints (Memory Complaint Scale), metacognition (Brief Metamemory and Meta concentration Scale), and breastfeeding self-efficacy (Breastfeeding Self-Efficacy Scale-Short Form) (Miranda et al. 2021).

The results revealed that 37% of women reported postpartum depression, 46% insomnia, 42% memory impairment, 60% low meta concentration, 50% low metamemory, and 23% low breastfeeding efficacy. Results also suggested that social isolation promoted postpartum depression and insomnia were reciprocally related, which compromised female cognition. This situation was aggravated in women during late postpartum, with older children and by low social support, and with non-exclusive breastfeeding being increased (Miranda et al. 2021).

This study also revealed higher rates of postpartum depression and insomnia compared with pre-pandemic studies. The authors pointed out the importance of Women's health being addressed by clinicians and researchers during the COVID-19 pandemic, as early evaluation and promotion of mental health will improve the health status of specific female populations (Miranda et al. 2021).

Another international study was conducted to evaluate the mental well-being of pregnant and breastfeeding women amid the pandemic. In total, 9041 women that were pregnant and breastfeeding women up to 3 months postpartum, answer the Edinburgh Depression Scale (EDS), the Generalized Anxiety Disorder seven-item scale (GAD-7), and the Perceived Stress Scale (PSS) (Ceulemans et al. 2021). This multinational study found high levels of depressive symptoms and generalized anxiety among pregnant and breastfeeding women during the COVID-19 outbreak. These findings support the importance of monitoring perinatal mental health during pandemics and perhaps overall safeguarding maternal and infant mental health (Ceulemans et al. 2021).

Other cross-sectional, multinational studies were conducted to examine factors linked to the mental health of women during the perinatal period and well-being during the COVID-19 pandemic, such as the one conducted between May and June 2020, with a total number of participants of 7562 individuals across 64 countries (Basu et al. 2021).

This study suggests that addressing the mental health needs of pregnant and postpartum women requires a population-based approach that goes beyond medical and mental healthcare requirements and address the unique challenges posed by the pandemic concerning specific concerns related to child well-being and access to medical care (Basu et al. 2021).

However, these last two studies summarized above, didn't approach the sleep quality and sleep challenges that these pregnant and new mothers faced as potential contributors to the mental health problems during the COVID-19 pandemic. Sleep disruption and insomnia increase the risk for depression and anxiety disorders independently in general populations. If these problems get amplified by pregnancy and continue in the postpartum period, sleep disturbance becomes a risk factor for maternal mental health. It has been shown that poor subjective sleep quality predicts depressive symptoms during pregnancy and postpartum (Goyal et al. 2007; Skouteris et al. 2008).

16.5.1 Therapeutic Strategies to Improve Sleep During Pregnancy and Postpartum

A healthy mother equals a healthy baby. Sleep is crucial for promoting all essential aspects of optimal human functioning. Starting sleep education and supporting the future mother and families during pregnancy can be extremely important to establish healthy rest and sleep habits which can continue for the rest of the baby's and family's life (De Carlo 2019). As shown through recent studies, this aspect became even more important facing the challenges of the COVID-19 pandemic.

On their own, pregnancy sleep issues are enough reason to address sleep and rest during pregnancy. This is critical due to the physiological and psychological impact of pregnancy on the mother, which is also felt by the unborn child. In addition, by not prioritizing rest during their days, expecting families might not understand the importance of sleep and how their sleep habits influence the unborn child and the newborn after delivery and other members of the family.

Addressing sleep issues before a baby arrives allows all family members to be involved and address their own needs, which can lead to the focus not to be only on the baby's disrupted sleep, but family's sleep and rest needs.

By supporting a woman during pregnancy and her family to face sleep challenges, the family has an opportunity to evaluate their quality of life and the choices they are making and there are high chances that the family is set up for more effective collaboration and understanding during the postpartum period (De Carlo 2019).

All these aspects can work together as the prevention of postpartum distress, depression, and anxiety. The review done by Sweet, and collaborators recently showed that current treatments for sleep disorders are based on non-pregnant populations and may not be appropriate for expecting women (Sweet et al. 2020). Moreover, disordered sleep is a more widespread and serious issue, and there are numerous obstetric and general health implications to cause clinical concern. This group also suggests that more research is needed on both electroencephalography architecture changes and treatment options for sleep problems in pregnant and postpartum women (Sweet et al. 2020).

Treatment of sleep problems during pregnancy has been shown to result in lower symptoms of postpartum depression (Khazaie et al. 2013), which has significant implications for the well-being of the family (Letourneau et al. 2013). Moreover, treatment of sleep disruptions during pregnancy may prove to be an important mechanism for prevention of further mental health problems during the postpartum period.

Given the potential negative implications of disrupted sleep in the perinatal period, the study conducted by Gueron-Sela, and collaborators suggested that future intervention studies that target sleep problems during pregnancy are required and important. This came after their main findings pointed out that mothers with shorter sleep durations during pregnancy (measured objectively with Actigraphy and sleep diaries) were more likely to belong to the high comorbidity or moderate symptoms classes compared to the low symptomology class (Gueron-Sela et al. 2021).

16.5.2 The Importance of Early Assessment

Several factors can affect sleep during pregnancy and postpartum. To establish the support required and therapeutic approaches, it is essential to investigate the aspects that are affecting sleep and how they are manifesting themselves.

An attentive and timely therapeutic approach reflecting the multifaceted interactions of anatomical, physiological, psychological, and hormonal alterations experienced by pregnant women is indispensable to prevent avertible maternal morbidity and adverse birth outcomes (Silvestri and Aricò 2019).

A multicenter retrospective study was conducted using clinical data of women who delivered at term between 2014 and 2018 at 12 maternity care hospitals in Japan. A total of 15,314 women were stratified into five groups according to their pre-pregnancy sleep duration: <6, 6–7, 7–8, 8–9, and ≥9 h (Matsuo et al. 2022). This study aimed to investigate the association between pre-pregnancy sleep duration and PPD. The investigators also evaluated whether the risk for PPD differs between primipara, and multipara women classified according to pre-pregnancy sleep duration. A regression analysis was conducted to determine whether pre-pregnancy sleep duration affects the Edinburgh Postnatal Depression Scale (EPDS) scores at 1-month postpartum (Matsuo et al. 2022).

A 1-h increase in sleep duration was associated with an approximately 14% reduction in the risk for high EPDS scores. The association between short sleep duration and high EPDS scores was more remarkable in multipara women than in primipara women. Short pre-pregnancy sleep duration was associated with high EPDS scores, and this association is more significant in multipara women than in primipara women.

These findings emphasize the importance of collecting information on pre-pregnancy sleep duration to identify women at high risk for PPD (Matsuo et al. 2022). Stress-related sleep disturbances during pregnancy are associated with increased odds of psychiatric disorders. Therefore, inquiring about stress-related sleep disturbances during antenatal care may be beneficial for identifying and caring for women at high risk of psychiatric disorders (Sanchez et al. 2020).

Moreover, the importance of assessing sleep quality prior to and during pregnancy is important to reduce potential risks associated with delivery and postpartum. The population-based prospective cohort study conducted by Paine and colleagues (Paine et al. 2020) on a group of 939 women revealed that women who reported poor quality sleep as measured by the General Sleep Disturbance Scale in later pregnancy had almost twice the odds of delivering via emergency cesarean than women with good sleep quality (OR = 1.98, 95% CI 1.18–3.31) (Paine et al. 2020).

Reported breathing pauses during sleep (OR = 3.27, 95% CI 1.38–7.74) or current snoring (OR = 1.65, 95% CI 1.00–2.72) were also independently associated with a higher likelihood of an emergency cesarean. Short sleep duration and leg twitching/jerking were not independently associated with emergency cesarean section in this study (Paine et al. 2020).

This study came to support that healthy sleep during pregnancy could be a novel intervention to reduce the risks associated with the emergency cesarean section. Research on the effectiveness of sleep interventions for reducing cesarean section risk is also recommended.

16.5.3 Sleep Hygiene for Pregnant Women

The cornerstone in the treatment of sleep problems in all populations is good sleep hygiene. These aspects become even more important during pregnancy, when sleep disturbance is already anticipated. Prioritizing, planning, and scheduling rest and sleep will support not only mothers' recovery, connection with the well-being of the newborn, encompassing their physical and mental health, as well as the sleep patterns of the newborn and the rest of the family.

Here are some suggestions to improve sleep quality and reduce insomnia during pregnancy:

- Utilize pregnancy sleep aids such as specialized pillows or eye masks.
- Create a cool, dark, and quiet environment in the bedroom, reserving the bed for sleeping and intimacy only.
- Make sleep a priority and establish a consistent bedtime routine. Schedule daytime naps earlier in the day to avoid disrupting night-time sleep.
- Engage in calming activities before bedtime, such as reading a book or taking a relaxing bath.
- Use a nightlight to facilitate falling back asleep after bathroom breaks.
- Avoid consuming caffeine, spicy foods, and heavy meals close to bedtime to minimize the risk of experiencing GERD.
- Create a technology-free environment in the bedroom and switch off electronic screens at least 1 h prior to going to bed.
- Engage in regular exercise and seek natural daylight exposure early in the day.
- Stay hydrated throughout the day, but reduce liquid intake before bed to minimize night-time bathroom visits.
- If unable to sleep, get out of bed and engage in another activity until feeling sleepy.
- Consider writing down thoughts in a journal or seeking support from your partner, friends, doctor, or childbirth classes if experiencing anxiety or worry.
- Process emotions by keeping a journal or using a worry worksheet.

Utilizing sleep aids such as specialized pillows or a supportive mattress topper can play a vital role in enhancing both the duration and quality of sleep, particularly in the third trimester of pregnancy. These aids provide comfort and help maintain proper posture for optimal sleep.

Taking the time to rest and promoting relaxation are also important aspects to promote healthy habits and reduce anxiety levels.

Even if pregnant individuals did not experience significant sleep disruptions during pregnancy, they may encounter sleeping difficulties during the postpartum period due to the challenges associated with caring for a newborn, postpartum mood disorders, and other factors that can impact women after delivery. Therefore, it is crucial to make efforts to optimize sleep during the postpartum period, which might include:

- **Rearrange the sleep schedule**: Newborns often have unpredictable sleep patterns. Mothers can adapt to their baby's sleep schedule and ensure they get sufficient rest by sleeping when their baby sleeps, both at night and during daytime naps. However, it is crucial to allow an appropriate amount of sleep pressure to build up during the day to prevent circadian rhythm disorders.
- **Share the workload**: Mothers who have the support of a partner or extended family in raising their children can take turns in sharing the responsibilities of caring for the infant. This arrangement allows for more opportunities to get adequate sleep.
- **Engage in a morning stroll:** Walking in the morning and being exposed to natural sunlight can help reset the body's circadian rhythm, which is naturally synchronized with the sun's rise and fall. Additionally, moderate exercise can contribute to better sleep quality the following night.
- **Avoid alcohol consumption:** Drinking alcohol can have a negative impact on sleep quality. While alcohol may initially promote faster sleep onset due to its sedative effects, it often leads to disrupted sleep later in the night.

The treatment of insomnia varies depending on an individual's health and medical background. However, cognitive-behavioral therapy for insomnia (CBT-I) can be effective in alleviating symptoms. Notably, CBT-I has no known side effects and is suitable for use by pregnant women and new mothers.

The American Academy of Sleep Medicine (AASM) suggests the use of multicomponent cognitive-behavioral therapy for insomnia (CBT-I) as a recommended treatment for both acute and chronic cases of insomnia, by launching clinical practice recommendations based on systematic review and evaluation of evidence (Edinger et al. 2021).

CBT-I involves identifying problematic or inaccurate thoughts and beliefs about sleep and replacing them with healthier attitudes. Specific components of cognitive-behavioral therapy may include:

Sleep education: Enhancing knowledge about the mechanisms of sleep and adopting strategies to improve sleep can empower individuals with insomnia and other sleep issues to approach their difficulties with a better understanding. Keeping a sleep diary also fosters awareness of sleep patterns.

Sleep hygiene: The term sleep hygiene refers to practices that improve sleep.

Stimulus control: Some people develop anxieties about sleeping after dealing with insomnia and other night-time disturbances, and they need to be reconditioned to sleep better. Stimulus control stresses the importance of only using a bed for sleep and intimacy, getting out of bed on nights when it's difficult to fall asleep, and setting an alarm for the same wake up time every day of the week.

Sleep restriction and compression: Based on the patient's sleep diary, a CBT-I practitioner may recommend curtailing the amount of time they spend in bed each night to help with the regulation of circadian rhythm and build up of sleep pressure.

Relaxation: Controlled breathing, meditation, and other relaxation techniques may help people unwind in the evening and fall back asleep if they wake up during the night.

Cognitive Therapy: invite patients to write down and point out potential solutions to worries and thoughts that might have a chance to disturb sleep.

These behavioral and cognitive techniques will help to decrease sleep anxiety and rumination prior to bed to change catastrophic thinking and beliefs about sleep and reduce excessive worry.

A very recent study evaluated the feasibility and efficacy of a scalable cognitive-behavioral therapy (CBT) sleep intervention tailored for the perinatal and postpartum periods (Bei et al. 2023). The participants who didn't present with severe medical or psychiatric conditions were randomly assigned in a 1:1 ratio to receive either cognitive-behavioral therapy (CBT) or a control intervention that matched in terms of attention and time. All participants received a telephone session lasting 1 h and automated multimedia emails starting from the third trimester and continuing for a duration of 6 months after giving birth. Various validated instruments were used to assess outcomes at different time points, including baseline at gestation week 30, pregnancy endpoint at week 35, and postpartum months 1.5, 3, 6, 12, and 24. The CBT intervention was well-received, with no reported negative effects. Data analysis revealed that compared to the control group, those who received CBT demonstrated lower severity of insomnia and sleep disturbances (the primary outcomes) at the pregnancy endpoint and 24 months postpartum (with p-values ≤ 0.001). Additionally, they experienced reduced sleep-related impairment at the pregnancy endpoint and at the 24-month postpartum follow-up (with p-values ranging from 0.01 to 0.05).Participants who exhibited higher levels of insomnia symptoms at the beginning of the study experienced significant benefits from cognitive-behavioral therapy (CBT) compared to the control group. These benefits included consistently lower insomnia symptoms throughout the first year after childbirth. However, there were no significant differences between the groups in terms of symptoms related to depression or anxiety (Bei et al. 2023). The findings of this study highlight the effectiveness of a scalable CBT sleep intervention in mitigating sleep disturbances during pregnancy and improving sleep quality up to 2 years after childbirth, particularly for individuals who experienced insomnia symptoms during pregnancy. Implementing this intervention into routine perinatal care shows promise (Bei et al. 2023).

Sleep disorders may worsen or present de novo during gestation. Managing a pregnant patient is complicated by the risk of teratogenicity, pharmacokinetic changes, and the dynamic nature of pregnancy. Although nonpharmacologic interventions are likely safest, they are often ineffective, and a patient is left dealing with frustrations of sleep disturbance, as well as the negative outcomes of poor sleep in pregnancy. Sleep disorders' management requires an understanding of pregnancy physiology, knowledge of the impact of a given condition on pregnancy or neonatal outcomes, and an ability to weigh the risk of exposure to an untreated, or poorly treated condition, against the risk of a given drug. In conjunction with pregnant patients and families, the options for therapy should be reviewed in the context of the impact of the condition on pregnancy and outcomes, while understanding that research on the impact of therapy on perinatal outcomes is needed (Miller et al. 2020). This article reviews the epidemiology of sleep disorders in pregnancy, general principles of prescribing in pregnancy and lactation, and safety surrounding therapeutic options in pregnancy.

In the case of severe or persistent insomnia, short-term sedatives may be considered after a discussion about potential risks and benefits. Although the tricyclic antidepressant amitriptyline is considered safe during pregnancy in the treatment of depression and can be used in lower doses to treat insomnia. However, it is considered as Class C in pregnancy (possible adverse fetal effects in humans). Other antidepressants used to treat insomnia are also classified as Class C and should be avoided. Benzodiazepines have been classified in pregnancy as Class D (human data show risk although benefits may outweigh these) or X (benefit unlikely to outweigh clear evidence of increased human risk). These are also likely to induce irritability, seizures, and abdominal distension (Miller et al. 2020).

In contrast to findings from animal studies, current evidence from clinical studies indicates that the use of melatonin during pregnancy and breastfeeding is likely safe for humans. A recent review highlights the importance of conducting clinical studies on sleep disorders, including the use of exogenous melatonin, during pregnancy and lactation. Clinical trials involving the administration of exogenous melatonin for other medical conditions during pregnancy and breastfeeding have not revealed significant safety concerns or adverse events (Vine et al. 2022).

A systematic review that investigated four studied sleep disturbances (OSA, snoring, extreme sleep duration, and poor sleep quality) during pregnancy revealed that these were associated with adverse perinatal outcomes, and suggestions were made that expectant mothers should be advised to practice healthy sleep hygiene measures (Yang et al. 2022).

A recent systematic review investigated 16 studies that included 1252 expecting mothers. Studies evaluated the efficacy of various interventions: cognitive-behavioral interventions for insomnia (4); pharmacotherapy (1); acupuncture (1); mindfulness and yoga (3); relaxation (5) and herbal medication (2). Only six were randomized controlled trials, and only four evaluated longitudinal outcomes. Preliminary support was found for all interventions, but the knowledge is still too limited by lack of evidence.

The authors pointed out the utmost urgency to perform high-quality randomized controlled trials for insomnia interventions during pregnancy and to implement effective programs in standard gynecological care (Bacaro et al. 2020).

16.5.4 Sleep Movement Disorders

Medical treatment of RLS during pregnancy is challenging considering the risks to the mother and fetus. Nevertheless, in certain instances, the condition may be sufficiently severe to necessitate medical intervention (Gupta et al. 2016).

Pharmacological therapies for RLS include dopaminergic, opiate, and benzodiazepine medications. During pregnancy, hormonal changes, and metabolic alterations, especially iron and folate deficiency are of major importance to act on (Manconi et al. 2012).

Only oxycodone and pergolide are classified as Class B (no evidence of risk in humans). However, pergolide has been associated with valvular health abnormalities in non-pregnant women. Therefore, should be avoided during pregnancy (Hedman et al. 2002). Only in extreme cases, when the iron-folate supplementation or other conservative approach do not provide enough support, short-term use of oxycodone or pergolide beyond the first trimester may be considered (Hedman et al. 2002).

Consensus clinical practice guidelines for the treatment of RLS during pregnancy and lactation indicate the importance of considering non-pharmacological treatments and simple reassurance whenever symptom severity allows it. Moderate-intensity exercise, yoga, massages, pneumatic compression devices, and treatment of OSA should always be attempted before initiating any type of pharmacological therapy (Picchietti et al. 2015).

16.5.5 Sleep Disorder Breathing

SDB treatment recommendations during pregnancy underscore the importance of sleep hygiene, including lateral positioning, adequate bed-timing, and avoidance of alcohol and other sedatives. On the contrary, oral appliances, surgical procedures, and supplemental night oxygen are not recommended during the gestational period.

No extensive reviews concerning the role of positive airway pressure ventilatory therapy during pregnancy exist to date. Edwards was among the first to recommend continuous positive airway pressure (CPAP) therapy to reduce nocturnal blood pressure in pre-eclampsia (Edwards et al. 2000). Later studies conducted by Guilleminault et al. confirmed a superior control of blood pressure with CPAP in women suffering from chronic hypertension and OSA (Guilleminault et al. 2007). CPAP treatment is recommended for SDB with AHI ≥ 15 (level A) in the general population. However, several authors broaden this recommendation to include hypertensive pregnant women with clinical symptoms and an AHI >5, despite a prospective longitudinal study reporting that nasal CPAP was insufficient to prevent negative pregnancy outcomes (Guilleminault et al. 2007).

Given that numerous GERD-related issues tend to worsen during night-time, certain lifestyle adjustments concentrate on recommendations for managing sleep with GERD.

- Avoiding late meals to prevent lying down with a full stomach can reduce the chances of reflux.
- Sleep on the left side. Multiple research studies have found that lying on the left side is the best sleeping position for people with GERD (Person et al. 2015)
- Improve sleep hygiene
- Elevate the head of the bed: Tilting the upper part of the bed by a minimum of six inches can help reduce reflux symptoms while lying down.

Medications are available for treating GERD and may be required if lifestyle modifications alone do not consistently alleviate symptoms. Over-the-counter medications, such as antacids, can provide temporary relief but may have limited effectiveness for many individuals. Additional medications, including proton pump inhibitors (PPIs) and H2 blockers, are also utilized as treatment options, aim to reduce the production of stomach acid. However, they can have side effects and their use should be closely assessed.

References

Allen RP, Walters AS, Montplaisir J et al (2005) Restless legs syndrome prevalence and impact. Arch Intern Med 165(11):1286. https://doi.org/10.1001/archinte.165.11.1286

Bacaro V, Benz F, Pappaccogli A et al (2020) Interventions for sleep problems during pregnancy: a systematic review. Sleep Med Rev 50:101234. https://doi.org/10.1016/j.smrv.2019.101234

Baratte-Beebe KR, Lee K (1999) Sources of midsleep awakenings in childbearing women. Clin Nurs Res 8(4):386–397. https://doi.org/10.1177/10547739922158377

Basu A, Kim HH, Basaldua R et al (2021) A cross-national study of factors associated with women's perinatal mental health and wellbeing during the COVID-19 pandemic. PLoS One 16(4):e0249780. https://doi.org/10.1371/journal.pone.0249780

Beebe KR, Lee KA (2007) Sleep disturbance in late pregnancy and early labor. J Perinat Neonatal Nurs 21(2):103–108. https://doi.org/10.1097/01.JPN.0000270626.66369.26

Bei B, Coo Calcagni S, Milgrom J, Trinder J (2012) Day-to-day alteration of 24-hour sleep pattern immediately before and after giving birth. Sleep Biol Rhythms 10(3):212–221. https://doi.org/10.1111/j.1479-8425.2012.00563.x

Bei B, Pinnington DM, Quin N et al (2023) Improving perinatal sleep via a scalable cognitive behavioural intervention: findings from a randomised controlled trial from pregnancy to 2 years postpartum. Psychol Med 53(2):513–523. https://doi.org/10.1017/S0033291721001860. Published online July 7, 2021

Bourjeily G, Ankner G, Mohsenin V (2011) Sleep-disordered breathing in pregnancy. Clin Chest Med 32(1):175–189. https://doi.org/10.1016/j.ccm.2010.11.003

Buysse DJ (2013) Insomnia. JAMA 309(7):706. https://doi.org/10.1001/jama.2013.193

Cain SW, Dennison CF, Zeitzer JM et al (2010) Sex differences in phase angle of entrainment and melatonin amplitude in humans. J Biol Rhythm 25(4):288–296. https://doi.org/10.1177/0748730410374943

Ceulemans M, Foulon V, Ngo E et al (2021) Mental health status of pregnant and breastfeeding women during the COVID-19 pandemic—a multinational cross-sectional study. Acta Obstet Gynecol Scand 100(7):1219–1229. https://doi.org/10.1111/aogs.14092

Chen BY, He QY (2007) A multi-center study on the association between sleep apnea and prevalence of hypertension. Zhonghua Jie He He Hu Xi Za Zhi 30(12):894–897

Chen SJ, Shi L, Bao YP et al (2018) Prevalence of restless legs syndrome during pregnancy: a systematic review and meta-analysis. Sleep Med Rev 40:43–54. https://doi.org/10.1016/j.smrv.2017.10.003

Darvishi N, Daneshkhah A, Khaledi-Paveh B et al (2020) The prevalence of Restless Legs Syndrome/Willis-ekbom disease (RLS/WED) in the third trimester of pregnancy: a systematic review. BMC Neurol 20(1):132. https://doi.org/10.1186/s12883-020-01709-0

De Carlo M (2019) Awakening through sleep: a transformational and spiritual guide for pregnancy, adult and child sleep. Baby Planner

Dørheim SK, Bondevik GT, Eberhard-Gran M, Bjorvatn B (2009) Sleep, and depression in postpartum women: a population-based study. Sleep 32(7). https://doi.org/10.1093/sleep/32.7.847

Edinger JD, Arnedt JT, Bertisch SM et al (2021) Behavioral and psychological treatments for chronic insomnia disorder in adults: an American Academy of Sleep Medicine clinical practice guideline. J Clin Sleep Med 17(2):255–262. https://doi.org/10.5664/jcsm.8986

Edwards N, Sullivan CE (2008) Sleep-disordered breathing in pregnancy. Sleep Med Clin 3(1): 81–95. https://doi.org/10.1016/j.jsmc.2007.10.010

Edwards N, Blyton DM, Kirjavainen T, Kesby GJ, Sullivan CE (2000) Nasal continuous positive airway pressure reduces sleep-induced blood pressure increments in preeclampsia. Am J Respir Crit Care Med 162(1):252–257. https://doi.org/10.1164/ajrccm.162.1.9905006

Esposito G, Chiaffarino F, Odelli V, Romiti L, di Martino M, Parazzini F (2017) Restless legs syndrome in pregnancy and risk of gestational hypertension in a low-risk population. Ital J Gynaecol Obstet 29(4). https://doi.org/10.14660/2385-0868-79

Facco FL, Parker CB, Reddy UM et al (2017) Association between sleep-disordered breathing and hypertensive disorders of pregnancy and gestational diabetes mellitus. Obstet Gynecol 129(1): 31–41. https://doi.org/10.1097/AOG.0000000000001805

Fung AM, Wilson DL, Barnes M, Walker SP (2012) Obstructive sleep apnea and pregnancy: the effect on perinatal outcomes. J Perinatol 32(6):399–406. https://doi.org/10.1038/jp.2012.14

Gallaher KGH, Slyepchenko A, Frey BN, Urstad K, Dørheim SK (2018) The role of circadian rhythms in postpartum sleep and mood. Sleep Med Clin 13(3). https://doi.org/10.1016/j.jsmc.2018.04.006

Goyal D, Gay CL, Lee KA (2007) Patterns of sleep disruption and depressive symptoms in new mothers. J Perinat Neonatal Nurs 21(2):123–129. https://doi.org/10.1097/01.JPN.0000270629.58746.96

Gueron-Sela N, Shahar G, Volkovich E, Tikotzky L (2021) Prenatal maternal sleep and trajectories of postpartum depression and anxiety symptoms. J Sleep Res 30(4). https://doi.org/10.1111/jsr.13258

Guilleminault C, Palombini L, Poyares D, Takaoka S, Huynh NTL, El-Sayed Y (2007) Pre-eclampsia, and nasal CPAP: part 1. Early intervention with nasal CPAP in pregnant women with risk-factors for pre-eclampsia: preliminary findings. Sleep Med 9(1):9–14. https://doi.org/10.1016/j.sleep.2007.04.020

Gupta R, Dhyani M, Kendzerska T et al (2016) Restless legs syndrome and pregnancy: prevalence, possible pathophysiological mechanisms, and treatment. Acta Neurol Scand 133(5):320–329. https://doi.org/10.1111/ane.12520

Hajak G (2001) Epidemiology of severe insomnia and its consequences in Germany. Eur Arch Psychiatry Clin Neurosci 251(2):49–56. https://doi.org/10.1007/s004060170052

Hedman C, Pohjasvaara T, Tolonen U, Suhonen-Malm AS, Myllylä VV (2002) Effects of pregnancy on mothers' sleep. Sleep Med 3(1):37–42. https://doi.org/10.1016/S1389-9457(01)00130-7

Hertz G, Fast A, Feinsilver SH, Albertario CL, Schulman H, Fein AM (1992) Sleep in normal late pregnancy. Sleep 15(3):246–251. https://doi.org/10.1093/sleep/15.3.246

Izci-Balserak B, Keenan BT, Corbitt C, Staley B, Perlis M, Pien GW (2018) Changes in sleep characteristics and breathing parameters during sleep in early and late pregnancy. J Clin Sleep Med 14(07):1161–1168. https://doi.org/10.5664/jcsm.7216

Kang MJ, Matsumoto K, Shinkoda H, Mishima M, Seo YJ (2002) Longitudinal study for sleep-wake behaviours of mothers from pre-partum to post-partum using actigraph and sleep logs. Psychiatry Clin Neurosci 56(3):251–252. https://doi.org/10.1046/j.1440-1819.2002.00992.x

Khazaie H, Ghadami MR, Knight DC, Emamian F, Tahmasian M (2013) Insomnia treatment in the third trimester of pregnancy reduces postpartum depression symptoms: a randomized clinical trial. Psychiatry Res 210(3):901–905. https://doi.org/10.1016/j.psychres.2013.08.017

Kim Y, Lee YJ, Park JS et al (2018) Associations between obstructive sleep apnea severity and endoscopically proven gastroesophageal reflux disease. Sleep Breath 22(1):85–90. https://doi.org/10.1007/s11325-017-1533-2

Lee K (2000) Parity and sleep patterns during and after pregnancy. Obstet Gynecol 95(1):14–18. https://doi.org/10.1016/S0029-7844(99)00486-X

Lee J, Shen L, Neemia D, Fulgoni C, Manber R, Bei B (2018) Sleep complaints, regardless of insomnia, are associated with worse sleep and mental health during pregnancy. J Sleep Res 27: e69_12766. https://doi.org/10.1111/jsr.69_12766

Letourneau NL, Tramonte L, Willms JD (2013) Maternal depression, family functioning and children's longitudinal development. J Pediatr Nurs 28(3):223–234. https://doi.org/10.1016/j.pedn.2012.07.014

Liu G, Li L, Zhang J et al (2016) Restless legs syndrome and pregnancy or delivery complications in China: a representative survey. Sleep Med 17:158–162. https://doi.org/10.1016/j.sleep.2015.02.541

Ma S, Shang X, Guo Y, Liu G, Yang J, Xue R (2015) Restless legs syndrome and hypertension in Chinese pregnant women. Neurol Sci 36(6):877–881. https://doi.org/10.1007/s10072-015-2094-4

Maghami M, Shariatpanahi SP, Habibi D et al (2021) Sleep disorders during pregnancy and postpartum depression: a systematic review and meta-analysis. Int J Dev Neurosci 81(6): 469–478. https://doi.org/10.1002/jdn.10118

Manber R, Armitage R (1999) Erratum: sex, steroids, and sleep: a review. Sleep 5:540–555; Sleep 2000;23(2)

Manconi M, Ulfberg J, Berger K et al (2012) When gender matters: restless legs syndrome. Report of the "RLS and woman" workshop endorsed by the European RLS Study Group. Sleep Med Rev 16(4):297–307. https://doi.org/10.1016/j.smrv.2011.08.006

Matsuo S, Ushida T, Iitani Y et al (2022) Pre-pregnancy sleep duration and postpartum depression: a multicenter study in Japan. Arch Womens Ment Health 25(1):181–189. https://doi.org/10.1007/s00737-021-01136-1. Published online July 13, 2021

Miller MA, Mehta N, Clark-Bilodeau C, Bourjeily G (2020) Sleep pharmacotherapy for common sleep disorders in pregnancy and lactation. Chest 157(1):184–197. https://doi.org/10.1016/j.chest.2019.09.026

Mindell JA, Jacobson BJ (2000) Sleep disturbances during pregnancy. J Obstet Gynecol Neonatal Nurs 29(6):590–597. https://doi.org/10.1111/j.1552-6909.2000.tb02072.x

Mindell JA, Cook RA, Nikolovski J (2015) Sleep patterns and sleep disturbances across pregnancy. Sleep Med 16(4):483–488. https://doi.org/10.1016/j.sleep.2014.12.006

Miranda AR, Scotta AV, Cortez MV, Soria EA (2021) Triggering of postpartum depression and insomnia with cognitive impairment in Argentinian women during the pandemic COVID-19 social isolation in relation to reproductive and health factors. Midwifery 102:103072. https://doi.org/10.1016/j.midw.2021.103072

Morin CM, Blais F, Savard J (2002) Are changes in beliefs and attitudes about sleep related to sleep improvements in the treatment of insomnia? Behav Res Ther 40(7):741–752. https://doi.org/10.1016/S0005-7967(01)00055-9

Morin CM, Belleville G, Bélanger L, Ivers H (2011) The insomnia severity index: psychometric indicators to detect insomnia cases and evaluate treatment response. Sleep 34(5):601–608. https://doi.org/10.1093/sleep/34.5.601

Neyal A, Senel GB, Aslan R et al (2015) A prospective study of Willis–Ekbom disease/restless legs syndrome during and after pregnancy. Sleep Med 16(9):1036–1040. https://doi.org/10.1016/j.sleep.2015.01.026
Okun ML, Luther JF, Wisniewski SR, Sit D, Prairie BA, Wisner KL (2012) Disturbed sleep, a novel risk factor for preterm birth? J Women's Health 21(1):54–60. https://doi.org/10.1089/jwh.2010.2670
Oyieng'o DO, Kirwa K, Tong I, Martin S, Antonio Rojas-suarez J, Bourjeily G (2016) Restless legs symptoms and pregnancy and neonatal outcomes. Clin Ther 38(2):256–264. https://doi.org/10.1016/j.clinthera.2015.11.021
Paine SJ, Signal TL, Sweeney B et al (2020) Maternal sleep disturbances in late pregnancy and the association with emergency caesarean section: a prospective cohort study. Sleep Health 6(1): 65–70. https://doi.org/10.1016/j.sleh.2019.11.004
Person E, Rife C, Freeman J, Clark A, Castell DO (2015) A novel sleep positioning device reduces gastroesophageal reflux. J Clin Gastroenterol 49(8):655–659. https://doi.org/10.1097/MCG.0000000000000359
Picchietti DL, Hensley JG, Bainbridge JL et al (2015) Consensus clinical practice guidelines for the diagnosis and treatment of restless legs syndrome/Willis-Ekbom disease during pregnancy and lactation. Sleep Med Rev 22:64–77. https://doi.org/10.1016/j.smrv.2014.10.009
Pien GW, Pack AI, Jackson N, Maislin G, Macones GA, Schwab RJ (2014) Risk factors for sleep-disordered breathing in pregnancy. Thorax 69(4):371–377. https://doi.org/10.1136/thoraxjnl-2012-202718
Porkka-Heiskanen T, Saaresranta T, Polo-Kantola P (2014) Physiological basis of sleep: gender differences in sleep. In: Bassetti C, McNicholas W, Paunio T, Peigneux P (eds) Sleep medicine textbook ESRS, 1st edn, pp 83–94
Redhead K, Griffin C, Hillman D et al (2017) Sleep apnoea and symptoms of depression and anxiety during pregnancy. J Sleep Res 31–31. https://doi.org/10.1111/jsr.67_12618
Redman CW, Sargent IL (2005) Latest advances in understanding preeclampsia. Science 308(5728):1592–1594. https://doi.org/10.1126/science.1111726
Román-Gálvez RM, Amezcua-Prieto C, Salcedo-Bellido I, Martínez-Galiano JM, Khan KS, Bueno-Cavanillas A (2018) Factors associated with insomnia in pregnancy: a prospective Cohort Study. Eur J Obstet Gynecol Reprod Biol 221:70–75. https://doi.org/10.1016/j.ejogrb.2017.12.007
Sanchez SE, Friedman LE, Rondon MB, Drake CL, Williams MA, Gelaye B (2020) Association of stress-related sleep disturbance with psychiatric symptoms among pregnant women. Sleep Med 70:27–32. https://doi.org/10.1016/j.sleep.2020.02.007
Sarberg M, Bladh M, Svanborg E, Josefsson A (2016) Postpartum depressive symptoms and its association to daytime sleepiness and restless legs during pregnancy. BMC Pregnancy Childbirth 16(1):137. https://doi.org/10.1186/s12884-016-0917-9
Sedov ID, Tomfohr-Madsen LM (2021) Trajectories of insomnia symptoms and associations with mood and anxiety from early pregnancy to the postpartum. Behav Sleep Med 19(3):395–406. https://doi.org/10.1080/15402002.2020.1771339
Sedov ID, Anderson NJ, Dhillon AK, Tomfohr-Madsen LM (2021) Insomnia symptoms during pregnancy: a meta-analysis. J Sleep Res 30(1). https://doi.org/10.1111/jsr.13207
Shaker R, Castell DO, Schoenfeld PS, Spechler SJ (2003) Nighttime heartburn is an under-appreciated clinical problem that impacts sleep and daytime function: the results of a Gallup survey conducted on behalf of the American Gastroenterological Association. Am J Gastroenterol 98(7):1487–1493. https://doi.org/10.1111/j.1572-0241.2003.07531.x
Sharma SK, Nehra A, Sinha S et al (2016) Sleep disorders in pregnancy and their association with pregnancy outcomes: a prospective observational study. Sleep Breath 20(1):87–93. https://doi.org/10.1007/s11325-015-1188-9
Sibai B, Dekker G, Kupferminc M (2005) Pre-eclampsia. Lancet 365(9461):785–799. https://doi.org/10.1016/S0140-6736(05)17987-2

Silvestri R, Aricò I (2019) Sleep disorders in pregnancy. Sleep Sci 12:3. https://doi.org/10.5935/1984-0063.20190098

Skouteris H, Germano C, Wertheim EH, Paxton SJ, Milgrom J (2008) Sleep quality and depression during pregnancy: a prospective study. J Sleep Res 17(2):217–220. https://doi.org/10.1111/j.1365-2869.2008.00655.x

Steiger A (2002) Sleep and the hypothalamo–pituitary–adrenocortical system. Sleep Med Rev 6(2):125–138. https://doi.org/10.1053/smrv.2001.0159

Suzuki K, Ohida T, Sone T et al (2003) The prevalence of restless legs syndrome among pregnant women in Japan and the relationship between restless legs syndrome and sleep problems. Sleep 26(6):673–677. https://doi.org/10.1093/sleep/26.6.673

Sweet L, Arjyal S, Kuller JA, Dotters-Katz S (2020) A review of sleep architecture and sleep changes during pregnancy. Obstet Gynecol Surv 75(4):253–262. https://doi.org/10.1097/OGX.0000000000000770

Tomfohr LM, Buliga E, Letourneau NL, Campbell TS, Giesbrecht GF (2015) Trajectories of sleep quality and associations with mood during the perinatal period. Sleep 38(8):1237–1245. https://doi.org/10.5665/sleep.4900

Uhlig BL, Sand T, Ødegård SS, Hagen K (2014) Prevalence, and associated factors of DSM-V insomnia in Norway: the Nord-Trøndelag Health Study (HUNT 3). Sleep Med 15(6):708–713. https://doi.org/10.1016/j.sleep.2014.01.018

Vela MF, Kramer JR, Richardson PA, Dodge R, El-Serag HB (2014) Poor sleep quality and obstructive sleep apnea in patients with GERD and Barrett's esophagus. Neurogastroenterol Motil 26(3):346–352. https://doi.org/10.1111/nmo.12265

Vine T, Brown GM, Frey BN (2022) Melatonin use during pregnancy and lactation: a scoping review of human studies. Braz J Psychiatry 44(3):342–348. https://doi.org/10.1590/1516-4446-2021-2156. Published online November 1, 2021

Ward BA (2010) In: Hrayr A, Schuman C (eds) Clinical handbook of insomnia: insomnia during pregnancy, 2nd edn

Wesström J, Skalkidou A, Manconi M, Fulda S, Sundström-Poromaa I (2014) Pre-pregnancy restless legs syndrome (Willis-Ekbom disease) is associated with perinatal depression. J Clin Sleep Med 10(05):527–533. https://doi.org/10.5664/jcsm.3704

Wilson DL, Walker SP, Fung AM et al (2018) Sleep-disordered breathing in hypertensive disorders of pregnancy: a BMI-matched study. J Sleep Res 27(5):e12656. https://doi.org/10.1111/jsr.12656

Woodley J, Smith S (2006) Safety behaviors and dysfunctional beliefs about sleep: testing a cognitive model of the maintenance of insomnia. J Psychosom Res 60(6):551–557. https://doi.org/10.1016/j.jpsychores.2006.03.002

Yang Z, Zhu Z, Wang C, Zhang F, Zeng H (2022) Association between adverse perinatal outcomes and sleep disturbances during pregnancy: a systematic review and meta-analysis. J Matern Fetal Neonatal Med 35(1):166–174. https://doi.org/10.1080/14767058.2020.1711727

Young T, Peppard PE, Taheri S (2005) Excess weight and sleep-disordered breathing. J Appl Physiol 99(4):1592–1599. https://doi.org/10.1152/japplphysiol.00587.2005

Chapter 17
Updates in Pediatric Sleep Apnea

Cláudio D'Elia, Chris Landon, and Miguel Meira e Cruz

Abstract Sleep-disordered breathing encompasses a group of highly impacting conditions with a significant prevalence among the pediatric population. Although manifested in a wide spectrum of clinical presentations, upper airway increased resistance is often a common finding which therefore contributes to its deleterious role on several systems and functions. Challenging diagnostic algorithms and novel therapeutic options are relevant aspects that should be validated and established under an evidence-based matrix applied to a real-world scenario to improve the standards of care and reduce the negative consequences of a dangerous disturber of sleep from infancy to adolescence. This chapter aims to clarify the factors subsiding impaired respiratory function during sleep and provide an updated view of the mechanisms, clinical manifestations, and treatment options available for this range of pediatric ages.

Keywords Sleep-disordered breathing · Pediatric population · Upper airway resistance · Diagnostic algorithms · Treatment options

17.1 Introduction

Obstructive sleep apnea is the most clinically impacting form of the sleep-disordered breathing spectrum. During sleep, the disruption of normal respiratory patterns and ventilation in children causes several health issues and significantly affects learning,

---

C. D'Elia
Department of Pediatrics, Lusíadas Hospital, Lisbon, Portugal

European Sleep Center, Lisbon, Portugal

C. Landon
Family Practice and Pediatrics, University of California, Los Angeles, CA, USA

M. Meira e Cruz (✉)
European Sleep Center, Lisbon, Portugal

Cardiovascular Center, University of Lisbon, Lisbon School of Medicine, Lisbon, Portugal
e-mail: mcruz@medicina.ulisboa.pt

A. S. BaHammam, M. Hunasikatti (eds.), *Sleep Apnea Frontiers*, Progress in Sleep Research, https://doi.org/10.1007/978-981-99-7901-1_17

academic performance, and quality of life. Recognition of the condition is imperative and requires an adequate assessment with an active search for signs, symptoms, and behaviors related to normal sleep deviation. Data on the appropriate sequence of interventions is still scarce and standards of care are debatable. The ongoing medical science is, however, looking to improve each of the algorithmic steps, from diagnosis to treatment pursuing successful approaches allowing to reduce the burden of OSA as well as its deleterious consequences. Knowledge of epidemiological aspects will provide basics for improving screening strategies and performing accurate diagnosis to achieve optimized results.

17.2 Epidemiology

For the practicing clinician sleep problems are a common parental complaint, occurring in up to 25% of children over the course of infancy, childhood, and adolescence. Presentations of significance pointing toward obstructive sleep apnea can include learning and behavioral problems, cardiovascular complications, and impaired growth (including failure to thrive). Population estimates of obstructive sleep apnea in children suggest it is most common between ages 2 and 6 years, with incidence ranging between 1 and 5% of children (Marcus et al. 2012). The major risk factors in otherwise healthy children are **adenotonsillar hypertrophy** and **obesity** (BMI > 97th percentile for age and gender). Findings of obesity and adenotonsillar hypertrophy can be mitigated by age and anatomic considerations.

The clinician must be aware of additional risk factors, keeping in mind medical, neurological, skeletal, or dental conditions that reduce upper airway size, affect the neural control of the upper airway, or impact the collapsibility of the upper airway. In **Cerebral Palsy** the Pediatric Sleep Questionnaire was abnormal in 58% with concomitant epilepsy increasing positivity to 67%, compared to population values of 27%, while medical review for symptoms suggested only 8.2% (Garcia et al. 2016). In **Down Syndrome** estimates of pediatric abnormality range from 50 to 100%, progressing to 100% in adulthood. In a group of unselected patients with down syndrome, 66% were found to have OSA on full night polysomnogram (PSG), and about half of these children presented with severe disease, even among those children with no history of snoring or witnessed apneas, the majority (53%) were found to have OSA on PSG and that traditional risk factors for OSA such as tonsil size or body mass index (BMI) *z*-score were not associated with the risk of OSA among a large cohort of children with Down Syndrome (Maris et al. 2016). **Craniofacial anomalies (e.g., retrognathia, micrognathia, midface hypoplasia)** were found in 15.5% of pediatric OSA patients in a population-based case control study (Lam et al. 2010), while children with Apert, Crouzon, or Pfeiffer syndromes, which all have significant maxillary hypoplasia as a key feature, reported an estimated prevalence of OSA of 53% in this population based on symptom report on a mail-in questionnaire (Pijpers et al. 2004). **Children born preterm** compared to term were significantly more likely to be diagnosed with sleep apnea (<32 weeks versus term) hazard ratio 2.74 this remained even after adjustment for known confounding variables. Children born small for gestational age were not at increased risk of sleep apnea compared to children born appropriate for gestational age, hazard ratio of 0.95 (Raynes-Greenow et al. 2012). **Muscular dystrophy or other neuromuscular disorders** are frequently associated with sleep-disordered breathing (SDB) abnormalities, although the cumulative prevalence is not well known and probably underestimated. Between 27–62% of children and 36–53% of adults with neuromuscular disorders have sleep-disordered breathing, depending on the type of neuromuscular disorder involved, the definition of respiratory impairment, and the tools used to measure it (Arens and Muzumdar 2010). The prevalence of SDB in **myelomeningocele** (MMC) has been reported between 62 and 81%; the prevalence of moderate to severe OSA in this population has been reported to be as high as 20–31% (Walker Jr and Perez 2021). Obstructive events are more likely to occur during a rapid eye movement (REM) sleep state. Efforts to quantify the incidence of central apneas are based on small series and case reports and likely represent under-recognition/reporting. In children with myelomeningocele, risk factors for sleep-related disordered breathing include higher spinal cord lesions, brain stem dysfunction resulting from decreased corticomedullary subarachnoid space, very abnormal Chiari malformations, pulmonary function abnormalities (obstructive lung disease and restrictive lung disease), and disorders of upper airway maintenance. Obesity is an independent risk factor for SDB in all children with an estimated prevalence of 13–78% (Walker Jr and Perez 2021). The combination of higher obesity rates in individuals with MMC and an abnormal brainstem respiratory brainstem control center secondary to their Chiari malformation places this population at increased risk. Obstructive sleep apnea (OSA) has been observed in individuals with **achondroplasia** and is associated with compression of the brain stem, although overnight sleep studies showed a reduction in respiratory disturbance index (RDI) and movements/arousals after treatment to relieve upper airway obstruction. Improvement of respiratory indices was associated with an increased proportion of slow-wave sleep and a decrease in stage 1–2 sleep (Waters et al. 1995). The prevalence of OSA across studies in **Prader–Willi syndrome** (PWS) was 79.91% (Sedky et al. 2014). Among youths with OSA, 53.07% had mild OSA, 22.35% moderate OSA, and 24.58% severe OSA. Adenotonsillectomy was associated with improvement in OSA for most children with PWS. However, residual OSA was present in most of the cases post-surgery (Sedky et al. 2014). Forty patients out of 42 **mucopolysaccharidoses** (MPS) (e.g., Hunter syndrome or Hurler syndrome) children tested with PSG had OSA (95%) (Gönüldaş et al. 2014). Adenoid grade, Mallampati grade, restricted mouth opening, and rate of difficult intubation were significantly different among MPS subtypes. There was no significant difference between MPS subtypes according to tonsil grade and OSA severity. Most MPS patients have airway obstruction and OSA due to adenotonsillar hypertrophy and most benefit from adenotonsillectomy, after which OSA significantly improves (Gönüldaş et al. 2014). **Orthodontic problems** (e.g., high narrow hard palate, overlapping incisors, cross bite) are significantly associated with OSA (Galeotti et al. 2018). The prevalence of malocclusions in OSA children without genetic syndromes was 89.9%

compared to 60.6% in the control group. Factors independently associated with OSA compared to the control group were posterior crossbite, reduced overbite increased overbite, and increased overjet (Galeotti et al. 2018). A high prevalence of OSA among family members, suggested that a **family history of OSA** has a role apart from the influence of obesity. Several small family-based studies have focused on defining the genetic basis of OSA. In a case series, non-overweight relatives of patients with OSA were found to have more daytime sleepiness, snoring, apneas, and arousal compared with matched controls (Gharibeh and Mehra 2010). In the Cleveland Family Study, which is the largest, assembled familial aggregation study investigating OSA, sleep apnea was identified to be more prevalent in relatives of index probands of OSA (21%) compared with neighborhood control subjects (12%) (Gharibeh and Mehra 2010). Data from twin studies showed that the concordance of snoring, a symptom of OSA, is higher among monozygotic twins than dizygotic. The heritability of OSA is partially explained by intermediate, genetically determined risk factors. Obesity, which is a known risk factor for OSA, is partially under genetic control. Other risk factors which predispose to OSA that are genetically determined include craniofacial morphologic characteristics and ventilatory control mechanisms (Gharibeh and Mehra 2010).

17.3 Pathophysiology

OSA is caused by complete or partial collapse during sleeping, of one or multiple segments of the extrathoracic airway, including the nasal vault, oral cavity, pharynx, larynx, and trachea. Although interactively related, the pathophysiological factors involved in OSA can be broadly and arbitrarily divided into anatomical factors that effectively reduce airway caliber and those that promote increased upper airway collapsibility (Tan et al. 2013). A clear anatomic factor cannot always be identified and several contributing factors may coexist to a greater or lesser degree (Arens and Marcus 2004; Gozal et al. 2020).

Key pathophysiologic possible causes include (1) an anatomically compromised or collapsible upper airway (high passive critical closing pressure of the upper airway [Pcrit]); (2) inadequate responsiveness of the upper airway dilator muscles during sleep (minimal increase in EMG activity to negative pharyngeal pressure); (3) waking up prematurely to airway narrowing (a low respiratory arousal threshold); and (4) having an oversensitive ventilatory control system (high loop gain) (Eckert et al. 2013).

17.3.1 Dysregulation of the Upper Airway Neuromotor System During Sleep

The upper airway has important physiologic functions such as respiration, swallowing, speech and vibration, and local immunity. The pharynx is a collapsible segment that performs these functions under the balanced control of dilator and constrictor muscles (Chang and Chae 2010). Neuromuscular factors for upper airway patency are affected by central ventilatory drive, arousal responses to respiratory occlusion (arousal threshold), and upper airway reflexes (Chang and Chae 2010).

The respiratory cycle is controlled by an extensive network of neurons within the brainstem, which is dependent upon unspecific and specific drives like hypoxia or hypercapnia. The typical changes in this network and its drives during the transition from wakefulness to sleep and within the various sleep states favor destabilization of breathing during sleep (Schäfer 2006). During sleep in a normal child there is a modest increase in upper airway resistance and alterations in central respiratory drive, both resulting in a decrease in nocturnal ventilation, a slight drop in oxygen saturation, and a rise in end-tidal PCO_2. Brief central pauses are common in children, but obstructive apneas are never normal in children (Marcus et al. 1992).

The upper airway is a complex, multifunctional, dynamic neuromechanical system. Maintaining the openness of the airway during breathing necessitates constant coordination between neural and mechanical factors, and this can vary depending on the individual's posture. Failure to continuously recruit and coordinate dilator muscles to counterbalance the forces that act to close the airway results in hypopneas or apneas (Bilston and Gandevia 2014).

During wakefulness and sleep, the pharyngeal dilators help to maintain pharyngeal patency by tonic contraction and/or phasic contractions, thereby helping to stiffen the upper airway prior to and during inspiration. At sleep onset, airway muscle activity is reduced, and ventilatory variability increases. Airway collapse is offset by pharyngeal dilator activity in response to hypercapnia and negative luminal pressure (Katz and D'Ambrosio 2008). In children with sleep-related breathing disorders, this response does not completely relieve the upper airway obstruction although, in many children with OSA, this is usually sufficient to maintain a partially open pharyngeal airway for lengthy periods of sleep without experiencing discrete obstructive apneic events.

The upper airway is more collapsible in lighter sleep and REM sleep compared to deeper SWS (Carberry et al. 2016). During REM sleep, paroxysmal reductions in pharyngeal dilator activity related to central REM sleep processes likely account for the predisposition toward an increase in resistance and obstructive events (Katz and D'Ambrosio 2008; Ayappa and Rapoport 2003). The withdrawal of excitatory noradrenergic and serotonergic inputs to upper airway motor neurons during REM sleep further reduces pharyngeal muscle activity and substantially increases the propensity for upper airway collapse (Mokhlesi and Punjabi 2012). Obstructive sleep events in children with OSA have an inverse pattern concerning the sleep phase, compared with that observed in adults with OSA. The REMS-predominant disease is by far the most usual form of OSA occupying more than 80% of the cases. In adults, NREMS-predominant disease occupies approximately 80% or more of the totality of patients being evaluated for OSA (Spruyt and Gozal 2012). The pathophysiology of the increased variability of pharyngeal dilator muscle activity during Stage 2 sleep is postulated to arise from alterations in mechanoreceptor/chemoreceptor control (Katz and White 2004).

The pharyngeal collapsible segment is controlled by at least 20 sets of paired muscles, which can alter its size, shape, and dynamics, and behaves like a Starling resistor (Ayappa and Rapoport 2003). Collapse occurs when the pressure surrounding the collapsible segment of the upper airway, known as critical tissue pressure (Pcrit), becomes greater than the pressure within the collapsible segment of the airway and when upstream pressure is lower than Pcrit (Arens and Marcus 2004). Pcrit is, thus, an objective measure of airway collapsibility and was found to be higher (i.e., less negative) in children with OSA than in those with primary snoring and correlated with the AHI (Tan et al. 2013).

Intense local and systemic inflammation is present in OSA patients. In the upper airway, this process may promote oropharyngeal inspiratory muscle dysfunction and amplify upper airway narrowing and collapsibility thereby worsening the frequency and duration of apneas during sleep (Hatipoğlu and Rubinstein 2003).

The pediatric airway is more resistant to collapse compared to the adult airway; children snore less than adults and have fewer obstructive apneas (Katz and D'Ambrosio 2008; Marcus et al. 1999a). The upper airway compensatory responses to subatmospheric pressure loading decreases with age. Obese adolescents with OSAS manifest blunted upper airway neuromuscular compensatory responses to decreases in upper airway pressure during sleep. In contrast, obese adolescents without OSAS maintain protective upper airway reflexes during adolescent development (Huang et al. 2012).

The pattern of upper airway muscle recruitment in children with OSAS may be different from that in adults; children may have greater upper airway muscle activation, thereby preventing total airway occlusion (Marcus et al. 1999a).

17.3.2 Arousal Threshold

Arousal from sleep immediately opens the airway and normalizes gas exchange abnormalities. However, arousal may also be considered an adverse epiphenomenon that potentiates obstructive cycling and interferes with sleep homeostasis (Younes 2004). With a decreased arousal threshold, there is insufficient time for respiratory stimuli to engage the pharyngeal muscles, reopen the airway prior to awakening, and disrupt breathing stability. Accordingly, a decreased respiratory arousal threshold is considered a contributor to OSA pathogenesis. It was hypothesized that repetitive arousal from sleep led to hypocapnia and subsequent dilator muscle hypotonia and apneas.

However, recent clinical and experimental studies have not demonstrated low upper airway dilator muscle activity following arousal-induced hypocapnia (Cori et al. 2017; Malhotra and Jordan 2016). Inconsistences between the hypothesized adverse effects of arousal and prior research may be due to methodological issues.

The arousal response in prepubertal children with OSA was observed at a significantly higher inspiratory resistive load and hypercapnia than did controls (Marcus et al. 1999b, 1998). The etiology of the increased arousal threshold to respiratory stimuli has not been established and may be a secondary phenomenon.

The arousal threshold was particularly high during REM sleep, which is when most pediatric OSA occurs. This lack of arousal in children with OSA, and presumed lack of a ventilatory compensation to loading may account for the preservation of sleep architecture and contribute to the occurrence of prolonged obstructive hypoventilation (prolonged periods of partial upper airway obstruction associated with hypercapnia) (Marcus et al. 1999b). Arousal responses to acoustic nonrespiratory stimuli seems to be normal in children with OSA (Moreira et al. 2005).

17.3.3 Unstable Ventilatory Control

At sleep onset, the wakefulness drive is lost, and ventilation is primarily governed by the metabolic chemoreflex control system (Deacon-Diaz and Malhotra 2018). The contribution of ventilatory control to the pathogenesis of OSA in children is so far unknown (Domany et al. 2019).

Loop gain (LG) is a dimensionless value quantifying the stability of the ventilatory control system; it represents the sensitivity of the feedback loop controlling ventilation and is defined as the size of a corrective ventilatory response relative to the size of the ventilatory disturbance that elicits the correction (Bokov et al. 2020).

Components of LG in the system of ventilatory regulation are the controller gain (CG), which refers to the extent of the ventilatory response to specific variations in hypercapnia or hypoxia, regulated by chemoreceptors, and the plant gain (PG), which indicates the impact of the ventilatory response on arterial oxygen and carbon dioxide levels. Unstable ventilatory control is characterized by high LG and may contribute to cyclical airway obstruction by promoting airway collapse during periods of a low ventilatory drive (Deacon and Catcheside 2015).

It was shown that children with moderate to severe OSA had not an increased LG as compared to those without OSA or with mild OSA. These children with OSA had a higher PG compared to children without OSA, and PG was positively associated with the severity of OSA (Domany et al. 2019; Bokov et al. 2020).

High PG, acting in concert with a highly collapsible upper airway and low arousal threshold, may contribute significantly toward increasing the severity of OSA, even when controller gain is low (Khoo et al. 2020). Raised PG might exacerbate OSA severity through a feedback instability pathway that is different from the classical pathway that involves high chemoreflex loop gain (Khoo et al. 2020).

Factors such as lung volume, pulmonary blood flow, and the ability to eliminate CO2 measured by the difference between the inspired and expired CO2 are major determinants of PG. Accordingly, an abnormally elevated PG may result from abnormal lung volumes, a reduced ability to eliminate CO2 adequately such as in ventilation/perfusion mismatch (VQM), and decreased pulmonary blood flow (Domany et al. 2019).

In most OSA patients the main factor affecting plant gain is functional residual capacity. Reductions in lung stores of either CO2 or O2 will increase plant gain because smaller tidal volumes will produce greater fluctuations in alveolar gas tensions (Deacon-Diaz and Malhotra 2018). PG and OSA severity may increase with certain kinds of VQM where the effective gas-exchanging lung volume is reduced, such as what may exist in patients with asthma or subjects with developmental lung abnormalities (e.g., alveolar simplification), such as in down syndrome (Khoo et al. 2020).

The reversibility of the elevated PG in children with OSA after adenotonsillectomy and with improvement in OSA suggests that obstructive breathing during sleep might be responsible for the increase in PG (Domany et al. 2019). There is no evidence to suggest that lung volumes or plant gain are inherently different in OSA patients independent of other factors known to alter lung volume, such as obesity (Deacon-Diaz and Malhotra 2018).

17.3.4 Anatomic Factors

The anatomic predisposing factors for OSA differ over the lifespan. In infants, high nasal resistance, reduced airway stiffness, highly compliant chest wall, spontaneous neck flexion, airway edema from gastroesophageal reflux disease, laryngomalacia, altered facial skeletal development, glossoptosis, macroglossia are important risk factors (Arens et al. 2021). Adenotonsillar hypertrophy (AH) and obesity are the main risk factors in childhood and adolescence, respectively (Arens and Marcus 2004).

Compared with controls, children with OSA have smaller upper airway cross-sectional area (Arens et al. 2005) and this is noted in all age groups, probably predisposing to airway collapse during sleep (Arens and Marcus 2004).

External factors impact the size of the upper airway, particularly when it is in a retropalatal or retroglossal position. Three of these factors are particularly prominent: fat deposits (related to the body mass index), craniofacial features (related to genetic and functional factors), and hypertrophied tissues, in part related to local inflammation. Genetic and environmental factors can both have an influence on these external factors (Huang and Guilleminault 2017).

Fluctuations in the airway area during tidal breathing are significantly greater in children with OSA, maybe due to increased airway compliance although resistive pressure loading is another probable explanation (Arens et al. 2005).

Children with syndromic dysmorphology involving craniofacial synostosis, midface hypoplasia, retro positioning of the mandible or maxilla, and glossoptosis frequently have OSA. These morphologic alterations in genetic syndromes, such as Pierre Robin, are well-recognized risk factors, but the etiology is often multifactorial, and many children have multilevel airway obstruction. In the down syndrome phenotype, upper airway obstruction may be associated with a combination of factors such as midfacial hypoplasia, AH, relative macroglossia, enlarged or swollen nasal turbinates, poor muscle tone, and elevated BMI (Simpson et al. 2018).

The contribution of skeletal abnormalities to the development of OSA in otherwise normal children is controversial (Katz and D'Ambrosio 2008). No significant lower face skeletal differences were found between young children with OSA and matched controls (Arens et al. 2001), although cephalometric studies in children with OSA frequently report narrower maxilla, mandibular retrognathia, longer lower facial height, and caudal placement of the hyoid bone (Katz and D'Ambrosio 2008). These craniofacial skeletal abnormalities may contribute to the development of OSA in some otherwise nonsyndromic children and also are the effect of craniofacial growth modification in mouth breathing (MB) children (Zheng et al. 2020).

The low and anterior position of the tongue related to mouth breathing results in a lack of internal pressure, leading to a reduction of transversal growth of the upper arch with the development of lateral and posterior crossbite. On the other hand, mouth breathing is associated with decreased nose prominence and width dimensions compared to normal children. These facial characteristics could lead to a reduced upper airway space resulting in obstructive apnea events (Galeotti et al. 2018).

The etiology of MB may be multifactorial and attributable to anatomic factors, including narrow airways, AH, nasal septal deviation, nasal polyps, respiratory allergies, nasal turbinate hypertrophy, and sleep position (Zheng et al. 2020).

Inherited abnormalities of the craniofacial structure appear to explain at least a portion of the familial clustering of OSAS, and the genetic basis is suggested by twin and family studies (Casale et al. 2009).

AH substantially raises pharyngeal resistance and ultimately results in episodic airway collapse and complex interactions between the anatomical features and other elements such as upper airway tone and ventilatory drive. AH is the most common cause of OSA in children but a rare cause of OSA in adults.

It was demonstrated that tonsil size does not always correlate well with baseline OSA severity (Nolan and Brietzke 2011; Tang et al. 2015). In young children, using magnetic resonance imaging, irrespective of the relationship with the tonsillar pillars, has shown that the true tonsil size was positively correlated to OSAS severity (Arens et al. 2001).

The lack of a clear association between subjective tonsil size and OSA disease severity may be attributed to the difficulty in accurately estimating tonsil size due to the position of the tonsil in the fossa (endophytic tonsils) or difficulty in accurately estimating the anterior-posterior dimension of the tonsil (Tang et al. 2015).

The adenoids, located on the roof of the nasopharynx, normally enlarge in size during infancy and childhood and subsequently progressively reduce in size during adolescence and adulthood. In normal children, the airway size grows proportionately with the soft tissues surrounding it. Children with OSA have larger adenoids, tonsils, and soft palates compared with normal control subjects which decrease the pharyngeal airway (Arens et al. 2001).

The disproportionate proliferation of the adenoids and tonsils occurs more frequently in some groups of children, including those with allergic rhinitis, asthma, recurrent upper airway respiratory infections, and those exposure to environmental tobacco smoke (Evcimik et al. 2015).

There is now ample data confirming that OSA associated with obesity is highly prevalent in children and adolescents. The association between obesity and OSA emerges from two sets of observations; the first is the observed high prevalence of OSA among obese children and adolescents, and the second is the higher proportion of children with OSA who are obese. Thus, it appears that both conditions can coexist and yet potentiate the adverse impact of each. It is believed that the prevalence of OSA among obese children and adolescents can be as high as 60% (Narang and Mathew 2012). Increased deposition of fat in the parapharyngeal fat pads near and within the soft palate contributes to airway obstruction. Obese children with OSA may also have AH and surgical removal of enlarged tonsils and adenoids in obese children with OSA results in a marked improvement of the AHI but with high rates of residual OSA (Katz and D'Ambrosio 2008).

17.4 Assessment and Diagnosis in Pediatric Respiratory Sleep Medicine

Over the past decades, children presenting with sleep-related respiratory impairment revealing snoring and large tonsils became less frequent in specialized sleep centers since these cases are often screened and followed by the primary care providers with privilege access to ENT surgeon referral for surgery. Meanwhile, several symptoms more often presented to the specialist are also contextualized in such respiratory-based pathology even though it could happen to appear in a more subtle presentation. In these circumstances and depending on the specific age group, may be those major complaints like "can hardly sleep," "restless sleep" or "is sleepy during the day" suggest other problems demanding a careful differential diagnosis and an accurate assessment of possible comorbid conditions.

Therefore, a variety of questionnaires were developed to evaluate children with impaired sleep-wake cycles and disturbed sleep. Although not related to sleep-disordered breathing, it could be relevant to perform on these patients with nocturnal or sleep-related respiratory difficulties. Yet, some of them are more prone to be used as screeners by primary care practitioners, while others are preferent directed to research. As such useful assessment tools are very well described elsewhere, in this chapter we only cite a few of those questionnaires more widely used and with potential additional value for assessment and diagnosis of sleep-related breathing disturbance in pediatric ages.

17.5 Children's Sleep Habits Questionnaire (CSHQ)

Developed by Judith Owens, from Boston, this 45 items questionnaire to be answered by adult caregivers, this questionnaire covers the most prevalent sleep disorders in children. CSHQ allows to assessment of bedtime behavior and sleep onset, sleep duration, the effect of anxiety in sleep and frequent night wakings, sleep-related breathing disorders, parasomnias, and morning waking/daytime sleepiness. Parents/caregivers are asked to fill out the Likert-type scale about the child's typical behaviors in recent weeks and assign one of the three options for each item. A higher score indicates a more disturbed sleep (Owens et al. 2000).

17.6 Pediatric Daytime Sleepiness Scale (PDSS)

These eight items self-reported questionnaire was developed by Christopher Drake and is validated for children between 11 and 15 years old and is used to measure daytime sleepiness. This is provided in a Likert scale format with five options per item with a higher score suggesting more daytime sleepiness (Drake et al. 2003).

17.7 Adolescent Sleep Habits Survey

This two-version questionnaire developed by Judith Owens is a relevant tool with the aim of summarizing an appreciable amount of information on demographics, medical history, and sleep habits as well as schedules and related sleep behaviors. It can be applied for children from 4 to 12 grade and the complete version consist of 61–65 questions to be answered in about 25 min. There is no need for calculation and the user should be familiar with general sleep disorders to proceed with recommending further assessment as needed (Shahid et al. 2012).

17.8 BEARS Sleep Screening Tool

Is an easy-to-do pediatric sleep screener to be used preferably in the primary care setting. Developed by Judith Owens, the BEARS has five items surpassing bedtime issues, excessive daytime sleepiness, night awakenings, regularity of sleep and sleep duration, and snoring, where parents/caregivers are asked to point out if there are any sleep problems referred for each item. When positively answered, further assessment should be required. No scoring is needed but the interpreter should have a basic knowledge of sleep medicine (Owens and Dalzell 2005).

17.9 Obstructive Sleep Apnea-18 (OSA-18)

An 18-item (five subscales) questionnaire developed by Ramon Arturo Franco, covering five distinct aspects (five items), like sleep disturbance, physical symptoms, emotional symptoms, daytime function, and caregivers' concerns. It serves to infer the impact of pediatric OSA on quality of life. It is validated for children ranging from 6 months to 12 years and takes up to 10 min to be completed. Each item is cored using a Likert-type scoring system and the caregiver is called to point, from 1 to 7, the frequency of a list of symptoms for the last 4 weeks. A total symptom score of less than 60 indicates a small impact on health-related quality of life. A moderate (60–80) and severe impact (more than 60) would be inferred accordingly (Franco et al. 2000).

17.10 Obstructive Sleep Apnea-5 (OSA-5)

A short form of the OSA-18, proposed by Gillian Nixon for the 2–18 years old population. This questionnaire evaluates the frequency of symptoms during the last 4 weeks as well as it takes less than 5 min to completed. Total score rage from 0 to 15, with a score of 5 and above indicating OSA with an increasing score meaning a more likelihood of respiratory impairment during sleep (Soh et al. 2018).

17.11 Clinical Assessment and Featuring Pediatric OSA

Defining normal is not as straightforward as the literature seems to suggest. This is particularly true in pediatric sleep medicine. Not only the wide range of the standard deviations is common regarding sleep of younger age groups but also there is a lack of robust normative data on the different objective tests commonly assumed as standards. As such, substantial controversial issues remain over many decision algorithms on the base of different suggested criteria. This implies that a careful clinical approach is not only useful but mandatory to progress among diagnostic hypothesis and the final assumption as well as the associated therapeutic decisions.

Disruption of both internal and external factors which depend on human life may affect sleep in different complex ways allowing to the development of circadian clock impairments as well as sleep disturbances. The major complaints of pediatric patients presenting in a sleep clinic are insomnia and hypersomnia. Distinct mechanisms may account for: (a) daytime complaints, such as hypersomnia or fatigue often associated with behavioral, cognitive, and emotional problems, and (b) nighttime difficulties, such as insomnia or interrupted sleep together with restlessness, respiratory impairments, motor disturbances or parasomnias. The sleep history with a focus on a systematic sleep examination will allow us to isolate seven major diagnostic hypotheses and allow us to go further in a more oriented way through therapeutic management.

The diagnostic hypothesis at this stage will be oriented according to details obtained both from the history and clinical examination. Further support will be provided from the sleep lab and other tests in sleep-related breathing disorders, central disorders of hypersomnolence, and sleep-related motor disorders. For parasomnias, insomnia disorders and circadian rhythm sleep disorders, sleep logs, and actigraphy measurements would be preferred to confirm the diagnostic hypothesis.

Although less frequently in younger ages as compared to adulthood, comorbid conditions may occur and impact either diagnostic strategy or prognosis of sleep apnea in pediatrics. As such, it is important to consider a global view even if the focus should be directed to respiratory impairment during sleep.

Parents or friends of patients with SRBD will often report at least the occurrence of snoring, witness apneas, gasping, or choking during sleep or restless sleep. Night sweats and morning headaches are common complaints in those patients. The history of adenotonsillar hypertrophy in preschool or early school years is also common and comorbid conditions such as obesity, craniofacial abnormalities, and neuroendocrine disturbances may be crucial role players.

Crosstalk between different pathways and inadequate restorative processes will turn into evident signs of poor daytime performance. Psychobehavioral problems as well as developmental issues are also commonly concurrent and as so, these aspects must be valorized.

Several tools objectively assessing sleep parameters are correlated to the symptoms of inadequate sleep and serve as standards for diagnostic assessment in pediatric sleep medicine. From in-lab polysomnography to unattended sleep studies, oximetry, spectral analysis of the EEG signal, actigraphy measures, autonomic assessments, and sound analysis, there are plenty of proposals. Several wearable devices and software's were developed in recent years providing a substantial number of options. However adequate validation and data from carefully taken clinical studies should be considered when managing the available tools either for diagnosis or for screening purposes.

OSA in children may present in a wide spectrum of clinical features. Snoring, observed increased effort to breath and respiratory pauses, snorting and gasping, nighttime sweats, restless sleep, and enuresis. This is often associated with particular sleeping positions and occurs together with some habits usually observed during wakefulness like mouth breathing (also more likely to occur at night in these patients), recurrent infections in the upper airway tract, increased fatigue during daytime, attention deficit problems, hyperactivity (majorly contributing to a poor school performance), increased aggressivity, hearing and speech problems, increased pain complaints (headaches and others), and excessive daytime sleepiness.

The differential diagnosis for pediatric OSA may be helped by an exhaustive and systematic evaluation of symptoms. Inspiratory sounds are suggestive of snoring should be considered face to snoring presenting as an isolated symptom or in the context of upper airway obstructive events. It is, however, crucial to be aware of other sounds such as stridor and catathrenia which may sometimes confound conclusions. On the other hand, and to a certain extent, respiratory pauses during sleep could be normal in children. As so, it is important to differentiate those silences from those accompanying central sleep apnea or OSA. Choking or gasping must be adequately assessed and not be confused with sleep-related laryngospasm or sleep-related reflux. OSA-associated daytime impairment in a wide range of pediatric ages would present differently. But other causes could rely on its background as obesity hypoventilation syndrome, hypersomnia, or insufficient sleep syndrome. Also, motor events often occurring with OSA as periodic limb movements should be adequately differentiated from those related to restless leg syndrome, Finally, in the context of the complaints, history, and objective clinical evaluation, parasomnias should be considered in these patients also.

17.12 Polysomnography (PSG)

PSG is the gold standard for sleep-related disorders either in adulthood or pediatric age. The standard overnight PSG assesses sleep parameters utilizing a multi-channel recording and was already detailed in another chapter of this book. Meanwhile, the PSG montage may be adequate to each specific patient according to their specific condition and suspected disorder. On the other hand, although with a great clinical value, standard PSG is linked to several challenging factors to be overwhelmed (Marcus et al. 2012). Yet, although with some important limitations, in recent years, home sleep apnea tests (HSATs) for children are becoming more interesting in the general clinical setting. With the need of counting on significant discrepancies when measuring with PSG or HSATs, later ones may have to be very well considered when applied to the pediatric population, particularly to children and young adolescents. Therefore, ambulatory unattended polysomnography (H-PSG) which has been involved in such a controversial debate for years, has the advantage of being easier to perform than standard PSG, and not only evaluate SDB but also a numerous other sleep disorders (Chesson Jr et al. 2003).

Indeed, the future of pediatric sleep assessment would probably account for simplifications regarding important technical aspects that would allow for a model of parents-assisted remote monitoring preserving the major standards for clinical integration of sleep evaluation in children and young adolescents.

17.13 Overnight Pulse Oximetry

Overnight pulse oximetry allows the relatively noninvasive estimation of peripheral blood hemoglobin oxygen saturations (SpO2) and forms part of the typical polysomnogram (PSG) for investigation of OSA (Terrill 2020). Compared to PSG, oximetry is simple, inexpensive, and readily available (Chan et al. 2019).

In clinical practice, obtained data are summarized with simplified indices: the oxygen desaturation index (ODI) that calculates the number of desaturation events (also referred to as dips or transient hypoxemic events) per hour which drop 3% (ODI3) or 4% (ODI4) below baseline levels; the proportion of total recording time with SpO2 $\leq$ 90% (T90); and other simple time-series statistics. When used as part of a PSG, oximetry contributes to the scoring of hypopneas, and consequently, the apnea–hypopnea index (AHI) that summarizes the severity of sleep-disordered breathing as the number of apneas and hypopneas per hour. While such simplified indices are convenient, they likely fail to capture important pathophysiological characteristics (Terrill 2020).

Normative oximetry parameters in healthy children 6 months to 12 years using oximetry motion-resistant technology able to exclude motion artifact and with short averaging times was published recently (Table 17.1) (Ong et al. 2021).

Based on this data, normality was defined as ODI 3% <7/h, to some extent higher than in previous studies that found in healthy children 2.2/h (95th centile ODI4%) (Urschitz et al. 2003) and 1.2/h (2–10 years of age, 90th centile ODI3%) (Scholle et al. 2011).

Nixon et al. observed a correlation between higher oximetry scores and elevated apnea–hypopnea index (AHI), increased desaturation index, lower minimum arterial oxygen saturation (SaO2) levels, and a higher respiratory arousal index (Nixon et al. 2004). Based on these findings, the McGill oximetry score was conceived. The scoring system comprises four severity levels (McGill 1–4), which are determined

Table 17.1 Descriptive statistics for oximetry desaturation events and indices for the entire study group ($n = 66$): number of oxygen desaturations >4% and >3% from baseline per hour (ODI4% and ODI3%), delta index 12 s (DI12s), and percentage time with saturations below 92% and 90% (Ong et al. 2021)

Parameter	Mean	SD	Mean 95% CI	Median	Median 95% CI	90th centile	95th centile	97.5th centile
ODI4%	1.14	0.84	0.93–1.34	0.92	0.73–1.15	2.60	3.00	3.38
ODI3%	2.90	1.65	2.50–3.29	2.58	1.96–3.10	5.65	6.43	7.06
ODI12s	0.35	0.09	0.33–0.38	0.34	0.31–0.38	0.48	0.53	0.57
Time < 92%	0.02	0.03	0.01–0.03	0.00	0.00–0.01	0.05	0.11	0.15
Time < 90%	0.005	0.015	0.002–0.009	0.00	0.00–0.00	0.01	0.05	0.07

Table 17.2 The McGill Oximetry Scoring System (MOS) (Nixon et al. 2004)

Oximetry score	Criteria					Recommendation
		No. of drops in SaO_2 <90%	No. of drops in SaO_2 <85%	No. of drops in SaO_2 <80%		
1	Normal study/inconclusive for OSA	<3	0	0	Baseline: stable (<3 clusters of desaturation) and >95%	Additional evaluation of breathing during sleep required to rule out OSA
2	OSA, mild	>3	3	0	Three or more clusters of desaturation events	Recommend T&A on the waiting list
3	OSA, moderate	>3	>3	3	Three or more clusters of desaturation events	Recommend surgery within 2 week
4	OSA, severe	>3	>3	>3	Three or more clusters of desaturation events	Recommend urgent surgery (within days)

by the quantity and severity of desaturations observed in an overnight pulse oximetry recording (Table 17.2). A score higher than one was proposed as an indication of obstructive sleep apnea (OSA). They suggested that prioritizing operative intervention based on oximetry score can shorten the diagnostic and treatment process for those with more severe diseases. However, oximetry is inconclusive in many children with suspected OSA and cannot replace PSG for the definitive diagnosis of OSA (Nixon et al. 2004).

In agreement with other studies (Chan et al. 2019; Chuanprasitkul et al. 2021; Jonas et al. 2020), it was shown that nocturnal oximetry may not be used for a definite diagnosis of OSA in obese children (Van Eyck et al. 2015). Although it has good specificity, with a positive result most likely indicating a positive PSG result, an absence of abnormal oximetry results does not definitively exclude a diagnosis of OSA, as the sensitivity of this test is only moderate (Van Eyck et al. 2015). In the context of a child suspected of having OSA, Brouillette et al. reported that positive nocturnal oximetry results have a minimum positive predictive value of 97% for OSA, whereas negative oximetry findings cannot completely rule out the possibility of OSA (Brouillette et al. 2000).

OSA by full night polysomnography was demonstrated in 42% of children who had a normal or inconclusive McGill oximetry score (MOS 1) (Horwood et al. 2014). Despite the low negative predictive value, the utility of nocturnal oximetry as a particularly effective tool for identifying cases of severe OSA was verified given that the clinical values for this group, such as the AHI, desaturation index, and oxygen saturation nadir, designate these children as having OSA in the mild to moderate range and without clinically significant desaturations (Horwood et al. 2014). Additionally, in another study, the specificity of MOS 3 and 4 in diagnosing moderate/severe OSA on PSG was 100% (Jonas et al. 2020).

MOS scores over different nights may have only been fair to a moderate agreement. In a recent study of children, 38% screened positive for OSA on the first night. When three nights of oximetry were done 59% screened positive on at least one of the nights suggesting that one night of oximetry screening may not be sufficient to screen for OSA (Galway et al. 2021).

Efforts to develop novel strategies for the analysis of oximetry data have been heavily concentrated on the development of tools for the automated (or semi-automated) diagnosis of OSA (Terrill 2020).

17.14 Pediatric OSA Treatment

Due to its complexity, OSAS management requires a multidisciplinary approach by different specialist teams to make an early diagnosis and a correct treatment plan. Several specialists are engaged in the treatment of pediatric OSA such as pediatricians, otorhinolaryngologists, pediatric dentists, orthodontists, neurologists, nutritionists, physiotherapists, and speech therapists (Giuca et al. 2021).

Although AH is the most common risk factor for Pediatric OSA, obesity has become an increasingly important risk factor. Prevalent, especially in early childhood (6–9 years) and adolescence (Bue et al. 2020). Therefore, nutritional education and physical exercise/training may be a part of the OSA therapeutic plan (Giuca et al. 2021).

The size and location of tonsils and adenoids are influenced by genetics, infection, and inflammation. Although tonsils that look big may contribute to a reduction in the diameter of the pathways, there is no clear linear correlation between enlarged tonsils and adenoids with the severity of OSA. Therefore, even the "small" tonsils (e.g., within the tonsillar pillars) may be clinically significant and cause airway obstruction during sleep. Adenoids also contribute to the reduction in airway size and are better visualized by nasal endoscopy or magnetic resonance image processing and anatomic measurements (Arens and Marcus 2004; Arens et al. 2001).

Children with OSA experience a combination of oxidative stress, inflammation, autonomic activation, endothelial dysfunction, increased aggregation platelet count, metabolic dysregulation, and interruption of sleep homeostasis (O'Driscoll et al. 2009). OSA is a common disorder in children and those at risk must be identified, studied, and treated promptly because untreated OSA can be responsible for cardiovascular, metabolic, and neurocognitive morbidities, with the potential to induce deficits, sometimes nonreversible, given its persistence in a critical period of physical and neuropsychic development (Bue et al. 2020). Treatment of OSA should be viewed as rather urgent in children, particularly when daytime symptoms such as hyperactivity, inattention, somnolence, or poor school performance are detected by either parents or school teachers (Gozal et al. 2021).

Treatment improves behavior, attention, quality of life, neurocognitive functioning, enuresis, parasomnias, and restless sleep and reversing of associated cardiovascular sequelae (Chinnadurai et al. 2017).

In addition to adenotonsillectomy, the treatment of OSA in children may include the use of continuous positive airway pressure (CPAP), weight loss, oral devices, and pharmacological treatment.

The decision to initiate treatment and choice of treatment depend upon the child's age, clinical symptoms (e.g., nighttime sleep problems or daytime dysfunction), presence of comorbidities (especially underlying genetic, craniofacial, neuromuscular disorders), risk factors (e.g., obesity, crowded oropharynx), and results of PSG, if performed.

17.15 Medical Therapy

Inflammation is consistently present in adenotonsillar tissues and upper airway tissues in children with OSA and treatment with corticosteroids and leukotriene receptor antagonists may reverse the underlying processes and resolve or mitigate OSA (Gozal et al. 2021; Goldbart and Tal 2008).

Airway inflammation is thought to be caused by the local, repeated, mechanical trauma related to the vibratory intermittent airway occlusion typical of OSA. Another potential mechanism mediating local and systemic inflammation involves the intermittent nocturnal hypoxemia that may, through the phenomenon of the ischemia-reperfusion injury, produce oxygen-free radicals (Goldbart and Tal 2008).

Corticosteroids are particularly effective in alleviating the magnitude of the inflammatory processes in the airway via their binding to corticosteroid receptors and their translocation to the nucleus where they operate via transcriptional, epigenetic, and post-translational regulatory processes (Gozal et al. 2021). The expression of cysteinyl-leukotriene receptors is markedly enhanced in the tonsillar and adenoid tissues of children with OSA and are predominantly located in T cell lymphocytes (both CD4+ helper T cells and even more so in CD8+ cytotoxic T cell lymphocytes). In vitro studies using mixed cell cultures of either adenoids or tonsils showed that the addition of leukotriene D4 to the media resulted in increased cellular proliferation and release of pro-inflammatory cytokines. When such cultures were treated with leukotriene receptor antagonists or with corticosteroids, dose-dependent reductions in proliferation and cytokine release occurred (Gozal et al. 2021).

Anti-inflammatory agents, especially topical nasal spray corticosteroids, are suggested to have a potential role in reversing adenotonsillar enlargement (Chan et al. 2015).

The initial randomized controlled trial (RCT) examining the effectiveness of nasal fluticasone for a duration of 6 weeks in children diagnosed with OSA who were waiting for AT was conducted in 2001, with a substantial decrease of the AHI in the fluticasone group (Brouillette et al. 2001). In a larger RCT, intranasal budesonide for 6 weeks in PSG-diagnosed children with mild OSA resulted in significant improvements in both polysomnographic measures (sleep latency, slow-wave sleep, and rapid eye movement sleep) and in the magnitude of respiratory disturbance (apnea/hypopnea index, nadir pulse oxygen saturation). Lateral X-rays of the upper airway showed that the size of the adenoids had also declined (Kheirandish-Gozal and Gozal 2008). In a more recent RCT of intranasal mometasone furoate (MF) for 4 months versus placebo in children aged 6–18 years with mild OSA, AHI, and ODI improved significantly in the MF group only (Chan et al. 2015).

However, the available evidence regarding the effectiveness of intranasal corticosteroids in treating OSA in children is insufficient. The level of certainty regarding the benefits on the primary outcome, including the apnea–hypopnea index (AHI) and respiratory arousal index, is considered low due to imprecise estimates and variations among studies (Kuhle et al. 2020). A large number of studies have been published on intranasal corticosteroids, but the majority did not assess the children using PSG (Gozal et al. 2021).

After the initial study in 2005, which showed improvements in AHI in children with mild OSA in an open-label trial (Goldbart et al. 2005), two double-blind, RCTs have been conducted and revealed that oral montelukast therapy for 12–16 weeks improves nocturnal symptoms, reduces the size of adenotonsillar tissues, and significantly reduces PSG-derived measures in the context of mild OSA (Goldbart et al. 2012; Kheirandish-Gozal et al. 2016).

In a review conducted in 2016, examining the VigiBase database analysis of psychiatric disorders in children linked to the use of Montelukast database, age-related neuropsychiatric adverse drug events were observed. Infants and children were found to develop sleep disturbances, while adolescents were more prone to experiencing depression/anxiety and psychotic reactions (Aldea Perona et al. 2016). The emergence of side effects related to montelukast raises a red flag that should prompt caution when deciding to use this compound (Gozal et al. 2021).

Studies comparing head-to-head intranasal corticosteroids and oral montelukast revealed rather similar improvements in children with OSA. Based on these results, both therapies seem to be of comparable efficacy (Gozal et al. 2021).

Overall, beneficial effects occurred in $>80\%$ of 752 children with combination therapy with intranasal corticosteroids and oral montelukast. In a subgroup of 445 patients who underwent follow-up polysomnography, 62% demonstrated improvement and normalization of sleep findings, while 17.1% experienced either no improvement or a worsening of their OSA. Among those who did not respond positively, older children (aged >7 years) and obese children (with a BMI z-score > 1.65) were notably more likely to be nonresponders (Kheirandish-Gozal et al. 2014).

Bacterial involvement in AH was first suggested based on the increase in lymphocytes in tonsils and adenoids and later based on a greater number of bacterial pathogens such as *Haemophilus influenzae* and other β-lactamase-producing bacteria in enlarged tonsils and adenoids (Don et al. 2005). Some studies have indicated the efficacy of broad-spectrum antibiotics to improve the symptoms and complications of AH and OSA. However, the results of a meta-analysis, which analyzed various forms of treatment for OSA, found statistical insignificance for all comparisons related to antimicrobial therapy (Lin et al. 2020). Evidence of the usefulness of using broad-spectrum antibiotics to treat pediatric OSA is limited.

Children with mild or moderate OSA and nasal obstruction due to adenoid hypertrophy, especially those with seasonal allergies, may be candidates for treatment with intranasal corticosteroids or leukotriene antagonist therapy. These anti-inflammatories are typically considered as an alternative to adenotonsillectomy, or as a therapy during the watchful waiting period before other interventions. If this therapeutic option is selected, an administration of at least 4 weeks of the drug must be performed to decide whether it will be useful to maintain the therapy for a longer period.

17.16 Adenotonsillectomy (AT)

If a child meets diagnostic criteria for OSA, clinical examination consistent with AH, and no contraindications to surgery, adenotonsillectomy should be recommended as the first line of treatment (Marcus et al. 2012). Certain children may present with multiple degrees of airway obstruction and might necessitate further assessment and treatment (Bitners and Arens 2020).

A variety of surgical techniques and instruments such as "cold dissection, electric monopolar cautery, bipolar cautery, harmonic scalpel, bipolar radiofrequency (coblation), and PEAK PlasmaBlade (Medtronic)" has been used for AH surgery. Various methods and devices have been employed to decrease post-operative pain and intra-operative bleeding, increase surgical speed, and decrease the risk of post-surgical hemorrhage (Sakarya et al. 2017). Adenoidectomy can be performed using suction diathermy ablation or curettage adenoidectomy (Sakarya et al. 2017).

A multicenter randomized controlled trial (Childhood Adenotonsillectomy Trial—CHAT) examined the outcome of early adenotonsillectomy compared with a conservative approach of watchful waiting without surgical intervention concerning neurobehavioral symptoms, quality of life, and changes in polysomnography (Marcus et al. 2013). The study population included 464 school-aged children (5–9 years old) with mild sleep apnea, moderate OSA, or severe OSA determined by polysomnography. More than 50% of subjects were African American and nearly 50% were overweight or obese. The primary outcome, attention, and executive function on the Developmental Neuropsychological Assessment did not show any significant difference in the early tonsillectomy group compared with the watchful waiting group. However, compared with the watchful waiting strategy, there was a significant improvement in behavior, reduction of OSA symptoms, normalization of polysomnography findings, and quality of life (evaluated by the pediatric quality of life inventory) in children undergoing early adenotonsillectomy. Although the percentage of subjects experiencing normalization of polysomnography after early adenotonsillectomy was higher than that in the watchful waiting group (79.5% vs. 46.5%), it should be noted that almost half of the subjects in the non-surgical intervention group demonstrated normalization of polysomnography at the end of the 7-month follow-up period. This improvement in the non-intervention group was attributed to the growth of the developing airway leading to a proportional reduction in airway obstruction because of AH (Marcus et al. 2013).

A recent systematic review suggests that AT treatment of OSA in children can improve behavioral symptoms, cognitive outcomes, and sleep-related quality of life. The clear improvement of patients after AT provides evidence for a cause-and-effect relationship between sleep breathing disorders and various negative results on behavioral, cognitive, and mental health (Di Mauro et al. 2021).

Using a multicenter study design consisting of eight pediatric sleep centers across the USA and Europe that routinely conduct preoperative and post-operative nocturnal polysomnography, the outcomes of adenotonsillectomy for OSA in a large cohort of children were examined and revealed a high frequency of residual OSA. Approximately 50% of included children were obese and only 27.2% had complete resolution of OSA (Bhattacharjee et al. 2010). There is compelling evidence that AT is not uniformly effective in curing OSA in children and that in the context of increasing obesity rates in children, the ability of AT to eradicate OSA needs to be viewed with great skepticism (Bhattacharjee et al. 2010).

The impact of adenotonsillectomy in children with severe OSA without comorbidities was reviewed (Galluzzi and Garavello 2021). AT significantly reduces apneas in otherwise healthy children with severe OSA. All the studies included documented a significant reduction of AHI or RDI values after adenotonsillectomy, the improving varied from 57.7 to 93.3% (Galluzzi and Garavello 2021). Nevertheless, the number of children with severe OSA having persistent apnea after surgery is high. Considering AHI $\geq$5, 30 to 55% of children still have apneas and with AHI $\geq$1 the percentage increases from 60 to 90.6% (Galluzzi and Garavello 2021).

Various risk factors for persistent OSA in otherwise healthy children with severe OSA have been investigated with contradictory data (Galluzzi and Garavello 2021). Such factors include small adenoids, abnormal facial morphology, narrower epipharyngeal airspace, more poorly developed mandibular protrusions, OSA severity, age, obesity, male sex, allergy, and asthma. Overall, to date, the demographic characteristics and the polysomnographic criteria that will predict residual OSA in otherwise healthy children after AT have not been completely established. This may be reflected in the unknown etiology of OSA which is probably due to multifactorial causes (Galluzzi and Garavello 2021).

Tonsillotomies instead of tonsillectomy are preferred by many surgeons since it has lower post-operative complication rates such as hemorrhage, less post-surgical pain, and shorter recovery times (Tan et al. 2016). there are no significant differences in the resolution of OSA symptoms, quality of life, and post-operative immune function between the two surgical procedures. However, the risk ratio of OSA recurrence was 3.33 in favor of tonsillectomy (Wang et al. 2015). Current data are insufficient to recommend one surgical technique over the other. Children undergoing tonsillotomy should be monitored carefully long-term to ensure that OSA symptoms do not recur (Tan et al. 2016).

Following AT, it is important to monitor all children for any lingering symptoms, and in cases of persistent disease, repeat polysomnography may be conducted to identify the need for ongoing management, including positive airway pressure ventilation and potentially additional surgical interventions if necessary (Bitners and Arens 2020).

17.17 Continuous Positive Airway Pressure (CPAP)

Treatment with CPAP is provided through an electronic device that maintains constant positive air pressure through a nasal or face mask, with effects of the mechanical endoprosthesis of the airways and improvement of functional residual capacity. The goals of CPAP therapy are to ensure unobstructed breathing, good sleep quality, improved daytime function, and in some cases, rest from the work of breathing by overcoming the critical closing pressure of the pharynx. It has no restrictions or recommendations regarding the age of application.

For patients with specific surgical contraindications, craniofacial abnormalities, minimal adenotonsillar tissue, persistent OSA after adenotonsillectomy, or for those who prefer non-surgical alternatives, CPAP therapy is an option (Gao et al. 2021). It may also be indicated as a bridge to surgery or orthodontic procedures (craniofacial alterations, severe adenoid, and tonsil hypertrophy with repercussions on growth and/or cardiac function). However, unlike adenotonsillectomy, which is a single procedure that, in most cases, results in significant resolution or improvement, CPAP, in many cases, will need to be used indefinitely (Section on Pediatric Pulmonology et al. 2002).

CPAP has been increasingly used in the treatment of pediatric OSA unresolved by surgical airway intervention. It is not clear what proportion of children can be successfully withdrawn from CPAP and the factors that affect this discontinuation.

In an Australian cohort, reasons for discontinuing CPAP included post-surgery improvement in the airway, weight loss, and improvement in symptoms and/or PSG parameters over time. It was concluded that the natural course of pediatric OSA is not clear, especially in severe and treatment-resistant cases (King et al. 2019). Children on CPAP should be monitored regularly for maintenance of symptoms and the need for positive pressure support, as a small proportion can be successfully removed from CPAP over time.

Initial CPAP prescriptions are made by specialists in sleep medicine reference centers, ideally integrated into multidisciplinary consultations that ensure the necessary diagnostic workup for therapeutic decisions.

The pressure levels needed vary between individuals; therefore, CPAP should be titrated in the sleep laboratory prior to devise prescription and periodically readjusted thereafter. The auto-CPAP machines are programmed to adjust CPAP pressure based on the detection of apnea, hypopnea, flow limitation, snoring, or free-flowing air. When successful, this precludes the need for CPAP titration in the laboratory. The pressure applied is set within a range determined by the prescriber The long-term use of CPAP may cause craniofacial abnormalities; thus, regular evaluation is recommended (Gao et al. 2021).

Objective monitoring of adherence, by means of software data analysis, is essential. If adherence is less than ideal, measures to improve, such as the behavioral modification or treatment of side effects (nasal symptoms, optic irritation, and skin damage) should be instituted. If these measures are ineffective, the institution of alternative treatments should be considered (Marcus et al. 2012).

CPAP should be used only in subjects with obstructive sleep apnea in the absence of hypoventilation. Bilevel-positive airway pressure (BiPaP) may be used instead of CPAP for patients with significant hypoventilation (elevated $PaCO_2$) or clinically significant central apneas (Arens and Muzumdar 2010).

CPAP is generally better tolerated in older children. Behavioral or desensitization techniques may be necessary to help young children or older children with learning or behavioral difficulties become accustomed to using the device.

17.18 Odontoesthomatological Management

Odontostomatological care providers are likely to be an important part of the sleep medicine team either in adults or pediatrics (Meira et al. 2020). Although from a traditional perspective, oral health professional does not make the initial diagnostic judgment, they can contribute to identifying pediatric OSA as they examine children and adolescents more frequently than pediatricians or family doctors. Furthermore, orofacial and craniomandibular-specific features may be strongly associated with some particular subsets of OSA in these age groups (Stauffer et al. 2018). Still, associations between OSA and orofacial characteristics are bidirectional, with morphometric changes predisposing to an impaired respiratory function during sleep and, breathing difficulties or changes in normal respiratory pattern, determining skeletal and soft tissue developmental challenges which perpetuate in a vicious cycle of unless there is an interventive measure to stop this negative feedback loop. Even though as children get older, other features prevail as etiological determinants for the development and worsening of OSA, this relationship remains of high relevance in a wide range of pediatric ages.

It became clear from several epidemiological and clinical studies, that phenotypic rather than anatomic markers regarding craniofacial features are linked to pediatric OSA. These markers are for instance long narrow fascia, convex profile, some degree of constriction within the palate, and anterior open bite. However, they seem to play only a minor role in other subsets of patients.

Therefore, as the positive predictive value of those markers is high, it would be important for dentists to adequately screen craniofacial bony and soft tissues as well as other general indicators such as body weight, presence of developmental changes, and syndromic features. Sometimes, together with a cautious examination of the medical and dental status, radiographs (panoramic and lateral) or tomographic images should strengthen diagnostic hypotheses and diagnostic accuracy.

Craniofacial characteristics can be valuated both in terms of clinical aspects and radiographic findings using cephalometry. Classically, increased height of the face is common in children with OSA as well as a reduction in maxillary width and retruded mandibular prominence. Other aspects have been supported by the evidence as characterizing pediatric OSA: a lower position of the hyoid bone, a reduction in SNB angle, supporting the retruded mandible, and a reduced Pg-NB as indicative of chin retrusion. Also, an initial screening for tonsils enlargement and adenoid hypertrophy as documented by imaging examination seems to be important in diagnosis and referral (Lee et al. 2020).

In patients with syndromes affecting the craniofacial area, these standards often have some distinctive markers which the dentist should be aware of. Orthodontic and orthopedic strategies can be put forward to achieve treatment success. Rapid maxillary expansion is indicated for taking care of transverse maxillary constriction or posterior crossbite and it can be used as a single option or as an adjunctive measure. Such effects are often potentiated by a previous adenotonsillectomy as well as the adequacy of healthy lifestyles.

17.19 Final Considerations

Pediatric OSA is a common and impacting condition with the potential to affect more than momentum in the children's life and have substantial influences on familial and social status of the patient. Health-related consequences require imperative and adequate managing to succeed. Screening solutions, diagnostic models, and therapeutic standards are challenges and a multi-professional approach should be required to successfully respond to the patient itself but also to the social and environmental determinants.

Future improvements on each of such dimensions will certainly confirm the advantage of a healthy, good sleeper human being.

References

Aldea Perona A, García-Sáiz M, Sanz Álvarez E (2016) Psychiatric disorders and montelukast in children: a disproportionality analysis of the VigiBase®. Drug Saf 39(1):69–78. https://doi.org/10.1007/s40264-015-0360-2

Arens R, Marcus CL (2004) Pathophysiology of upper airway obstruction: a developmental perspective. Sleep 27(5):997–1019. https://doi.org/10.1093/sleep/27.5.997

Arens R, Muzumdar H (2010) Sleep, sleep disordered breathing, and nocturnal hypoventilation in children with neuromuscular diseases. Paediatr Respir Rev 11(1):24–30. https://doi.org/10.1016/j.prrv.2009.10.003

Arens R, McDonough JM, Costarino AT et al (2001) Magnetic resonance imaging of the upper airway structure of children with obstructive sleep apnea syndrome. Am J Respir Crit Care Med 164(4):698–703. https://doi.org/10.1164/ajrccm.164.4.2101127

Arens R, Sin S, McDonough JM et al (2005) Changes in upper airway size during tidal breathing in children with obstructive sleep apnea syndrome. Am J Respir Crit Care Med 171(11): 1298–1304. https://doi.org/10.1164/rccm.200411-1597OC

Arens R, Sin S, Wootton DM (2021) Pathophysiology of obstructive sleep apnea syndrome in childhood. In: Gozal D, Kheirandish-Gozal L (eds) Pediatric sleep medicine, mechanisms and comprehensive guide to clinical evaluation and management, pp 437–457. https://doi.org/10.1007/978-3-030-65574-7_thirty-nine

Ayappa I, Rapoport DM (2003) The upper airway in sleep: physiology of the pharynx. Sleep Med Rev 7(1):9–33. https://doi.org/10.1053/smrv.2002.0238

Bhattacharjee R, Kheirandish-Gozal L, Spruyt K et al (2010) Adenotonsillectomy outcomes in treatment of obstructive sleep apnea in children: a multicenter retrospective study. Am J Respir Crit Care Med 182(5):676–683. https://doi.org/10.1164/rccm.200912-1930OC

Bilston LE, Gandevia SC (2014) Biomechanical properties of the human upper airway and their effect on its behavior during breathing and in obstructive sleep apnea. J Appl Physiol 116(3): 314–324. https://doi.org/10.1152/japplphysiol.00539.2013

Bitners AC, Arens R (2020) evaluation and management of children with obstructive sleep apnea syndrome. Lung 198(2):257–270. https://doi.org/10.1007/s00408-020-00342-5

Bokov P, Matrot B, Medjahdi N et al (2020) Cross-sectional study of loop gain abnormalities in childhood obstructive sleep apnea syndrome. Sleep Med 69:172–178. https://doi.org/10.1016/j.sleep.2020.01.023

Brouillette RT, Morielli A, Leimanis A, Waters KA, Luciano R, Ducharme FM (2000) Nocturnal pulse oximetry as an abbreviated testing modality for pediatric obstructive sleep apnea. Pediatrics 105(2):405–412. https://doi.org/10.1542/peds.105.2.405

Brouillette RT, Manoukian JJ, Ducharme FM et al (2001) Efficacy of fluticasone nasal spray for pediatric obstructive sleep apnea. J Pediatr 138(6):838–844. https://doi.org/10.1067/mpd.2001.114474

Bue AL, Salvaggio A, Insalaco G (2020) Obstructive sleep apnea in developmental age. A narrative review. International Classification of Sleep Disorders. Eur J Pediatr 179(3):357–365

Carberry JC, Jordan AS, White DP, Wellman A, Eckert DJ (2016) Upper airway collapsibility (Pcrit) and pharyngeal dilator muscle activity are sleep stage dependent. Sleep 39(3):511–521. https://doi.org/10.5665/sleep.5516

Casale M, Pappacena M, Rinaldi V, Bressi F, Baptista P, Salvinelli F (2009) Obstructive sleep apnea syndrome: from phenotype to genetic basis. Curr Genomics 10(2):119–126. https://doi.org/10.2174/138920209787846998

Chan CCK, Au CT, Lam HS, Lee DLY, Wing YK, Li AM (2015) Intranasal corticosteroids for mild childhood obstructive sleep apnea—a randomized, placebo-controlled study. Sleep Med 16(3): 358–363. https://doi.org/10.1016/j.sleep.2014.10.015

Chan W-S, Chan E-T, Ng D-K, Kwok K-L, Yip A-F, Leung S-Y (2019) McGill oximetry score to predict risk of obstructive sleep apnea in pediatric patients. Pediatr Respirol Crit Care Med 3(2): 36. https://doi.org/10.4103/prcm.prcm_7_19

Chang SJ, Chae KY (2010) Obstructive sleep apnea syndrome in children: epidemiology, pathophysiology, diagnosis, and sequelae. Korean J Pediatr 53(10):863–871. https://doi.org/10.3345/kjp.2010.53.10.863

Chesson AL Jr, Berry RB, Pack A, American Academy of Sleep Medicine, American Thoracic Society, American College of Chest Physicians (2003) Practice parameters for the use of portable monitoring devices in the investigation of suspected obstructive sleep apnea in adults. Sleep 26(7):907–913

Chinnadurai S, Jordan AK, Sathe NA, Fonnesbeck C, McPheeters ML, Francis DO (2017) Tonsillectomy for obstructive sleep-disordered breathing: a meta-analysis. Pediatrics 139(2): e20163491. https://doi.org/10.1542/peds.2016-3491

Chuanprasitkul C, Veeravigrom M, Sunkonkit K, Tansrirattanawong S, Sritippayawan S (2021) Incidence/predictors of pediatric obstructive sleep apnea with normal oximetry. Pediatr Int 63(11):1376–1380. https://doi.org/10.1111/ped.14663
Cori JM, Thornton T, O'Donoghue FJ et al (2017) Arousal-induced hypocapnia does not reduce genioglossus activity in obstructive sleep apnea. Sleep 40(6). https://doi.org/10.1093/sleep/zsx057
Deacon NL, Catcheside PG (2015) The role of high loop gain induced by intermittent hypoxia in the pathophysiology of obstructive sleep apnoea. Sleep Med Rev 22:3–14. https://doi.org/10.1016/j.smrv.2014.10.003
Deacon-Diaz N, Malhotra A (2018) Inherent vs. induced loop gain abnormalities in obstructive sleep apnea. Front Neurol 9:1–10. https://doi.org/10.3389/fneur.2018.00896
Di Mauro P, Cocuzza S, Maniaci A et al (2021) The effect of adenotonsillectomy on children's behavior and cognitive performance with obstructive sleep apnea syndrome: state of the art. Children (Basel) 8:921. https://doi.org/10.3390/children8100921
Domany KA, He Z, Nava-Guerra L et al (2019) The effect of adenotonsillectomy on ventilatory control in children with obstructive sleep apnea. Sleep 42(5):1–8. https://doi.org/10.1093/sleep/zsz045
Don DM, Goldstein NA, Crockett DM, Ward SD (2005) Antimicrobial therapy for children with adenotonsillar hypertrophy and obstructive sleep apnea: a prospective randomized trial comparing azithromycin vs placebo. Otolaryngol Head Neck Surg 133(4):562–568. https://doi.org/10.1016/j.otohns.2005.05.012
Drake C, Nickel C, Burduvali E, Roth T, Jefferson CPB (2003) The pediatric daytime sleepiness scale (PDSS): sleep habits and school outcomes in middle-school children. Sleep 26(4):455–458
Eckert DJ, White DP, Jordan AS, Malhotra A, Wellman A (2013) Defining phenotypic causes of obstructive sleep apnea: identification of novel therapeutic targets. Am J Respir Crit Care Med 188(8):996–1004. https://doi.org/10.1164/rccm.201303-0448OC
Evcimik MF, Dogru M, Cirik AA, Nepesov MI (2015) Adenoid hypertrophy in children with allergic disease and influential factors. Int J Pediatr Otorhinolaryngol 79(5):694–697. https://doi.org/10.1016/j.ijporl.2015.02.017
Franco RA, Rosenfeld RM, Rao M, York N (2000) First place—resident clinical science award 1999. Quality of life for children with obstructive sleep apnea. Otolaryngol Head Neck Surg 123:9–16. https://doi.org/10.1067/mhn.2000.105254. Published online 1999
Galeotti A, Festa P, Viarani V et al (2018) Prevalence of malocclusion in children with obstructive sleep apnoea. Orthod Craniofacial Res 21(4):242–247. https://doi.org/10.1111/ocr.12242
Galluzzi F, Garavello W (2021) Impact of adenotonsillectomy in children with severe obstructive sleep apnea: a systematic review. Auris Nasus Larynx 48(4):549–554. https://doi.org/10.1016/j.anl.2020.10.008
Galway NC, Maxwell B, Shields M, O'donoghue D (2021) Use of oximetry to screen for paediatric obstructive sleep apnoea: is one night enough and is 6 hours too much? Arch Dis Child 106(1):58–61. https://doi.org/10.1136/archdischild-2019-318559
Gao Z, Han D, Ni X et al (2021) Chinese guideline for the diagnosis and treatment of childhood obstructive sleep apnea (2020). Pediatr Invest 5(3):167–187. https://doi.org/10.1002/ped4.12284
Garcia J, Wical B, Wical W et al (2016) Obstructive sleep apnea in children with cerebral palsy and epilepsy. Dev Med Child Neurol 58(10):1057–1062. https://doi.org/10.1111/dmcn.13091
Gharibeh T, Mehra R (2010) Obstructive sleep apnea syndrome: natural history, diagnosis, and emerging treatment options. Nat Sci Sleep 2:233–255. https://doi.org/10.2147/NSS.S6844
Giuca MR, Carli E, Lardani L, Pasini M, Miceli M, Fambrini E (2021) Pediatric obstructive sleep apnea syndrome: emerging evidence and treatment approach. ScientificWorldJournal 2021:5591251. https://doi.org/10.1155/2021/5591251
Goldbart AD, Tal A (2008) State of the art series: sleep disordered breathing in children inflammation and sleep disordered breathing in children: a state-of-the-art review. Pediatr Pulmonol 43(12):1151–1160. https://doi.org/10.1002/ppul.20943

Goldbart AD, Goldman JL, Veling MC, Gozal D (2005) Leukotriene modifier therapy for mild sleep-disordered breathing in children. Am J Respir Crit Care Med 172(3):364–370. https://doi.org/10.1164/rccm.200408-1064OC

Goldbart AD, Greenberg-Dotan S, Tal A (2012) Montelukast for children with obstructive sleep apnea: a double-blind, placebo-controlled study. Pediatrics 130(3):1–6. https://doi.org/10.1542/peds.2012-0310

Gönüldaş B, Yilmaz T, Sivri HS et al (2014) Mucopolysaccharidosis: otolaryngologic findings, obstructive sleep apnea and accumulation of glucosaminoglycans in lymphatic tissue of the upper airway. Int J Pediatr Otorhinolaryngol 78(6):944–949. https://doi.org/10.1016/j.ijporl.2014.03.021

Gozal D, Tan HL, Kheirandish-Gozal L (2020) Treatment of obstructive sleep apnea in children: handling the unknown with precision. J Clin Med 9(3):31–34. https://doi.org/10.3390/jcm9030888

Gozal D, Ismail M, Brockmann PE (2021) Alternatives to surgery in children with mild OSA. World J Otorhinolaryngol Head Neck Surg 7(3):228–235. https://doi.org/10.1016/j.wjorl.2021.03.005

Hatipoğlu U, Rubinstein I (2003) Inflammation, and obstructive sleep apnea syndrome pathogenesis: a working hypothesis. Respiration 70(6):665–671. https://doi.org/10.1159/000075218

Horwood L, Brouillette RT, McGregor CD, Manoukian JJ, Constantin E (2014) Testing for pediatric obstructive sleep apnea when health care resources are rationed. JAMA Otolaryngol Head Neck Surg 140(7):616–623. https://doi.org/10.1001/jamaoto.2014.778

Huang YS, Guilleminault C (2017) Pediatric obstructive sleep apnea: where do we stand? Adv Otorhinolaryngol 80:136–144. https://doi.org/10.1159/000470885

Huang J, Pinto SJ, Yuan H et al (2012) Upper airway collapsibility and genioglossus activity in adolescents during sleep. Sleep 35(10):1345–1352. https://doi.org/10.5665/sleep.2110

Jonas C, Thavagnanam S, Blecher G, Thambipillay G, Teng AY (2020) Comparison of nocturnal pulse oximetry with polysomnography in children with sleep disordered breathing. Sleep Breath 24(2):703–707. https://doi.org/10.1007/s11325-019-01861-z

Katz ES, D'Ambrosio CM (2008) Pathophysiology of pediatric obstructive sleep apnea. Proc Am Thorac Soc 5(2):253–262. https://doi.org/10.1513/pats.200707-111MG

Katz ES, White DP (2004) Genioglossus activity during sleep in normal control subjects and children with obstructive sleep apnea. Am J Respir Crit Care Med 170(5):553–560. https://doi.org/10.1164/rccm.200403-262OC

Kheirandish-Gozal L, Gozal D (2008) Intranasal budesonide treatment for children with mild obstructive sleep apnea syndrome. Pediatrics 122(1):e149–e155. https://doi.org/10.1542/peds.2007-3398

Kheirandish-Gozal L, Bhattacharjee R, Bandla HPR, Gozal D (2014) Antiinflammatory therapy outcomes for mild OSA in children. Chest 146(1):88–95. https://doi.org/10.1378/chest.13-2288

Kheirandish-Gozal L, Bandla HPR, Gozal D (2016) Montelukast for children with obstructive sleep apnea: results of a double-blind, randomized, placebo-controlled trial. Ann Am Thorac Soc 13(10):1736–1741. https://doi.org/10.1513/AnnalsATS.201606-432OC

Khoo MCK, Hu WH, Amin R (2020) Effects of ventilation-perfusion mismatch on severity of obstructive sleep apnea: a modeling study. Proc Annu Int Conf IEEE Eng Med Biol Soc EMBS 2020(2):2792–2795. https://doi.org/10.1109/EMBC44109.2020.9175297

King Z, Josee-Leclerc M, Wales P, Masters IB, Kapur N (2019) Can CPAP therapy in pediatric OSA ever be stopped? J Clin Sleep Med 15(11):1609–1612. https://doi.org/10.5664/jcsm.8022

Kuhle S, Hoffmann DU, Mitra S, Urschitz MS (2020) Anti-inflammatory medications for obstructive sleep apnoea in children. Cochrane Database Syst Rev 2020(1):CD007074. https://doi.org/10.1002/14651858.CD007074.pub3

Lam DJ, Jensen CC, Mueller BA, Starr JR, Cunningham ML, Weaver EM (2010) Pediatric sleep apnea and craniofacial anomalies: a population-based case–control study. Laryngoscope 120(10):2098–2105. https://doi.org/10.1002/lary.21093.Pediatric

Lee Y, Huang Y, Chen I, Lin P, Chuang L (2020) ScienceDirect: craniofacial, dental arch morphology, and characteristics in preschool children with mild obstructive sleep apnea. J Dent Sci 15(2):193–199. https://doi.org/10.1016/j.jds.2019.09.005

Lin SY, Su YX, Wu YC, Chang JZC, Tu YK (2020) Management of paediatric obstructive sleep apnoea: a systematic review and network meta-analysis. Int J Paediatr Dent 30(2):156–170. https://doi.org/10.1111/ipd.12593

Malhotra A, Jordan A (2016) The importance of arousal in obstructive sleep apnea-updates from the American Thoracic Society 2016. J Thorac Dis 8(7):S542–S544. https://doi.org/10.21037/jtd.2016.06.81

Marcus CL, Omlin KJ, Basinki DJ et al (1992) Normal polysomnographic values for children and adolescents. Am Rev Respir Dis 146(5):1235–1239. https://doi.org/10.1164/ajrccm/146.5_Pt_1.1235

Marcus CL, Lutz J, Carroll JL, Bamford O (1998) Arousal and ventilatory responses during sleep in children with obstructive sleep apnea. J Appl Physiol 84(6):1926–1936. https://doi.org/10.1152/jappl.1998.84.6.1926

Marcus CL, Lutz J, Hamer A, Smith PL, Schwartz A (1999a) Developmental changes in response to subatmospheric pressure loading of the upper airway. J Appl Physiol 87(2):626–633. https://doi.org/10.1152/jappl.1999.87.2.626

Marcus CL, Moreira GA, Bamford O, Lutz J (1999b) Response to inspiratory resistive loading during sleep in normal children and children with obstructive apnea. J Appl Physiol 87(4):1448–1454. https://doi.org/10.1152/jappl.1999.87.4.1448

Marcus CL, Brooks LJ, Ward SD et al (2012) Diagnosis and management of childhood obstructive sleep apnea syndrome. Pediatrics 130(3). https://doi.org/10.1542/peds.2012-1672

Marcus CL, Moore RH, Rosen CL et al (2013) A randomized trial of adenotonsillectomy for childhood sleep apnea. N Engl J Med 368(25):2366–2376. https://doi.org/10.1056/nejmoa1215881

Maris M, Verhulst S, Wojciechowski M, Van De Heyning P, Boudewyns A (2016) Prevalence of obstructive sleep apnea in children with Down syndrome. Sleep 39(3). https://doi.org/10.5665/sleep.5554

Meira M, Estivill E, Kryger MH (2020) The dentist's role in sleep medicine: Why the Hesitation? J Dent Sleep Med 7(2):15331

Mokhlesi B, Punjabi NM (2012) "REM-related" obstructive sleep apnea: an epiphenomenon or a clinically important entity? Sleep 35(1):5–7. https://doi.org/10.5665/sleep.1570

Moreira GA, Tufik S, Nery LE et al (2005) Acoustic arousal responses in children with obstructive sleep apnea. Pediatr Pulmonol 40(4):300–305. https://doi.org/10.1002/ppul.20219

Narang I, Mathew JL (2012) Childhood obesity and obstructive sleep apnea. J Nutr Metab 2012: 134202. https://doi.org/10.1155/2012/134202

Nixon GM, Kermack AS, Davis GM, Manoukian JJ, Brown KA, Brouillette RT (2004) Planning adenotonsillectomy in children with obstructive sleep apnea: the role of overnight oximetry. Pediatrics 113(1 Pt 1). https://doi.org/10.1542/peds.113.1.e19

Nolan J, Brietzke SE (2011) Systematic review of pediatric tonsil size and polysomnogram-measured obstructive sleep apnea severity. Otolaryngol Head Neck Surg 144(6):844–850. https://doi.org/10.1177/0194599811400683

O'Driscoll DM, Foster AM, Ng ML et al (2009) Acute cardiovascular changes with obstructive events in children with sleep disordered breathing. Sleep 32(10):1265–1271. https://doi.org/10.1093/sleep/32.10.1265

Ong JWY, Williams D, Gavlak JC, Liddle N, Lowe P, Evans HJ (2021) Observational study to define reference ranges for the 3% oxygen desaturation index during sleep in healthy children under 12 years using oximetry motion-resistant technology. Arch Dis Child 106(6):583–586. https://doi.org/10.1136/archdischild-2020-320066. Published online 2020

Owens JA, Dalzell V (2005) Use of the 'BEARS' sleep screening tool in a pediatric residents' continuity clinic: a pilot study. Sleep Med 6:63–69. https://doi.org/10.1016/j.sleep.2004.07.015

Owens JA, Spirito A, McGuinn M (2000) The Children's Sleep Habits Questionnaire (CSHQ): psychometric properties of a survey instrument for school-aged children. Sleep 23(8): 1043–1051. https://doi.org/10.1093/sleep/23.8.1d

Pijpers M, Poels PJP, Vaandrager JM et al (2004) Undiagnosed obstructive sleep apnea syndrome in children with syndromal craniofacial synostosis. J Craniofac Surg 15(4):670–674. https://doi.org/10.1097/00001665-200407000-00026

Raynes-Greenow CH, Hadfield RM, Cistulli PA, Bowen J, Allen H, Roberts CL (2012) Sleep apnea in early childhood associated with preterm birth but not small for gestational age: a population-based record linkage study. Sleep 35(11):1475–1480. https://doi.org/10.5665/sleep.2192

Sakarya EU, Muluk NB, Sakalar EG et al (2017) Use of intranasal corticosteroids in adenotonsillar hypertrophy. J Laryngol Otol 131(5):384–390. https://doi.org/10.1017/S0022215117000408

Schäfer T (2006) Physiology of breathing during sleep. Sleep Apnea 35:21–28. https://doi.org/10.1159/000093139

Scholle S, Wiater A, Scholle HC (2011) Normative values of polysomnographic parameters in childhood and adolescence: cardiorespiratory parameters. Sleep Med 12(10):988–996. https://doi.org/10.1016/j.sleep.2011.05.006

Section on Pediatric Pulmonology, Subcommittee on Obstructive Sleep Apnea Syndrome, American Academy of Pediatrics (2002) Clinical practice guideline: diagnosis and management of childhood obstructive sleep apnea syndrome. Pediatrics 109(4):704–712. https://doi.org/10.1542/peds.109.4.704

Sedky K, Bennett DS, Pumariega A (2014) Prader Willi syndrome and obstructive sleep apnea: co-occurrence in the pediatric population. J Clin Sleep Med 10(4):403–409. https://doi.org/10.5664/jcsm.3616

Shahid A, Wilkinson K, Marcu S, Shapiro CME (2012) STOP, THAT and one hundred other sleep scales. Springer Science & Business Media, New York

Simpson R, Oyekan AA, Ehsan Z, Ingram DG (2018) Obstructive sleep apnea in patients with down syndrome: current perspectives. Nat Sci Sleep 10:287–293. https://doi.org/10.2147/NSS.S154723

Soh HJ, Rowe K, Davey MJ, Horne RSC, Gillian M (2018) The OSA-5: development and validation of a brief questionnaire screening tool for obstructive sleep apnea in children. Int J Pediatr Otorhinolaryngol 113:62–66. https://doi.org/10.1016/j.ijporl.2018.07.029. Published online 2018

Spruyt K, Gozal D (2012) REM and NREM sleep-state distribution of respiratory events in habitually snoring school-aged community children. Sleep Med 13(2):178–184. https://doi.org/10.1016/j.sleep.2011.10.025

Stauffer J, Okuji DM, Ii GCL, Bhattacharjee R, Whyte F (2018) A review of pediatric obstructive sleep apnea and the role of the dentist. J Dent Sleep Med 5(4):111–132

Tan HL, Gozal D, Kheirandish-Gozal L (2013) Obstructive sleep apnea in children: a critical update. Nat Sci Sleep 5:109–123. https://doi.org/10.2147/NSS.S51907

Tan HL, Kheirandish-Gozal L, Gozal D (2016) Obstructive sleep apnea in children: update on the recognition, treatment, and management of persistent disease. Expert Rev Respir Med 10(4): 431–439. https://doi.org/10.1586/17476348.2016.1163224

Tang A, Benke JR, Cohen AP, Ishman SL (2015) Influence of tonsillar size on OSA improvement in children undergoing adenotonsillectomy. Otolaryngol Head Neck Surg (United States) 153(2):281–285. https://doi.org/10.1177/0194599815583459

Terrill PI (2020) A review of approaches for analysing obstructive sleep apnoea-related patterns in pulse oximetry data. Respirology 25(5):475–485. https://doi.org/10.1111/resp.13635

Urschitz MS, Wolff J, Von Einem V, Urschitz-Duprat PM, Schlaud M, Poets CF (2003) Reference values for nocturnal home pulse oximetry during sleep in primary school children. Chest 123(1): 96–101. https://doi.org/10.1378/chest.123.1.96

Van Eyck A, Lambrechts C, Vanheeswijck L et al (2015) The role of nocturnal pulse oximetry in the screening for obstructive sleep apnea in obese children and adolescents. Sleep Med 16(11): 1409–1412. https://doi.org/10.1016/j.sleep.2015.07.023

Walker Jr WO, Perez IA (2021) Sleep related breathing disorders. Sleep Relat Breath Disord Guidel Spina Bifida Assoc. Accessed 12/10/2021. https://www.spinabifidaassociation.org/resource/sleep-related-disorders/

Wang H, Fu Y, Feng Y, Guan J, Yin S (2015) Tonsillectomy versus tonsillotomy for sleep-disordered breathing in children: a meta-analysis. PLoS One 10(3):1–11. https://doi.org/10.1371/journal.pone.0121500

Waters KA, Everett F, Sillence DO, Fagan ER, Sullivan CE (1995) Treatment of obstructive sleep apnea in achondroplasia: evaluation of sleep, breathing, and somatosensory-evoked potentials. Am J Med Genet 59(4):460–466. https://doi.org/10.1002/ajmg.1320590412

Younes M (2004) Role of arousals in the pathogenesis of obstructive sleep apnea. Am J Respir Crit Care Med 169(5):623–633. https://doi.org/10.1164/rccm.200307-1023oc

Zheng W, Zhang X, Dong J, He J (2020) Facial morphological characteristics of mouth breathers vs. nasal breathers: a systematic review and meta-analysis of lateral cephalometric data. Exp Ther Med 19(6):3738–3750. https://doi.org/10.3892/etm.2020.8611. Published online 2020

The manufacturer's authorised representative in the EU is Springer Nature Customer Service Centre GmbH, Europaplatz 3, 69115 Heidelberg, Germany. If you have any concerns regarding our products, please contact ProductSafety@springernature.com

Printed and bound by CPI Group (UK) Ltd, Croydon, CR0 4YY
01/10/2026
02225896-0008